ROYAL BC MUSEU

MARINE MAMMALS

OF BRITISH COLUMBIA

John K.B. Ford

Volume 6
The Mammals of British Columbia

John Ford

ROYAL BC MUSEUM
Victoria, Canada

Published by the Royal BC Museum, 675 Belleville Street,
Victoria, British Columbia, V8W 9W2, Canada.
www.royalbcmuseum.bc.ca

Printed in Canada.

MIX
Paper from
responsible sources
FSC® C016245

See page 454 for credits and copyright information on specific elements in this book.

Library and Archives Canada Cataloguing in Publication

Ford, John Kenneth Baker, 1955–, author
Marine mammals of British Columbia / John K.B. Ford.

(Royal BC Museum handbook)
(The mammals of British Columbia ; volume 6)
Includes bibliographical references and index.
ISBN 978-0-7726-6734-2 (pbk.)

1. Marine mammals – British Columbia. I. Royal BC Museum, issuing body II. Title. III. Series: Mammals of British Columbia; v. 6 IV. Series: Royal BC Museum handbook.

QL721.5.B7F67 2014 599.5'09711 C2014-904681-2

To my late mentor and friend, Michael A. Bigg (1939–90).

While working as marine-mammal scientist at the Pacific Biological Station in the 1970s and '80s, Mike Bigg began compiling material for a book on marine mammals aimed at readers keen on learning more about the natural history of these animals.
Sadly, Mike became gravely ill in the late 1980s and, realizing that he would not live to write his book, asked me to take on the project.
With the publication of this book, I can at long last feel like I've delivered on my commitment to Mike.
In many ways, I consider this to be his book as well.

CONTENTS

FOREWORD

The Royal BC Museum has been publishing handbooks since 1942. This series is dedicated to presenting information about groups of organisms that inhabit British Columbia (and adjacent areas) for readers who are not necessarily experts or even well-versed in the subject.

As with all our handbooks, countless hours of scientific research went into the writing of *Marine Mammals of British Columbia*. Dr John Ford has gleaned information from scientific journals and monographs, and condensed it into a more readable format for the wider audience. Like all our publications, this book contributes to the Royal BC Museum's mission: "to promote an understanding of the living landscapes and cultures of British Columbia and engage people in a dialogue about their future."

Marine Mammals of British Columbia is the sixth and final handbook in the Mammals of British Columbia series (see page 454 for the full list). It deals with some of our most familiar, yet poorly understood species. While there is plenty of published research on a few species of marine mammals, there is no detailed treatment focused on all species in our coastal waters. Most people are familiar with three of our baleen whales (Grey, Humpback and Blue whales), the Killer Whale (the largest dolphin) and our seals and sea lions; but many smaller dolphins and porpoises go unnoticed, especially those that inhabit areas farther from shore. How many of you have seen a Risso's Dolphin? (Don't feel bad. I haven't either.) This book goes a long way towards closing the knowledge gap about all the species residing in or passing through our waters.

Given the mountain of information and the size of the subjects represented in each species account, it is amazing that Dr Ford was able to keep this book to fewer than 500 pages. It's our largest handbook to date, but it remains a convenient and portable size that can fit in a daypack, tacklebox or dry-bag, and can be stashed within easy reach on a boat, canoe or kayak. I was thrilled to have the chance to review this handbook prior to publication, and since I have a background more with fishes, amphibians and reptiles, I had much to learn about our charismatic marine mammals. I anticipate that this book will remain a favourite in the personal libraries of British Columbians for many years to come. Read on, and have yourself a whale of a time.

Gavin F. Hanke,
Curator of Vertebrate Zoology,
Royal BC Museum

PREFACE

This is the final volume in a series of six handbooks revising the Royal BC Museum's Handbook 11, *The Mammals of British Columbia* by Ian McTaggart Cowan and Charles J. Guiguet (1965), which is out of print. As with the first five volumes in the series (*Bats of British Columbia*; *Opossums, Shrews and Moles of British Columbia*; *Hoofed Mammals of British Columbia*; *Rodents and Lagomorphs of British Columbia*; *Carnivores of British Columbia*), this handbook emphasizes identification, distribution and natural history. It's been almost 50 years since the publication of Cowan and Guiguet's classic volume, and substantial new information – both published and unpublished – has accumulated on the marine mammals of British Columbia. This handbook brings this information together to provide a comprehensive and up-to-date reference for naturalists, biologists, students, and anyone else interested in the mammals inhabiting BC's marine waters.

This volume departs from the taxonomic structure of the other handbooks in the series, because marine mammals do not constitute a discrete taxonomic group. Rather, they represent an array of species from different mammalian orders adapted to a primarily aquatic existence and relying on the marine environment for most or all of their life processes. Marine mammals in British Columbia belong to two very different lineages: order Cetacea, comprising the whales, dolphins and porpoises; and order Carnivora, which includes the pinnipeds (seals and sea lions) and the Sea Otter. Early in the planning for this series, it was decided that all marine mammals of BC be included in a single volume despite their phylogenetic differences. The Sea Otter, however, is also included in *Carnivores of British Columbia* (which covers terrestrial members of the order Carnivora) because most of its family, the Mustelidae, lives primarily on land.

In keeping with the museum's handbook format, specific sources of information are not attributed using in-text citations. Instead the most relevant literature is cited in Selected References at the end of each section or species account. In addition, I have attempted to ensure that researchers responsible for key information, particularly anything specific to marine mammals in BC, are identified along with when and where their studies were carried out or their observations made.

Studying marine mammals in their natural environment requires a team approach, perhaps more than the study of any other mammalian group. Observations at sea are made from many different kinds of vessels, as well as from aircraft. Small research boats usually have at least two people on board, and large ship surveys may involve more than a dozen researchers who work in shifts over long days of observation. Once back at the lab or office, data analysis often involves many more individuals with their own particular areas of expertise, from molecular genetics to underwater acoustics to statistical modelling. Accordingly, when describing personal observations and findings in this handbook, I typically use "we" rather than "I" to reflect these team efforts, particularly when referring to my co-workers and me in the Cetacean Research Program at the Pacific Biological Station (Fisheries and Oceans Canada, Nanaimo) or other colleagues with whom I have worked over the years (see Acknowledgements).

In assembling the species accounts and other sections of this handbook, I have compiled, synthesized and distilled current and historical data. So little is known about some of the rare species in BC waters that rendering information down to handbook size presented little challenge – some of the beaked whales, for example, are known only from examination of beach-cast carcasses or brief glimpses of the living animal at sea. For the well-known species, an overwhelming amount of information is available on their life history, behaviour and ecology, and I've had to make many difficult decisions about what to include and what to leave out. Hopefully, the information contained in this book will prove both useful and intriguing to a wide spectrum of readers interested in marine mammals off Canada's west coast.

GENERAL BIOLOGY

Introduction

Mammal species that spend all or a substantial portion of their life cycle in the marine environment are called marine mammals. But this term is only descriptive and does not refer to any distinct biological group or taxonomic classification. The main criterion by which mammals are judged to be "marine" is that they rely mostly or entirely on the aquatic environment for feeding, but this does not necessarily mean that they live entirely in salt water. For example, some species of dolphins have evolved to inhabit freshwater river systems such as the Amazon and Ganges and, although they probably could not survive in marine habitat for long, these dolphins (as well as a population of porpoises in the Yangtze River system of China) are usually lumped together with the other cetaceans as "marine mammals". An example in our region is a resident population of Harbour Seals in freshwater Harrison Lake on the southwestern BC mainland.

The dividing line between terrestrial mammals and marine mammals is also rather blurred. For example, the Polar Bear is considered a marine mammal by the international Society for Marine Mammalogy, but in Canada it is placed with terrestrial species under the jurisdictions of Environment Canada and relevant provincial and territorial governments (whereas all marine mammals are the responsibility of Fisheries and Oceans Canada). It could be argued that coastal populations of the Northern River Otter, which forage almost entirely in the ocean, should be considered marine mammals, but they are not because the species also lives in inland habitats (figure 1).

Figure 1. Northern River Otters (left) are frequently seen in BC's coastal marine waters, but these otters spend much time on land and are not considered marine mammals. They are often confused with Sea Otters (right), which live a wholly aquatic life and so are considered marine mammals. Photos: R. McDonell (left) and J. Towers.

The marine mammals of the world fall into three taxonomic orders in the class Mammalia: Cetacea★, which includes the whales, dolphins and porpoises; Carnivora, which includes the seals and sea lions, the Walrus and Sea Otter, as well as many species of terrestrial carnivores; and Sirenia, comprising the manatees and dugongs (also known as "sea cows"). The cetaceans are the most diverse of wholly aquatic mammals in terms of their morphological and physiological specializations, ecological and taxonomic diversity, and wide geographic distribution. Like cetaceans, sirenians are well-adapted to a completely aquatic existence, but they have not radiated as successfully into diverse types of marine habitat. There are just four extant species of sirenians and these are restricted to certain freshwater and nearshore marine regions in the tropics. Sirenians are unique among marine mammals in that they are entirely herbivorous, feeding on sea grasses and other aquatic plants. In our region, Steller's Sea Cow, a large sirenian that fed on kelp and other algae, was found in coastal waters of the Bering Sea and the eastern North Pacific as far south as Monterey Bay as recently as 19,000 years ago. But the last remnants of this species in the Commander Islands (Kamchatka, Russia) were hunted to extinction in the late 1700s. Of the carnivores, the seals, sea lions and Walrus – known as the pinnipeds – live primarily in the marine environment but must haul out on land or ice for at least some

---

★ A recently proposed taxonomic revision has the orders Cetacea and Artiodactyla (even-toed hoofed mammals) merged into a new order, called Cetartiodactyla. The Taxonomy and Nomenclature section has more details.

Figure 2. An extinct aetiocete whale, much like those discovered in Oligocene fossil deposits in the rocky beach near Carmanah Point, Vancouver Island. Illustration: C. Buell.

critical life processes such as breeding and moulting. Sea Otters are not radically different from their terrestrial cousins in the family Mustelidae – the weasels and allies – but are very competent in nearshore marine habitat and can live their entire lives without leaving the water.

These three major groups of marine mammals have separate evolutionary origins. The sirenians are quite ancient, having evolved from early African mammals at least 50 million years ago. The closest living relatives of sea cows are elephants (order Proboscidea) and hyraxes (order Hyracoidea). Fossils indicate that the earliest cetaceans are also very ancient, dating back more than 50 million years. Both fossil and molecular (DNA) evidence shows that cetaceans are most closely related to the artiodactyls – the even-toed hoofed mammals – and in particular to the hippopotamus lineage.

There were three major periods of radiation over the course of cetacean evolution, each involving rapid increases in ecological, functional and taxonomic diversity. The first was during the Eocene, 53 to 45 million years ago, which involved the most primitive cetaceans, the archaeocetes, moving from nearshore and riverine areas to fully oceanic habitat. At least 20 genera and 40 species of archaeocetes have been described from fossils. The next major phase of cetacean diversification took place during the Oligocene, about 35 to 25 million years ago. The Neoceti – an ancient group that gave rise directly to modern cetaceans – evolved during this period, and the archaeocetes died out. The two major groups of cetaceans, odontocetes (those with teeth) and mysticetes (those with baleen), diverged from a common ancestor early in the Oligocene. An unusual family of whales called aetiocetids had baleen and teeth (figure 2). These small whales (2–3 metres long) have been described from fossils collected among Oligocene rock formations found in only a few locations around the world. One such location is along Vancouver Island's West Coast Trail in Pacific Rim National Park. Here, near Carmanah Point light station, intertidal shelves of Oligocene rocks contain a rich bed of fossilized aetiocetid whales. Whale paleobiologist Nick Pyenson from the Smithsonian's National Museum of Natural History in Washington, DC, recently collected two specimens that should provide new insights on the evolution of early baleen whales.

The final period of marked cetacean radiation took place during the Miocene, about 15 to 12 million years ago. This period saw the appearance of the "modern" odontocetes and mysticetes, and the decline and extinction of ancient groups. By the start of the Pleistocene, about 2 million years ago, many of the current cetacean genera had appeared, but patterns of evolution during the following ice ages are poorly known due to a scanty fossil record. It is during this period that the "anti-tropical" divergence of cetacean species took place. Populations became isolated on each side of the tropics during warm-water periods and in many cases this led to speciation. Examples include the different northern and southern hemisphere species of *Lissodelphis* (right whale dolphins), *Berardius* (the southern Arnoux's and northern Baird's beaked whales) and *Eubalaena* (right whales).

Pinnipeds originated more recently than cetaceans and sirenians. There has long been controversy over whether the main groups of pinnipeds – the true seals (family Phocidae), eared seals (family Otariidae) and Walrus (family Odobenidae) – descended from two ancestral lineages (diphyletic) or share a common lineage (monophyletic). It was originally proposed that the Walrus and eared seals evolved from an ursid (bear-like) ancestor and true seals from a mustelid (weasels and allies) ancestor. But recent DNA evidence strongly supports a common origin from which all pinnipeds arose, probably an ursid-like carnivore ancestor in the North Pacific region about 35 to 30 million years ago. The otariids (from the Greek *otarion*, meaning "little ear") appear to have diverged most recently from this common pinniped lineage, about 12 to 11 million years ago.

Sea Otters and their close terrestrial otter relatives arose relatively recently from a mustelid carnivore ancestor about 5 million years ago. The Sea Otter itself is thought to have diverged from this lineage in the North Pacific during the Pleistocene, about 2 million years ago.

Today there are 125 known species of marine mammals in the world's oceans and seas (not including the Polar Bear) – 32 pinnipeds, 2 marine otters, 87 cetaceans and 4 sirenians. About 25 per cent of these, 31 species, are found in the waters of British Columbia. We are fortunate to have such a diversity of marine mammals in the province, and many can be easily observed in nearshore waters. In the following sections, I give a brief overview of the general biology of marine mammals with a focus on BC species. I also describe some of the oceanographic features that create diverse habitat for marine mammals in BC, the relationships between humans and marine mammals in this region, how we study these animals, and the management and conservation status of marine mammals in BC.

Selected References: Berta 2009, Berta and Deméré 2009, Berta and Sumich 1999, Estes et al. 2009, Fordyce 2009, Heyning and Lento 2002.

Form and Structure

When mammals invaded the aquatic environment after evolving for millions of years on land, they faced a number of major challenges. On land, gravity is the main physical force animals must contend with, but it is not significant in water. But water is 800 times denser and 60 times more viscous than air, making drag and buoyancy, as well as hydrostatic pressure at depth, the main physical forces to deal with. Water also conducts heat at a rate 24 times that of air, so keeping warm is also a major problem for mammals. All modern marine mammals have, to a greater or lesser extent, managed these challenges through evolutionary changes in body form and structure, as well as physiology. Arising from separate lineages, the four main groups of marine mammals – cetaceans, sirenians, pinnipeds and marine otters – have adapted independently to an aquatic life in different ways.

All marine mammals have body shapes that developed to reduce drag and turbulence in water. This is most evident in the cetaceans, which are the fastest and most hydrodynamically efficient of all the marine mammals. Cetaceans have very streamlined bodies, and their appendages are radically adapted for propulsion in water. The hindlimbs have been lost entirely and the forelimbs have been flattened to serve as flippers for steering (figure 3). Broad lateral projections from the tail, known as *flukes*, provide propulsion using an up-and-down motion (in contrast to side-to-side as in fishes). A dorsal fin is present on most species. Cetacean skin is smooth and hairless, and external projections that might cause drag have been lost (e.g., external earflaps) or moved internally (e.g., testes). The structure of

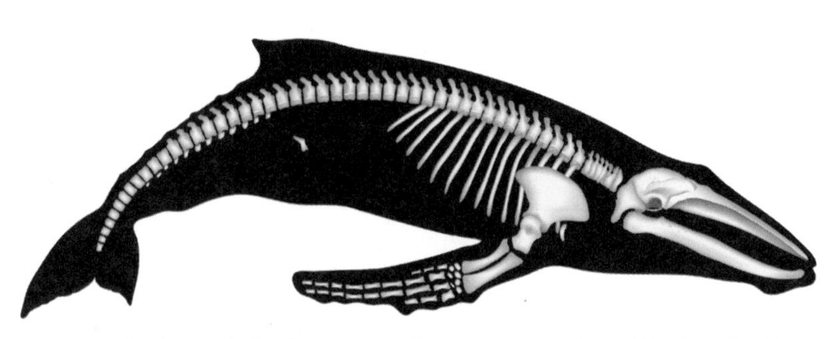

Figure 3. The basic skeletal anatomy of cetaceans, such as this Humpback Whale, is typical of land mammals, but the hindlimbs have been lost, the pelvis reduced to a small vestige detached from the vertebral column, and the forelimbs modified into flippers. Illustration: U. Gorter.

the cetacean head is also highly modified for efficient swimming and breathing at the water's surface. It is elongated and pointed to reduce drag, and the nostrils have moved from their typical mammalian position at the front of the head, or *rostrum*, to the top of the head to facilitate breathing without the animal having to slow down and raise its head above the surface.

Pinnipeds, while not entirely aquatic, have adaptations that allow them to exploit the marine environment efficiently but also return to land for breeding, nursing, resting and moulting. Like cetaceans, pinnipeds are morphologically adapted for efficient swimming, but not to the same extent. Their bodies are streamlined but they have retained the basic quadruped form of terrestrial mammals. In the true seals – phocids, also known as "earless" seals because they lack external earflaps (or *pinnae*), although they have good hearing – the hindlimbs are flattened into flippers, which are swept in side-to-side strokes to provide propulsion, while the fore flippers are used mainly for steering. Because the hind flippers can't be rotated forward, seals move somewhat awkwardly on land by undulating or wriggling their body and pulling themselves along with their fore flippers. In contrast, otariids – the sea lions and seals that have earflaps – propel themselves with broad strokes of their long fore flippers while the hind flippers serve mostly as rudders. On land otariids can raise their body off the ground using both the fore flippers and hind flippers, which can rotate forward to function somewhat like feet. Otariids are far more mobile on land than phocids, walking (or perhaps more accurately, waddling) by alternating movements of their fore flippers and hind flippers. When motivated, such as when chasing a rival on a breeding territory, otariids can lope across rocks at speeds rivalling humans for short distances.

Sea Otters are perhaps the least modified from the terrestrial mammalian form, even though they can live without ever leaving the water. The hindlimbs are long and wide, forming flippers similar to true seals, but sea otter forelimbs are not unlike the paws of other mustelids, and have no role in aquatic propulsion or steering. Instead, the front feet are used for grasping prey from the sea floor and manipulating it for feeding at the surface. Sea Otters are capable of locomotion on land, but they are limited to an ungainly shuffle rather than true walking. Being thus vulnerable to terrestrial predators, they seldom travel more than a few metres from the water's edge.

All marine mammals submerge themselves to feed or conduct other life processes, and they have a range of anatomical and physiological adaptations to facilitate diving. Some species, such as the beaked whales, sperm whales and elephant seals, are excellent divers and may descend to depths in excess of 1000 metres and stay down for an hour or longer. Others, such as the Sea Otter, have more modest diving abilities, rarely reaching more than 100 metres below the surface or dive durations of more than a few minutes. The most imposing physical challenge for deep-diving marine mammals is the great increase in pressure that accompanies increasing depth. At a depth of one metre, the pressure is 100 grams per square centimetre, but at 1000 metres it is more than 100 kilograms per square centimetre. Pressure effects are most profound on any air sacs or sinuses in the mammalian body, including the lungs. Deep-diving seals and whales have modifications that allow the lungs to collapse almost completely at depth. Their upper airways are strongly reinforced with cartilage and receive gases that are squeezed out of the lung's alveoli (air sacs) as it compresses, and a flexible rib cage allows the chest to compress as the lungs collapse.

Although cetaceans dive with full lungs, pinnipeds generally exhale about half of their lung volume before diving. Because the lungs collapse during deep dives, little or no gas exchange between the lungs and tissue takes place at depths greater than about 60 metres. This prevents the animals from getting the bends, since nitrogen would be forced into solution under pressure and would likely bubble out of the blood as the animal ascended.

Diving mammals exchange most or all oxygen while at or near the surface. To enable high levels of activity during prolonged submersion, marine mammals have exceptional abilities to store oxygen in their blood and tissue and to allocate this oxygen selectively to organs

Figure 4. A raft of California and Steller sea lions near Nanaimo. While they rest at the surface, the sea lions hold their flippers out of the water to conserve body heat. Photo: J. Ford, March 2013.

that need it the most. Marine mammals have about three times greater blood volume than land mammals by body size, about 1.5 times more hemoglobin in their blood to carry oxygen and about 10 times more myoglobin in their muscles, which is their main reservoir of oxygen. During a prolonged dive, the animal's heart rate slows to 20–50 per cent of that at the surface and blood flow to the skeletal muscles is cut to a trickle. Skeletal muscles continue to function aerobically using oxygen stored in the myoglobin, while oxygenated blood is preserved for the brain and heart. The blood flow to the brain is modulated by extensive bundles of arterioles and venules, known as the *rete mirable* (or "wonderful net"), which, when engorged, may also serve to prevent tissue compression in the head and neck region.

Marine mammals have also acquired morphological and physiological attributes to limit the loss of body heat in the cold marine environment. The bodies of cetaceans and most pinnipeds are encased in a thick layer of blubber, which serves as an insulating blanket as well as a storehouse of energy. Large body size, such as seen in the great whales, helps to reduce heat loss, since the ratio of surface area to body mass decreases with increasing size and rotundness. In sea lions and most species of seals, the layer of hair that covers most of their bodies plays little role in thermoregulation. Fur seals are an exception, as these otariids have an unusually dense and luxuriant pelage that provides considerable insulation. Sea Otters have no blubber and rely entirely on their fur for insulation, which is double that of fur seals. In fact, Sea Otter fur is the densest of any mammal, with more than 125,000 hairs per square centimetre. The insulation, though, is actually provided by a layer of air trapped in the fur that prevents water from reaching the skin. Sea Otters also compensate for heat loss to the ocean by having a comparatively high metabolic rate.

Both cetaceans and pinnipeds have efficient vascular counter-current heat-exchange systems that reduce heat loss in their flippers, flukes and dorsal fins, where there is little blubber or fur. Arteries carrying blood to the surfaces of these appendages are surrounded by veins returning blood to the heart, and in the process, the outgoing arterial blood warms the cooler returning venous blood, thereby retaining core body heat. Sea lions and fur seals may further reduce heat loss to the ocean by holding their flippers above the surface, where the rate of cooling is much slower than in the water (figure 4). Ironically, the excellent insulative properties of blubber can present a problem when these animals generate excessive body heat through exertion. In such cases, blood is shunted preferentially to the appendages, which serve as radiators to dump excess heat into the cold water.

Selected References: Berta and Sumich 1999, Kooyman 2009, Williams and Worthy 2002, Würsig et al. 2000.

Sensory Systems and Sound Production

Sensory systems allow animals to receive and interpret information from their surroundings. In the transition from a terrestrial to an aquatic environment, the sensory structures of marine mammals have evolved to detect and process signals in water, a very different medium than air. The greater density of water compared to that of air dramatically alters the propagation properties of light, sound and chemicals. This allows marine mammals to use sound to a greater extent than most terrestrial mammals, while the utility and importance of other signals, like chemicals, have diminished in the marine environment.

In water sound travels about 4.5 times faster and suffers less attenuation over distance than it does in air, making it particularly effective for communication. All marine mammals have special adaptations in the external and middle ear to accommodate rapid and extreme pressure changes associated with diving. But they retain an air-filled middle ear and have the same basic inner-ear configuration as terrestrial mammals. The pinnae – the external earflaps attached to the outside of the mammalian head – evolved to collect and amplify sounds in air but are ineffective in water. The density of water is similar to that of mammalian tissue, so sound passes easily into and through the body, except where it meets air cavities or sinuses and, to a lesser extent, bone. Because pinnae are ineffective in water and would cause hydrodynamic drag, they are absent in most marine mammals and reduced to small flaps in the otariids – the "eared" seals.

Adaptations for sound reception and production in water are most advanced in the cetaceans, particularly the odontocetes. Most whales, dolphins and porpoises are exceptionally sonic creatures, producing a wide range of sounds for social communication and, in the odontocetes, for echolocation of objects. The auditory meatus, or ear canal, of all cetaceans is reduced to a pinhole and blocked with tissue or waxy secretions, and sound reaches the middle ear through soft tissue pathways. In the toothed whales, dolphins and porpoises, sound passes into the lower jaw through a very thin section of bone called the *acoustic window* and is conducted by a special body of lipids in the mandible to the middle ear. This system provides excellent directional hearing at ultrasonic frequencies (about 20 to >100 kHz) and is associated with the use of echolocation. Lower frequency sounds (<20 kHz), such as the diversity of whistles and burst-pulse signals used for communication, enter the odontocete ear through tissues around the head. Audiograms of dolphins in captivity show that these odontocetes can hear over a wider range of frequencies than any other mammal. The sound production structures of odontocetes are also highly modified for an aquatic life. High-frequency echolocation clicks are produced in the nasal passages beneath the blowhole and then projected ahead of the animal by nasal air sacs and the forehead, or *melon*, of the head, which contains special lipids that further focus the sound energy. Whistles and burst-pulse sounds (resembling screeches, squawks and screams) are also produced primarily in the nasal passage.

The sophisticated echolocation system of toothed whales, dolphins and porpoises is likely their primary sensory mode. They rely on it for navigation and orientation as well as for discrimination of objects at close range. Echolocation is no doubt particularly helpful in turbid waters or at night, when vision is restricted, and different types of sounds are used for different purposes. For example, echolocation clicks rich in low-frequency energy and repeated at relatively low rates of about 1 to 30 per second are used mostly to scan surroundings acoustically, while rapid clicks (100 to more than 300 per second) with most energy at ultrasonic frequencies are for inspection of small objects at ranges of a body length or less. Field studies I have conducted with colleagues Whitlow Au and John Horne on Resident Killer Whale echolocation indicate that the whales should be able to detect a large Chinook Salmon, their primary prey, at a range of at least 100 metres in moderately noisy ambient conditions and even farther if it is quiet. Though echolocation provides odontocetes with an excellent navigation tool, it is not infallible. Misreading acoustic cues in shallow water is thought to be one of the reasons that some toothed whales strand en masse on shoaling beaches (figure 5).

Baleen whales lack the sophisticated sound reception structures of toothed whales and have ears best suited for receiving the low-frequency signals that they produce. Blue and Fin whales produce intense sounds at infrasonic frequencies of 10–20 Hz, below the hearing range of humans, which may allow communication at ranges of tens or even hundreds of kilometres. The low-pitched calls and songs of baleen whales are associated mostly with mate acquisition but may also serve as a long-range form of echolocation (although this has yet to be proven). In contrast to ondotocetes, mysticete whales produce sounds in the larynx, although the exact mechanism is not known. All cetaceans have no vocal cords.

Pinnipeds have sound reception and production structures that are more typical of terrestrial mammals, since these amphibious animals must hear and communicate in air as well as water. Unlike cetaceans, they do not appear to have any novel pathway for sound transmission to the middle ear, nor do they have elaborate adaptations for making sounds. Pinnipeds have well-developed rings of muscles

Figure 5. Killer Whales and other odontocetes use echolocation for orientation in their environment, but they occasionally make navigational errors. Two Transient (Bigg's) Killer Whales became stranded when swimming over sand flats during an ebb tide in Marcus Passage on BC's north coast. Several hours later when the tide rose again, the whales refloated themselves and swam off without any apparent ill effects. Photo: D. Davis, July 2011.

around the ear canal that may serve to close it while under water, but overall, their hearing sensitivity is better in water than in air. Pinniped sounds range from the barks, growls and roars of sea lions (which they make in both air and water) to the long, descending underwater trills used in breeding displays by certain species of Antarctic and Arctic seals. The mechanism for underwater sound generation in pinnipeds is not well known, but it may involve cycling air between the larynx and trachea. There is no evidence that pinnipeds use sound for echolocation.

The ears of Sea Otters are the most like those of land mammals. Although Sea Otters are often quiet, they can produce intense scream-like calls above the surface, especially for contact between mother and pup. They are not known to make sounds underwater, and their hearing sensitivity under water has not been tested.

Light attenuates rapidly with depth in water, particularly in turbid conditions typical of productive coastal regions. As a result, the vision of marine mammals is well adapted for low-light conditions, and in many ways, their eyes resemble those of nocturnal terrestrial mammals. Marine mammals have high densities of photoreceptors, with maximum sensitivities at the blue end of the spectrum, matching the wavelengths of light that best penetrate the ocean. They also have a *tapetum lucidum* – a layer of tissue behind the retina that reflects light back to the photoreceptors – that is the most well-developed of all mammals.

The refractive indices of light rays entering the eye are very different under water than in air, and therefore the lenses in marine mammal eyes are more like those of fish than land mammals. Cetacean lenses are almost spherical, which optimizes visual acuity under water, while pinniped lenses have a shape that is intermediate between cetaceans and terrestrial mammals. Since marine mammals have lenses designed for underwater vision, they would be expected to have severe myopia (short sightedness) in air, but they do not – their acuity in air is as good as in water. The mechanism by which they can see well above and below the surface is not well understood, but it may be through their ability to contract their pupil to a very small slit that functions like a pinhole camera, which has great depth of field. The visual acuity of cetaceans and pinnipeds is similar to that of a cat (on land). Sea Otters are also able to accommodate their vision for use in air and water – possibly better than any other vertebrate – by changing the shape of the eye lens. Their acuity, however, has not yet been tested. And while all marine mammals have both rod and cone retinas, it is not yet clear if they can detect colours.

Chemoreception through olfaction (smell) and gustation (taste) is not well studied in marine mammals, but generally its importance appears to have declined as the extent of aquatic adaptation has increased. Chemical communication in water is constrained by the very low rates chemicals can diffuse, which is about 10,000 times slower than in air. Odontocetes lack olfactory bulbs in their brain and associated nerves, so they apparently have no sense of smell. Olfaction has been little studied in mysticetes, but recent research has shown that Bowhead Whales have well-developed olfactory bulbs and likely have a good sense of smell. This may also be the case in other baleen whale species. Olfaction is reduced in pinnipeds compared to land mammals, but is still functional and may be important in, for example, mother–pup recognition and detection of female estrus by males. Sea Otters appear to have a well-developed sense of smell. Gustation in marine mammals is poorly known, overall. Experiments have shown that captive dolphins can distinguish all four taste groups assigned by humans – sour, bitter, sweet and salt – and sea lions can distinguish all of these except sweet.

Mechanoreception – the sense of touch – is, in contrast, very well developed in marine mammals. Cetacean skin has high tactile sensitivity, especially around the blowhole where detection of the air-water interface is critical. Pinnipeds have *vibrissae*, or whiskers, that are exceptionally sensitive, with a density of nerve receptors in the follicles of some species that is 10 times that of any land mammal. Vibrissae can be extended forward from the nose to inspect objects with levels of tactile discrimination that rival even the prehensile hands of some primates. Vibrissae may well be an indispensible sensory system for prey detection and capture. Sea Otters have well-developed vibrissae as well, and the excellent tactile sensitivity of their forepaws is important in foraging, manipulating prey for consumption and grooming.

Selected References: Thewissen 2009, Tyack and Miller 2002, Wartzok and Ketten 1999, Würsig et al. 2000.

Feeding Ecology

Over the course of evolutionary time, marine mammals have diversified to occupy an amazing variety of aquatic habitats – from inland river systems to the shallow intertidal marine zone to the high seas, from the ocean's surface to depths of more than 1000 metres, and from the tropics to ice-covered polar waters. In doing so, these animals have developed morphological and behavioural adaptations allowing them to exploit most vertebrate and invertebrate food sources in these ecosystems, with the exception of organisms that occupy the deepest parts of the oceans. As apex predators at the top of the food web or major consumers of low trophic-level resources such as zooplankton, marine mammals play important roles in shaping marine ecosystems.

Different groups of marine mammals are specialized to consume different types of food resources, with the cetaceans showing the greatest diversity in feeding ecology. Mysticetes – the baleen whales – are the largest marine mammals, yet they feed on some of the smallest organisms. Instead of teeth, they have multiple rows of baleen plates that hang from each side of the roof of the mouth. The fringes along the inside of these baleen plates, which are made of keratin like our hair and nails, function as a sieve to trap zooplankton and small schooling fishes. To feed in this way, baleen whales must strain great masses of water, and their large mouths and large bodies support this function.

Different baleen whales have different modes of feeding. The balaenopterids, which include Blue, Fin, Sei, Humpback and Common Minke whales, are primarily "gulpers" – they lunge at high-density patches of prey and engulf huge amounts of water from which they strain out the small krill (euphausiids) or schooling fishes. These whales all have relatively short, coarse baleen plates (figure 6) and pleated throats that expand during the lunge, allowing their mouths to engulf a water mass that can be equal to that of their own body. Right whales are "skimmers" or "ram feeders" that swim open mouthed through dense patches of zooplankton, mostly tiny copepods that

Figure 6. This Humpback Whale lunging for krill in Juan Perez Sound, Haida Gwaii, shows the mats of baleen in the roof of its mouth separated by the smooth pink palate. Photo: J. Ford, May 2003.

become trapped against the long, finely-fringed baleen (figure 7). Skimming can be done either at the surface or at depth. Although Sei Whales are mainly gulpers, they sometimes use a skim technique when feeding on copepods. Unlike other baleen whales, Grey Whales use a "sucking" technique to feed on amphipods and other small invertebrates on or buried in soft sediment on the sea floor or on zooplankton in the water column, especially near the bottom. Plumes of mud are often seen behind the whale as it swims to the surface following a feeding dive. Most baleen whales feed alone, but Humpback Whales are well known for the cooperative "bubble-net" feeding technique,

Figure 7. A North Pacific Right Whale killed off the west coast of Vancouver Island in July 1951 is ready for processing at the Coal Harbour whaling station. Long baleen plates hang from the roof of the mouth (which is held open by the unnaturally enlarged tongue, due to either decomposition or inflation by the whalers). Biologist Gordon C. Pike stands beside the whale . Photo: Cetacean Research Program, PBS.

which may involve a dozen or so individuals working together to corral and capture small schooling fish.

The toothed whales, dolphins and porpoises generally feed on larger prey than do mysticetes – mostly fishes and squid. Killer Whales are the only cetaceans that routinely prey on other marine mammals. Odontocetes typically capture and swallow prey individually rather than in batches. In species that have functional teeth (not all do), dentition is homodont – all teeth are of uniform size and shape – unlike the normally heterodont mammalian teeth, which are quite variable (e.g., incisors, canines, and molars). Odontocete teeth are typically conical with a sharp crown, and they are used for grasping and piercing prey, rather than chewing. They can be numerous – some dolphins have more than 200 in their long, slender jaws – but some species, including Risso's Dolphins and beaked whales, have few or no exposed teeth at all. These "toothless" odontocetes likely catch and swallow prey – primarily squid – simply by sucking them into their mouth. Beaked whales have muscular throats with deep grooves that allow the oral cavity to expand, presumably to create a strong suction for prey capture.

Some toothed whales and dolphins have developed innovative, cooperative techniques to catch prey. Groups of Common Bottlenose Dolphins, for example, may work together to drive schooling fish out of the water and onto muddy banks of estuaries. The dolphins then slide partially out of the water to pick up the wriggling stranded fish. Killer Whale populations in different regions of the world have developed foraging specializations that can involve quite sophisticated tactics. In northern Norway, groups of Killer Whales cooperate to encircle and concentrate schools of herring; some of them slap at the school with their tail flukes to stun the fish, then the whales pick up and consume the drifting, debilitated fish one by one. In the Antarctic, some Killer Whale groups that specialize in seal hunting work cooperatively to generate waves that wash seals off pans of ice so that they can be captured and shared. Killer Whales and other delphinids often ram their prey with their rostrum to subdue it before grasping it in their jaws (figure 8).

Like toothed cetaceans, pinnipeds feed mostly on fishes and squid, although crustaceans such as krill are important parts of the diets of some Arctic and Antarctic seals. Most pinnipeds are generalist predators, feeding on a wide variety of prey species. Yet in most cases the great majority of their diet is made up of only two or three key species that meet the bulk of the seal's energy needs. Fish-eating pinnipeds feed mostly on species that are 10 to 35 cm long. They

Figure 8. A Transient (Bigg's) Killer Whale ramming a Dall's Porpoise at high speed in Johnstone Strait. Photo: J. Towers, June 2010.

sometimes capture larger fish, but these must be broken up prior to consumption. Some of the larger otariids, such as Steller Sea Lions, occasionally feed on seal pups. Sea lions and the smaller true seals such as Harbour Seals are not deep divers and feed mostly in the top 100 metres of the water column within about 200 kilometres of the coast. Northern Elephant Seals and their southern hemisphere counterparts, Southern Elephant Seals, are superb divers, spending about 90 per cent of their time more than 300 metres under the surface in the deep open ocean, coming up to breathe for less than three minutes every 20 minutes, on average, for months at a time. Northern Fur Seals also spend the majority of their lives on the high seas far from land, but they are shallow divers, feeding at depths of less than about 200 metres. Pinnipeds may aggregate in areas with abundant prey, but they typically feed alone without any form of cooperation.

Sea Otters forage in rocky habitat and soft substrates in shallow nearshore waters. They feed mostly on sea urchins, clams, mussels, crabs, snails and other invertebrates that they pick up from the sea floor and bring to the surface to eat. Unlike other marine mammals, Sea Otters have flat-topped molars that are used to crush prey. They are among the few mammals to use tools, often bringing a rock to the surface to help break open a hard-shelled animal. The Sea Otter is known as a keystone species, playing a central role in rocky reef ecosystems by limiting the abundance of herbivorous invertebrates and thus allowing kelp and other algae to flourish.

Selected References: Bowen and Siniff 1999, Bowen et al. 2009.

Distribution and Migration

Marine mammals are a diverse group of animals, and so are their patterns of distribution and movements. Some have restricted ranges, while others are found throughout the world's oceans. Many species undertake annual migrations over vast distances, while others make only minor seasonal movements or none at all. Many factors can influence the distribution of marine mammals – the habitat features, including bathymetry, temperature and salinity; the abundance of prey, predators and competitors; demographic characteristics, such as age, sex and reproductive status; and disturbance from human activities. Overall, the most important determinants of marine mammal distribution and movements are the distribution of food and the need to reproduce and rear offspring.

There are several different large-scale patterns of marine mammal distribution. Species that are widespread in most or all of the world's oceans are considered *cosmopolitan*. Examples of this pattern include the Killer Whale and Humpback Whale. Species that are found only in high-latitude waters in either the northern or southern hemisphere have a *circumpolar* distribution. Many species of pinnipeds and cetaceans in the southern hemisphere exhibit this distribution pattern, but relatively few in the northern hemisphere do – examples are the Bowhead Whale and the Beluga. Species that occur only in warm waters on both sides of the equator are defined as *pantropical*. Species found close to shore or over continental shelf waters are *coastal*, while those found beyond the shelf edge in oceanic waters are *pelagic*. The Harbour Porpoise, for example, is restricted to coastal waters, but the distribution of Dall's Porpoise is both coastal and pelagic. Deepwater cetaceans such as the beaked whales are entirely pelagic and rarely occur in coastal waters. Northern Fur Seals have a pelagic distribution for most of the year, but move to coastal waters en route to their summer breeding rookeries.

Seasonal movements are common in many species of marine mammals and can be either migratory or non-migratory, although the distinction between the two is sometimes blurred. True migration involves long-distance return movements, usually for separate purposes; a common form of seasonal migration is between breeding locations and feeding grounds. Most baleen whales, for example, migrate between summer feeding grounds in cold, productive waters at high latitudes and breeding grounds in warm-temperate to tropical waters. Well-known to British Columbians is the migration of Grey Whales, which takes place close along the outer shores of the province (figure 9). Most Grey Whales feed during summer and fall in the Bering and Chukchi seas, swim 7500–10,000 km to breeding lagoons along the west coast of Baja California, Mexico, where they mate and give birth to their calves, then return northward in March and April. Another strongly migratory mysticete is the Humpback Whale, which feeds in cool-temperate to polar waters from spring through fall and migrates to breeding areas around oceanic island groups or along continental margins for winter. Baleen whales typically feed little if at all during migration or while on their winter grounds. It has long been thought that these lengthy journeys are undertaken so that calves can be born in warm waters, allowing them to dedicate energy gained from nursing toward growth rather than staying warm. Zooplankton

Figure 9. Two Grey Whales off Estevan Point on their northward spring migration along the west coast of Vancouver Island.
Photo: J. Ford, March 2011.

productivity in high-latitude waters declines in winter, so swimming thousands of kilometres to warm waters for the winter months, while involving a great outlay of energy, may not be much more costly than staying in cold waters with minimal feeding opportunities. This would help to explain why males and non-breeding females, rather than just pregnant females, also make these long journeys. Another factor driving migration might be to avoid the risk of predation on calves by mammal-hunting Killer Whales. Newborn calves are especially vulnerable, and there are generally fewer Killer Whales in subtropical and tropical waters than in higher latitudes.

But the timing of baleen whale migration as a whole can be fairly diffuse. Some whales head to their winter grounds early and return early, while others leave much later, and some individuals may choose not to migrate at all. As a result, northbound and southbound migrants can cross paths, and some whales can be present on feeding grounds throughout the year. In British Columbia this is the case for Grey Whales, Humpback Whales and Fin Whales. Individuals tend to return to the same feeding area each year following the spring migration and often have strong preferences for certain areas or locations within the population's overall feeding area. This site fidelity may result from whales becoming familiar with the place that they spent their first summer with their mother prior to weaning.

Some toothed cetaceans also make long seasonal migrations. Belugas and Narwhals, for example, spend winter in areas of loose ice cover then migrate north to high Arctic waters during summer as the solid pack ice breaks up. The Sperm Whale is another migratory species, although movement patterns vary by region as well as by age and sex, and those in some equatorial areas may not migrate at all. Among Sperm Whales that do migrate, schools consisting of females and their immature offspring often move toward fairly high latitudes during summer and return to warmer waters in winter, but adult males travel much farther into cold, polar waters. Killer Whales may make seasonal movements related to prey availability but generally do not undertake long migrations. An exception are Pack Ice (or "Type B") Killer Whales in the Antarctic, which feed on seals among the ice for much of the year but make periodic rapid excursions thousands of kilometres to the north, where they spend several weeks in warm (21–24°C) waters before returning. Why they make these long-distance movements is not clear – the movements do not appear strictly seasonal in their timing, and they may serve a physiological purpose such as skin maintenance.

Many odontocetes that do not make seasonal migrations have home ranges that can remain stable throughout the year or vary seasonally with prey availability. Depending on the social structure of the species, home ranges may be specific to individuals or groups but are not exclusive defended territories – individuals and groups typically have overlapping home ranges and may or may not interact when in the same area. A few species may have *diel* movement patterns – that is, they vary between day and night. Spinner Dolphins in Hawaii, for example, rest in nearshore bays during the day, where they are safe from shark attack, and move offshore at night to feed.

The most important factor dictating the movement patterns of pinnipeds is their periodic need to leave the ocean to moult or give birth and rear their offspring. Since most phocids breed on ice and must migrate due to its seasonality, they are generally more migratory than otariids, but some of them, such as Harbour Seals, are non-migratory and only make seasonal movements related to prey. Harbour Seals move onto haulouts each day to rest, and the timing of these movements is largely dictated by the tide. Among the most impressive migrations of any pinniped – or any mammal for that matter – are those of the Northern Elephant Seal. These seals undertake two long-range migrations per year, each of which is a round trip distance of up to 10,000 km. One migration is from high-seas feeding grounds to on-shore breeding rookeries for mating and giving birth, and the second is for moulting. Most otariids move seasonally to traditional breeding rookeries, but these are usually relatively short-range coastal movements rather than migrations. But some species, such as the Northern Fur Seal, do make long annual migrations from pelagic feeding areas to breeding rookeries on shore.

Sea Otters are non-migratory, and even their seasonal movements are modest. Generally, they tend to move into protected nearshore bays and inlets during the stormy winter months, and to outer, more exposed areas during summer. Sea Otters are unusual among carnivores and marine mammals in the small size of their home range, which varies with age and sex. Adult males tend to have the largest home ranges, at about 50 linear kilometres of coastline, and subadult females the smallest, at about 25 kilometres.

Selected References: Bowen and Siniff 1999, Bowen et al. 2009, Gaskin 1982, Reeves et al. 2002.

Life History and Reproduction

As with most aspects of marine mammal biology, features of their life history and reproduction have been modified from those typical of land mammals to enable an aquatic existence. The extent of these modifications varies in different groups, especially those that give birth in water. The large body sizes of most marine mammals are associated with life history characteristics that include long lifespan, slow growth, delayed sexual maturation, long gestation and reduced number of offspring per cycle – only a single offspring is the norm in all marine mammals. The long gestation is likely due to the harsh environment into which the neonate is born – most newborn marine mammals are precocial, especially cetaceans, which are born in the water and must be able to swim at birth. Although female marine mammals generally give birth to only one young at a time, they usually invest heavily in the maternal care of the pup or calf to promote its survival. In many species, there is a prolonged period of nursing and care, and bonds between offspring and mothers can sometimes last well into adulthood in odontocetes. In others, the period of lactation is brief but intense, with the mother transferring much of her fat reserves to the offspring through remarkably rich milk over the course of only a few days after birth.

As is typical of mammals, offspring are fed and cared for by their mothers, and males contribute little or nothing to parental care. Instead, males expend considerable effort in competing for females to breed, and polygyny – where successful competitors mate with multiple females – is a common mating strategy among marine mammals. Various kinds of competition can, therefore, be observed in marine mammals, including direct contests over females, attempts to endure longer than other males (e.g., holding territories the longest), attempts to entice and attract females, and attempts to outcompete other males by producing greater quantities of sperm, thereby increasing the chances of fathering more offspring. Large body size can be advantageous in species with male-male competition, and as a result, sexual selection has led to males often being considerably larger than females in many pinnipeds and some toothed whales. In species with non-combative mating systems, there is usually little difference in size between the sexes.

Among cetaceans, life history and reproduction differ greatly between the two major groups – the mysticetes and the odontocetes. In most of the baleen whales, a two-year reproductive cycle is closely tied to an annual migratory cycle – females ovulate and conceive in low latitudes and warm waters during winter, migrate to summer feeding grounds while pregnant, then return to low latitudes to give birth the next winter following 11 or 12 months of gestation. Mothers nurse their calves while on the winter grounds and during the migration back to high-latitude summer feeding grounds. Calves are weaned by the time the females return to the winter grounds and conceive once again. Blue, Fin, Sei, Humpback and Grey whales all tend to follow this two-year cycle, although a small proportion of Humpbacks conceive shortly after giving birth and produce another calf the following winter. Common Minke Whales may routinely breed annually, and Right Whales have a 3–4 year cycle. Baleen whales live long – Common Minke Whales up to 60 years, Fin Whales to over 90 and Bowhead Whales in the Arctic can live well into their 100s, perhaps to 150 years. Because female Fin and Minke whales become sexually mature as young as 5 years of age and since they continue having offspring throughout their life, they may have as many as 40 pregnancies in a lifetime.

The toothed whales, dolphins and porpoises have a wider range of life history and reproductive patterns than do the baleen whales. Most breed and calve seasonally, but unlike the mysticetes, this season differs greatly in some species, with births having a seasonal peak but taking place in any month of the year. Some odontocetes, such as the Harbour Porpoise, have an accelerated life history compared to other species. Harbour Porpoises mature at 3–4 years old, give birth each year following 10.5 to 11 months of gestation, and have an average lifespan of only 8–10 years. Killer Whales are at the opposite extreme – they mature slowly, have long intervals between calves, and live much longer. Females have their first calf at an average age of 14, and males, although they reach sexual maturity at around age 13, do not usually attain "social maturity" and actually father offspring until they are in their 20s. With a gestation of 17 months and a nursing period of about a year, the interval between calves is about 3 years. Females give birth to their last calf at about age 40 and can live on for decades as post-reproductive matriarchs that help keep their social groups together.

Unlike cetaceans, all pinnipeds typically leave the water to give birth, either on land or on ice. Breeding and pupping tend to be highly seasonal and synchronized, and many species that are usually dispersed widely over the oceans during the non-breeding season gather in the hundreds or thousands at traditional rookeries each year to give birth, nurse their pups and mate. Northern Fur Seals and Northern Elephant Seals, for example, migrate many thousands of kilometres to their rookeries. An annual birth and mating strategy requires a basic

12-month reproductive cycle, but seals and sea lions do not have a gestation of such duration. Instead, they have a *diapause* – a lag between mating and the beginning of gestation – that allows this annual synchrony of breeding. After mating, the fertilized ovum enters a suspended state for 2 to 4 months, then implants in the uterus and gestation begins, leading to birth 8 to 10 months later.

Most pinnipeds have a polygynous mating system, with males competing with other males for well-defined territories on breeding rookeries. They mate with females that choose to haul out and pup on these rookery territories (e.g., Steller Sea Lions), or they may defend clusters of females rather than clearly demarcated places on the rookery, mating with those females after they have pupped (e.g., Northern Elephant Seals). In some species, polygyny is extreme and males must establish and defend their territories or clusters of females for weeks or months at a time. During this period they stay on land and must rely solely on their fat reserves for nutrition and hydration. Long fasting periods favour large body sizes, which is why males of highly polygynous species are two to four times bigger than females. Males cannot compete effectively until they have attained their full adult size, and they generally do not become successful breeders until several years following puberty – and some never breed at all. Defending territories is an arduous task for a breeding male, and most do not live for many years after attaining breeding status – few make it into their twenties.

Pinniped pup care differs slightly between the otariids and phocids. In otariids, nursing mothers stay with their pups for about a week, then begin leaving their pups on shore for variable periods while they search for food at sea. Females mate immediately before their first foraging trip to sea. Otariids have protracted nursing periods that may continue for many months, long after the mother and pup have left the breeding rookeries. Most phocids, however, stay on land or ice and nurse their pups until weaning, which may occur 4 to 30 (or more) days after birth. They then mate with the attending male and abandon the pups, which must leave the rookery and fend for themselves. In Harbour Seals, the mother goes on feeding excursions while still nursing her pup.

Although Sea Otters are less adapted to an aquatic existence than many other marine mammals, they both mate and give birth in the water. Pups are precocious but unable to swim and must be carried by their mother for the first few weeks. Like most of the pinnipeds, Sea Otters are polygynous, with males defending territories in the water that overlap the ranges of several females with whom they mate. They also have a period of delayed implantation, but this plus gestation is only six to seven months long. There is a seasonal peak in breeding, but pups can be born at any time of year.

Selected References: Boness et al. 2002, Boyd et al. 1999.

Behaviour and Social Organization

From a very broad perspective, an animal's success in terms of natural selection hinges on its efficiency at turning food into offspring. Although morphology, physiology, genetics and many other factors play key roles in this process, an animal's behaviour is critical for it to function in its environment. Unlike morphology and physiology, behaviour can change in an animal's lifespan in response to changing environmental conditions, and this behavioural flexibility can enhance its ability to survive and reproduce. Marine mammals are highly diverse in their morphology, physiology and ecology, and even more diverse in their behaviour patterns and forms of social organization.

Our understanding of marine mammal behaviour is still quite limited. We know more about some species – especially the pinnipeds and Sea Otter – because they're more accessible to shore-based observers. But we generally know less about cetacean behaviour, because these animals spend most of their time under water and boat-based observers have only brief opportunities to view them when they surface to breathe. In fact, some of the beaked whales, which are deep-diving pelagic species, have rarely or never been seen alive let alone studied, so their behaviour patterns and social organization are truly unknown. Until quite recently, our understanding of the behaviour of cetaceans was based mostly on whalers' accounts or came from studies of the smaller species held in captivity. But the past few decades have seen significant advances in knowledge from dedicated field studies of free-ranging cetaceans, often aided by techniques such as photo-identification of individuals and the use of data-logging instruments temporarily affixed to the animals to collect information on diving and travel behaviour (see Studying Marine Mammals).

Most cetaceans are social, but the extent of sociality and types of societies vary greatly. The odontocetes tend to be more socially complex than the mysticetes, usually living in groups (known as schools, pods or herds) of varying size and composition. Porpoises are some of the least social odontocetes; they are usually observed in small mating associations during the breeding season and as female/calf pairs.

Aggregations of porpoises may form in good feeding areas, but these appear to be ephemeral. Most dolphins, on the other hand, tend to live in relatively large groups composed of individuals drawn from an even larger socially connected network of animals in an area. Group membership may change from day to day, and the groups may be segregated by age and sex, with all-male groups travelling apart from females and dependent young. Pelagic dolphins on the high seas often form giant schools of many thousands of individuals, but these are dynamic aggregations with little long-term structure. Killer Whales, the largest of the dolphins, can have some of the most stable societies known in mammals. Resident Killer Whales in British Columbia live in cohesive matrilineal kin groups composed of a matriarch and her living descendants over two to five generations (figure 10). Although different matrilines may travel together for varying periods of time, each is made up only of individuals that have been born into the group and remain affiliated for life. Sperm Whales have a complex society that is similar in many ways to that of African Elephants. Females live in clusters of matrilineal groups composed of females, their calves and subadult offspring, while adult males are largely solitary and spend much of their lives in higher latitude waters. These adult males periodically return to the range of the female clusters, roving from group to group to mate.

In contrast, the social organization of mysticetes is characterized by small unstable groups. On their feeding grounds, baleen whales are usually solitary but may periodically travel or forage together in temporary groupings of a few individuals. Associations are generally not persistent across years, and no long-term bonds between females and their offspring are evident post-weaning. Most mysticetes undertake long migrations to winter grounds where they engage in more frequent social interactions. For example, male Humpback Whales join together in "surface active groups" to compete for access to receptive females. These groups may involve up to a dozen or so males that chase a single estrous female, ramming, butting and thrashing at each other aggressively to determine dominance and access to the female.

Group living in cetaceans can have a variety of benefits. Cooperative foraging, for example, can increase the success rate of prey capture such that individuals do better in a group than they would by foraging on their own (see Feeding Ecology). Some dolphin schools cooperate by herding fish schools to the surface where they are more easily caught, and mammal-hunting Transient (or Bigg's) Killer Whales cooperate while chasing or corralling high-speed prey, or by taking turns keeping a Harbour Seal trapped underwater until it must head for

Figure 10. Resident Killer Whales have a stable matrilineal social structure. This photo shows the Northern Resident I27 matriline in Caamaño Sound: the matriarch I27 (second from left), born in 1974; her adult son I77 (to her right), born in 1997; her adult daughter I63 (to her left), born in 1990; and her youngest offspring, I107 (at the rear), born in 2004.
Photo: G. Ellis, July 2013.

the surface to breathe, at which point it is captured and shared. Prey sharing among close kin is also important in matrilines of salmon-eating Resident Killer Whales, even though prey is usually captured by individuals. Humpback Whales do not usually form stable associations, but whales may work together in tightly coordinated foraging groups to "bubble-net feed" on schooling fish. The same individuals may cooperate in such feeding groups from year to year, although there is no evidence that they are close relatives.

Predation pressure is an important factor that promotes group living in many terrestrial mammals, and this may also be the case with some cetaceans. Killer Whales are potential predators of all other marine mammals in the ocean, while sharks are also a potential threat to the smaller species and to calves of large whales. Marine mammals that live in groups can enhance their vigilance for predator detection, allowing some to rest while others remain alert. Individuals can also benefit from group defence or from the confusion experienced by a predator when presented with multiple potential prey. Dolphins have been observed to mob approaching sharks to drive them away, and Sperm Whales establish tight protective perimeters around calves when under attack by Killer Whales. The slower species of baleen whales, such as Humpback Whales and Right Whales, also group together to defend themselves from Killer Whales, but the fast-swimming *Balaenoptera* whales – the Blue, Fin, Sei, Bryde's and Minke whales – rely on high-speed flight for escape rather than on group defence.

Other important benefits of long-term, stable kin groupings include communal care of offspring and the long-term transmission of information. In some odontocete species, nursing continues long after the calf needs milk for nutrition. Sperm Whales have been observed still nursing offspring up to 13 years of age, and Short-finned Pilot Whales have been seen nursing young up to 16 years. In the latter species, old post-reproductive females have been found to continue lactating for up to 15 years after giving birth to their last calf. In these species, older siblings may "babysit" calves, allowing their mothers to forage. Long-term kin associations also allow the transmission of important information across generations through mimicry and learning such things as group-specific vocal dialects and specialized foraging behaviour. Young animals born into kin groups can therefore benefit from the experience and skills that have developed in the lineage over many years. Transmission of information by learning has led to behavioural traditions in species such as Killer Whales and Sperm Whales that arguably constitute a form of culture. For example, the distinct diets of different Killer Whale lineages in British Columbia waters have likely evolved as cultural traditions rather than being determined by genetic differences.

Compared to the complex associations and organization seen in odontocete societies, the behaviour patterns and social structure of pinnipeds appear relatively simple. Seals and sea lions do not form highly structured groups, and there is no evidence of any persistent associations of animals beyond the mother/pup bond during nursing. Rather, behaviour patterns are strongly associated with the annual pupping and mating cycle that brings pinnipeds together on traditional breeding rookeries or in aggregations on sea ice. While reproduction in pinnipeds does not generally involve complex communication and behavioural relationships, competitive males determine dominance through a wide variety of aggressive physical and vocal displays, which can escalate to violent fights in some species if disputes are not resolved through these displays. Other species use displays to attract rather than compete. Harbour Seals, which form territories and mate in the water instead of on land, perform a variety of visual and vocal displays to attract females and repel rival males.

During the non-breeding season, pinnipeds are mostly alone when at sea and are focused on finding and catching food. Many species in BC waters, such as California Sea Lions, Steller Sea Lions and Harbour Seals, regularly haul out on rocks or beaches to rest between bouts of foraging. They are loosely gregarious on haulouts, and densities of animals can be quite high. Most interactions on popular haulouts are related to conflicts over space. If no convenient haulouts are available, Steller and California sea lions will rest in tightly packed rafts at the surface (figure 4), with individuals holding a flipper in the air to reduce heat loss to the cold water, while Harbour Seals will rest on the bottom in shallow water. Some species, such as Northern Elephant Seals and Northern Fur Seals, spend the non-breeding portion of their annual cycle at sea, only hauling out on rookeries to breed and, in the case of Elephant Seals, for their annual moult.

Although relatively little modified from the more terrestrial members of the family Mustelidae, Sea Otters are less dependent on land than most pinnipeds (some pinnipeds haul out mainly or only on sea ice and therefore are essentially free of any dependence on land) and can function well without ever hauling out. They are not known to form complex or persistent social relationships. Males form aquatic territories during the breeding season, and females in estrus associate only with a territory-holding male. Mating behaviour is vigorous and often violent, with the male grasping the female's nose in his teeth while copulating. Females often have scarred noses from wounds incurred during mating.

Foraging occupies much of the daily routine of Sea Otters, and foraging behaviour can be individually specialized as a result of learning or mimicry, usually of the otter's mother. When not foraging, Sea Otters are gregarious and join together in rafts at the surface, where they spend much time grooming their fur and that of their pups as well as resting. Rafts are usually sexually segregated, with males rafting separately from females and their young. Certain locations are used repeatedly for rafting, and these are likely chosen based on qualities such as shelter from rough seas and proximity to good foraging grounds.

Selected References: Connor 2002, Ford and Reeves 2008, Wells et al. 1999.

MARINE MAMMALS IN BRITISH COLUMBIA

The coast of British Columbia has complex geography and oceanography, which together create a diversity of habitats for marine mammals. This habitat diversity and the high biological productivity of BC coastal waters are the main reasons it is home to such a rich assemblage of marine mammals – 25 per cent of all marine mammal species that currently exist in the world have been recorded in BC waters. In the following sections, I describe the ocean conditions and marine ecosystems that support this diversity, as well as the distribution of marine mammals with respect to 12 distinct marine ecological areas ("ecosections") defined by the BC Ministry of Environment. I also outline the historical and contemporary exploitation of marine mammals in BC, and their current management and conservation.

Oceanography and Marine Ecosystems

The west coast of continental Canada is formed by the Coast Mountain Range, a chain of peaks that rise to more than 2000 metres above sea level within 70 km of the ocean. This mountain range is riven by many narrow fjords and two major river systems, the Fraser and Skeena. Lying offshore from the continental coast is another chain of mountains that forms the islands of the Southeast Alaska panhandle, Haida Gwaii, Vancouver Island and Washington's Olympic Peninsula. Known as the Insular Range, this mountain chain is broken by

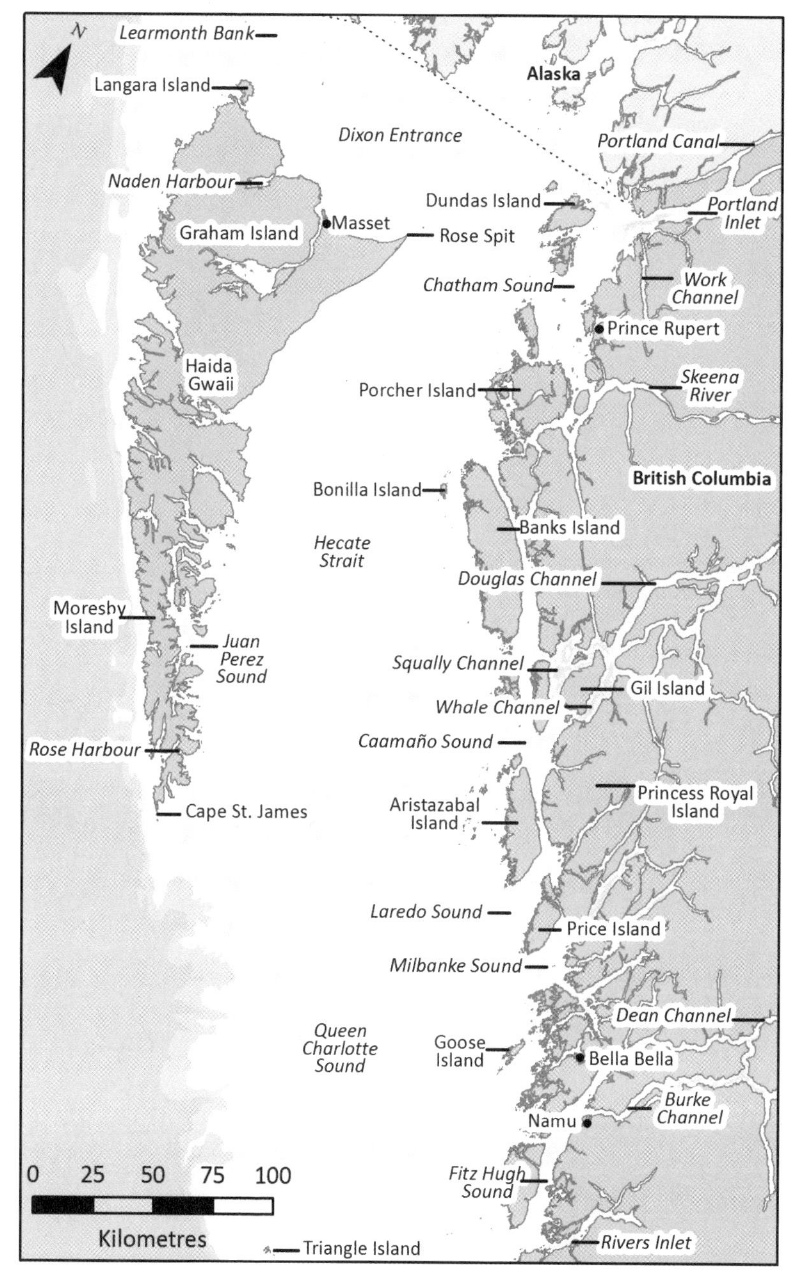

Figure 11. Major geographic features and place names for BC's north coast.

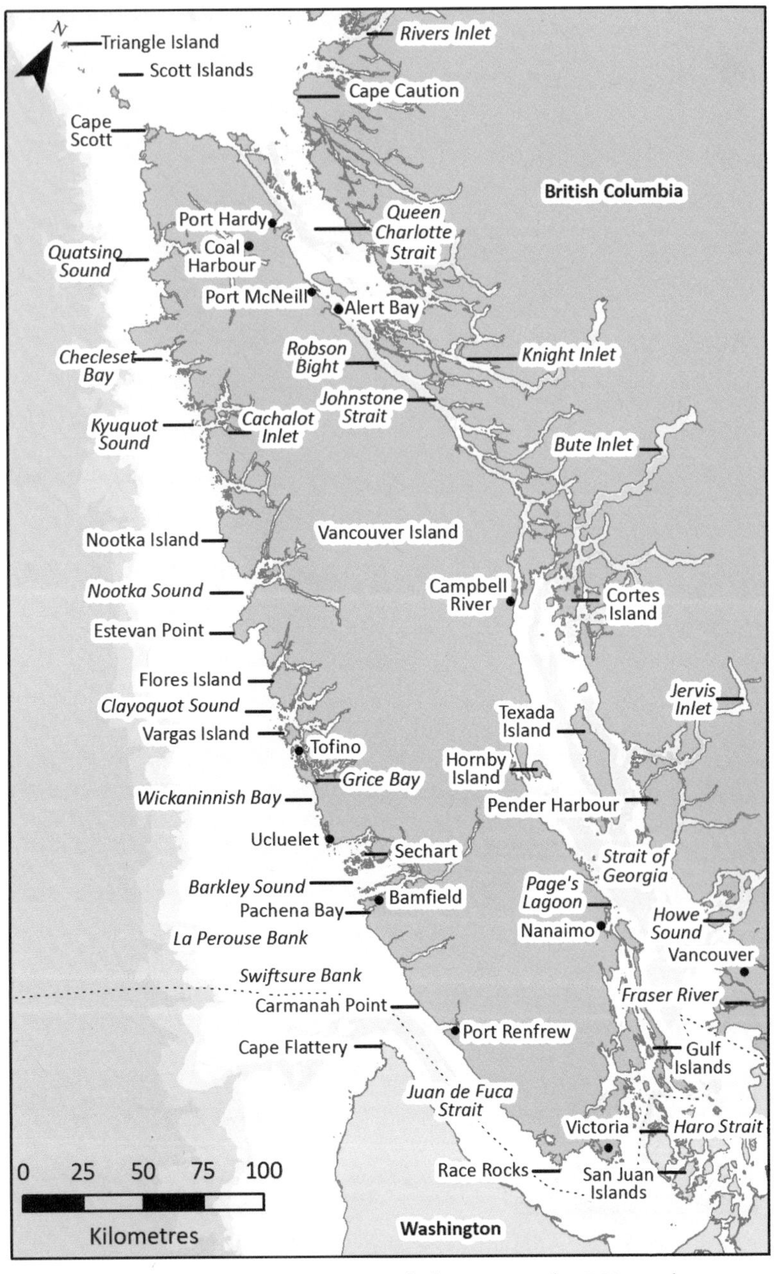

Figure 12. Major geographic features and place names for BC's south coast.

three major low-lying marine depressions: Dixon Entrance, between Southeast Alaska and northern Haida Gwaii; Queen Charlotte Sound, between southern Haida Gwaii and northern Vancouver Island; and Juan de Fuca Strait, between southern Vancouver Island and the Olympic Peninsula (figures 11 and 12). These three water bodies connect the oceanic waters of the eastern North Pacific to a labyrinth of inside passages, straits and channels formed by the roughly 40,000 islands that lie off the continental coast of BC. Although the straight-line distance between the southern tip of Vancouver Island and the Alaska border at the head of Portland Canal is only about 1000 km, the deep fjords and complicated island shorelines result in a total BC coastline length of over 25,000 km.

All of the islands off Canada's west coast lie on the continental shelf, with Vancouver Island and Haida Gwaii being closest to the shelf edge. The width of the shelf along the outer coast of these islands varies considerably. Off southern Vancouver Island, the shelf is about 80 km wide, but tapers toward the northern end of the island, where it is only about 20 km wide. The trend is reversed off the west coast of Haida Gwaii. The shelf is very narrow at the southern end – depths plummet to over 1000 metres within 5 km of shore – but it broadens to 30 km wide near the northern end of the islands.

The edge of the continental shelf is a very steep slope with depths that typically drop quickly from about 200 metres to 2000–3000 metres over a distance of 20–40 km, and then more gradually to about 4000 metres in the outer portion of Canada's Exclusive Economic Zone (EEZ), which extends 370 km (200 nautical miles) offshore. Approximately three-quarters of Canada's territorial waters off the Pacific coast (by area, not volume) is beyond the continental shelf. The continental slope has about 30 steep-walled canyons carved into it between Cape Flattery and Cape St James at the southern tip of Haida Gwaii. These were created by tongues of glaciers that extended over the shelf during the last ice age, or by fracturing along faults in the Earth's crust.

The oceanography of the west coast is influenced by the Subarctic Current, which originates off the coast of Asia and flows eastward across the North Pacific toward North America (figure 13). This broad current is very slow, typically only about 0.1–0.2 knots (about 2–4 metres per hour), and it takes two to five years for a body of water to complete the transit to the west coast of Canada. As the current approaches the BC coast from the west, it divides into two branches while still well offshore: a northern branch curves to the northeast to

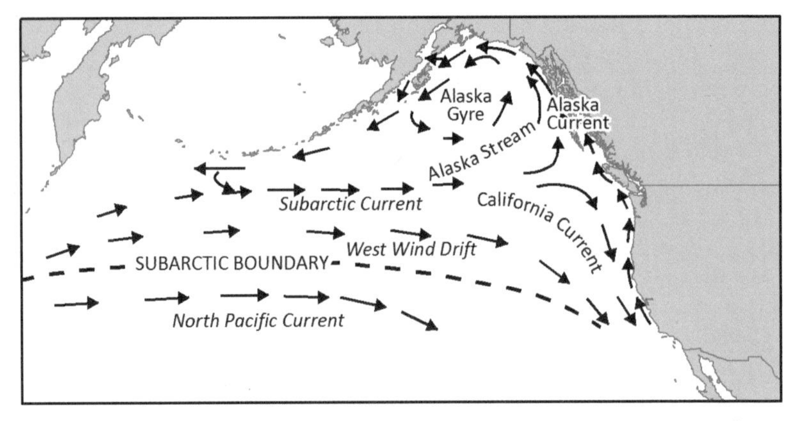

Figure 13. The prevailing surface currents in the central and eastern North Pacific Ocean. Illustration based on Thomson 1981.

form the Alaska Current, and a southern branch turns southeastward to form the California Current. The latitude at which these currents diverge is generally off Vancouver Island but varies with the seasons.

The BC Ministry of Environment has developed an ecoregion classification system for the province that defines distinct ecosystems geographically at five levels, from broad and general (ecodomains) to narrow and specific (ecosections). The marine environment off the British Columbia coast is divided into 12 ecosections, shown in figure 14. The distribution of marine mammals found regularly in BC waters, either currently or historically, is summarized according to their occurrence in each of these 12 ecosections in Table 1.

The two largest ecosections comprise the deep, oceanic waters from the 200 nautical mile EEZ boundary in the west to the edge of the continental slope, defined by the 2000-metre depth contour, in the east. This area is split roughly in half into the Subarctic Pacific ecosection to the north and the Transitional Pacific ecosection to the south. The Subarctic Pacific ecosection is dominated by the Alaska Current, which generally flows to the north throughout the year, while the Transitional Pacific ecosection is an area of variable currents that change seasonally as the Subarctic Current splits at different latitudes. In winter the Transitional Pacific area is dominated by the northward flowing Alaska Current and in summer by the southward flowing California Current (see figure 13). These two ocean ecosections are habitat for the deep-diving cetaceans such as Sperm Whales and beaked whales, as well as the pelagic species of baleen whales, particularly

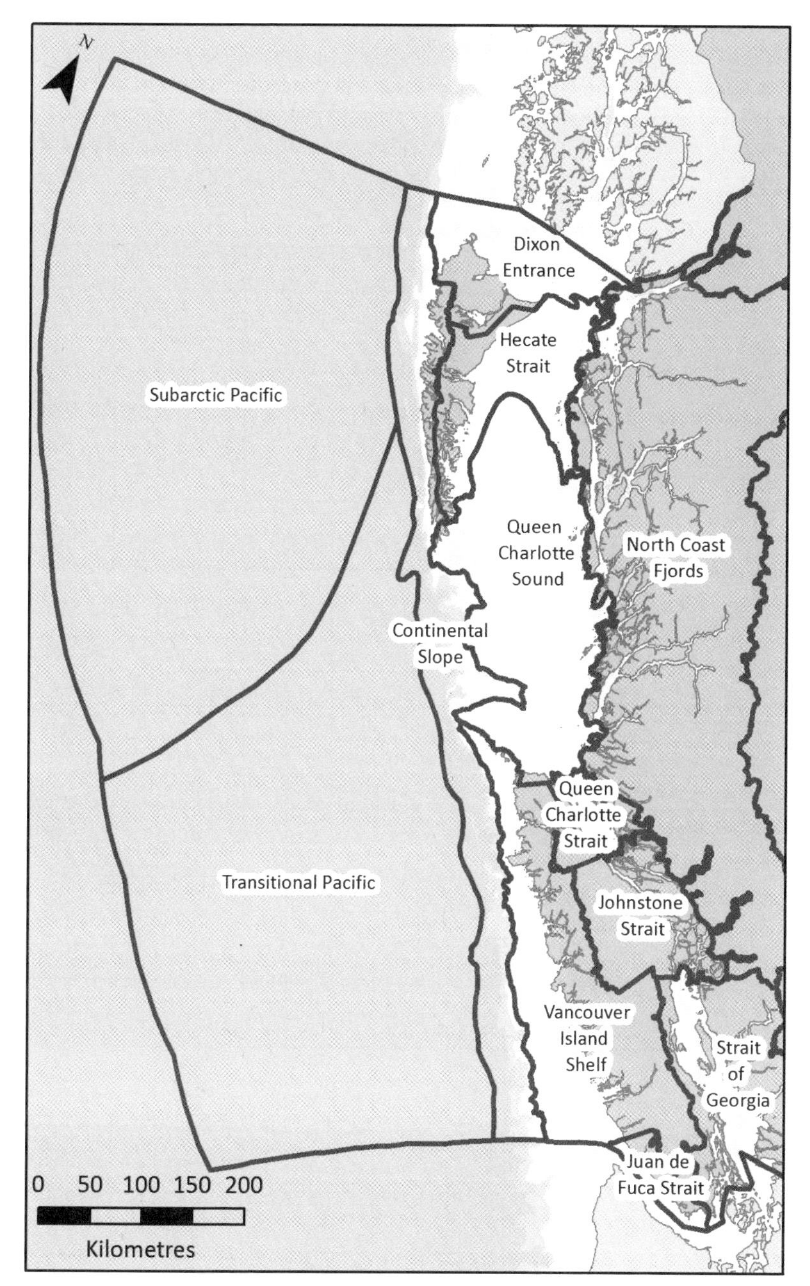

Figure 14. The 12 marine ecosections off BC's coast.

Table 1. Distribution of the 25 species of marine mammals that regularly occur (or did in the past) in British Columbia waters according to the 12 marine ecosections. Accidental or vagrant ("extralimital") species or occurrences are not included. P = regularly present, (P) = regularly present historically, (p) = minor or occasional presence, ? = possible or probable but unconfirmed presence in particular ecosection.

SPECIES / ECOSECTIONS	Subarctic Pacific	Transitional Pacific	Continental Slope	Vancouver Island Shelf	Juan de Fuca Strait	Strait of Georgia	Johnstone Strait	Queen Charlotte Strait	Queen Charlotte Sound	Hecate Strait	Dixon Entrance	North Coast Fjords
North Pacific Right Whale	(P)	(P)	(p)									
Grey Whale	P	P	P	P	P	(p)	(p)	(p)	P	P	P	(p)
Common Minke Whale	(p)	(p)	P	P	P	P	P	P	P	P	P	P
Sei Whale	P	P	(P)						(p)			
Blue Whale	P	P	P						(p)		(p)	
Fin Whale	P	P	P	P	(p)			(p)	P	P	P	
Humpback Whale	P	P	P	P	P	P	P	P	P	P	P	P
Sperm Whale	P	P	P						P			
Baird's Beaked Whale	P	P	P						P			
Hubbs' Beaked Whale	?	?	?						?			
Stejneger's Beaked Whale	?	?	?						?			
Cuvier's Beaked Whale	P	P	P	P					?	(p)		
Short-finned Pilot Whale		P	?									
Risso's Dolphin	(p)	P	P	P		(p)	(p)	(p)	(p)	P	(p)	
Pac. White-sided Dolphin	P	P	P	P	P	P	P	P	P	P	P	P
N. Right Whale Dolphin	P	P	P	(p)								
Killer Whale	P	P	P	P	P	P	P	P	P	P	P	P
Harbour Porpoise			(p)	P	P	P	P	P	P	P	P	P
Dall's Porpoise	P	P	P	P	P	P	P	P	P	P	P	P
Northern Fur Seal	P	P	P	P	P	(p)	(p)	(p)	P	P	P	P
Steller Sea Lion	P	P	P	P	P	P	P	P	P	P	P	P
California Sea Lion	(p)	(p)	P	P	P	P	P	(p)	(p)	(p)	(p)	(p)
Northern Elephant Seal	P	P	P	P	P	(p)	(p)	(p)	P	P	P	P
Harbour Seal			(p)	P	P	P	P	P	P	P	P	P
Sea Otter			(P)	P	(p)	(p)	(p)	P	P	P	(P)	P

Blue, Fin, Sei and North Pacific Right whales. Migratory Northern Fur Seals and Northern Elephant Seals also inhabit this oceanic habitat, as do Dall's Porpoises and Pacific White-sided Dolphins.

Moving east, the Continental Slope ecosection is a relatively narrow band bounded by the seaward 2000-metre depth contour and the 200-metre contour, which marks the edge of the continental shelf. The Continental Slope is an area with seasonal wind-driven upwelling of cold, nutrient-rich water that leads to high productivity and concentrations of zooplankton and forage fish that attract marine mammals. Canyons along the slope create localized eddies that also cause upwelling which promotes plankton production, and these eddies serve to further concentrate larger zooplankton. Deepwater species typical of the Subarctic Pacific and Transitional Pacific ecosections as well as species such as Steller Sea Lions and California Sea Lions can be found in the Continental Slope ecosection.

Off the west coast of Vancouver Island, from the Scott Islands to the entrance to Juan de Fuca Strait, lies the Vancouver Island Shelf ecosection. This is an area of strong tidal currents and upwelling, with upwelled water from several large canyons spilling over the shallow La Pérouse and Swiftsure banks off southwestern Vancouver Island. Here, a further influx of nutrients from estuarine water flows out of Juan de Fuca Strait. These banks are some of the most biologically productive areas along the BC coast, and high densities of marine mammals are typical there.

The waters over the inner continental shelf along the BC coast are divided into eight ecosections. Starting in the south, the Juan de Fuca Strait ecosection is a fairly deep, highly tidal trough that links outer coast waters with the Strait of Georgia ecosection, and it is an important migratory corridor for fish such as salmon and Pacific Herring. The Strait of Georgia ecosection is a large, relatively shallow inland sea between the southern half of Vancouver Island and the Lower Mainland. About three-quarters of BC's human population resides within 100 km or so of this ecosection. This is a biologically rich area supplied by nutrients from Juan de Fuca Strait in the south, Johnstone Strait in the north, and the Fraser River, which has a major influence on the ocean conditions of the region. The Strait of Georgia is an important migratory route for salmon headed to spawning grounds in the Fraser River system and also is an important spawning area for south BC coast Pacific Herring stocks. The Strait of Georgia supports densities of Harbour Seals that are about double those in other parts of the BC coast as well as large numbers of Steller and California sea lions in winter and spring. Harbour Porpoises are also found in the shallower areas of this ecosection. Portions of both the Strait of Georgia and Juan de Fuca ecosections are designated as critical habitat for salmon-eating Southern Resident Killer Whales under the Species at Risk Act.

North of the Strait of Georgia is the Johnstone Strait ecosection. Johnstone Strait separates the northern half of Vancouver Island from the mainland coast and adjacent clusters of islands. It is deep, steep-sided and narrow, typically only 2.5–4.5 km wide over its 100+ km length. It has vigorous tidal mixing of cold water that enters from several passes at its western terminus near the village of Alert Bay. Johnstone Strait is an important migratory route for salmon returning to the Fraser River and other rivers in the Strait of Georgia region. Most of Johnstone Strait is designated as critical habitat for salmon-feeding Northern Resident Killer Whales. The Robson Bight (Michael Bigg) Ecological Reserve protects the waters of Robson Bight and nearby rubbing beaches for use by Killer Whales.

West and north of Johnstone Strait, the inside passage widens to form the relatively shallow Queen Charlotte Strait ecosection, which further expands and deepens into the large Queen Charlotte Sound ecosection north of Vancouver Island. Queen Charlotte Sound is a large area open to the outer ocean between Vancouver Island and Haida Gwaii. It has complex bathymetry, with three broad troughs of 350–400 metre depths that run shoreward from the shelf break and various shallower banks between them. The largest is Moresby Trough, which extends from south of Cape St James diagonally about 270 km to the northeast, forming a deep underwater valley along the northeastern side of Queen Charlotte Sound. A combination of pelagic and nearshore marine mammal species can be found in Queen Charlotte Sound.

The Hecate Strait ecosection, which separates Haida Gwaii from the continental mainland and adjacent islands, is relatively shallow, especially over an extensive area in the northwestern part of the Strait, where depths are less than 50 metres. High densities of cetaceans are often seen in deeper waters along the east side of Hecate Strait, including Fin Whales, which usually prefer deep waters beyond the shelf break. North of Hecate Strait and extending to the west is the Dixon Entrance ecosection. This is a broad, relatively deep marine depression that separates Haida Gwaii from southeastern Alaska. At its western end, Dixon Entrance is split into two 400-metre deep channels by the shallow Learmonth Bank. There is much tidal activity in this area, and it is particularly rich in marine mammals, especially around Langara Island at the northwestern corner of Haida Gwaii. Both Hecate Strait and Dixon Entrance are the main corridors for northbound Grey

Whales migrating to their summer feeding grounds in the Bering Sea.

Finally, the North Coast Fjords ecosection comprises the inshore waters of BC's northern mainland coast, characterized by many long, narrow, fjord-like channels. Several of these fjords, including Douglas, Burke and Dean channels, extend more than 50 km into the Coast Mountains. The fjords are typically steep-sided with depths of 200 to 500 metres, and were scoured during past ice ages by glaciers that have since retreated (though glaciers are still present near the heads of Knight and Bute inlets). A variety of marine mammals can be found in these mainland inlets, including Humpback Whales, which are particularly common in summer and fall in Douglas Channel and adjacent channels on the north coast.

Variation in weather systems and prevailing wind patterns off the BC coast can cause marked changes in currents and upwellings that drive primary production (phytoplankton), and this can affect the distribution and abundance of zooplankton and forage fish both within and between years. This in turn can cause significant fluctuations in the distribution patterns of low trophic-level consumers, such as the baleen whales, within a feeding season and across years. Species such as Humpback Whales or Fin Whales may be common on one feeding area for several years in a row, then virtually absent the next. Other large-scale changes in ocean climate can also have a major effect on marine mammals. A periodic warm-water event known as El Niño, which occurs at irregular intervals of two to seven years, can cause a reduction in zooplankton and forage-fish production and survival. These effects are typically more severe off California than they are in BC waters. Warmer waters during El Niño may also result in a northward shift in the distribution of both marine mammals and prey species off the west coast, such as Pacific Mackerel and Humboldt Squid. A cool-water event known as La Niña, which often occurs between El Niños, can have the opposite effect, increasing ocean productivity and enhancing marine mammal survival. A longer-period cycle in ocean temperatures called the Pacific Decadal Oscillation (PDO) can also affect marine productivity and marine mammal distribution off the BC coast. The PDO is an alternating cycle of warm and cool ocean temperature conditions that occurs on a 20–30 year time scale, much longer than intervals between El Niños and La Niñas. When the PDO shifts from one temperature phase to another, it may trigger an ocean "regime shift", which is a broad change in the species composition and abundance of marine invertebrates and fish that persists for decades. Conditions during the cool phase of the PDO are favourable for zooplankton production, and there is some evidence that Blue Whales may occur more frequently off the BC coast during such periods of lower ocean temperatures.

Selected References: BC Marine Conservation Analysis 2011, Demarchi 2011, Gregr and Coyle 2009, Thomson 1981, Ware and McFarlane 1989, Zacharias et al. 1998.

History of Exploitation

Humans have exploited marine mammals in BC waters continuously for at least 10,000 years. First Nations people inhabiting coastal areas throughout the province hunted pinnipeds, small cetaceans and Sea Otters, and some groups, particularly the Nuu-chah-nulth of Vancouver Island's west coast were also proficient hunters of large whales. Marine mammals were extremely important to the economies of some coastal First Nations and less so to others. Hunting marine mammals required specialized skills, training and equipment that were based on traditions passed on across many generations. Elaborate rituals involving physical and spiritual preparation prior to and during hunts were typical among First Nations regardless of the species being pursued.

Along the east coast of Vancouver Island, the Kwakwa̱ka̱'wakw in the north and Coast Salish in the south had hunting traditions that targeted Harbour Seals, Harbour Porpoises and Steller Sea Lions, and they also pursued Dall's Porpoises and Pacific White-sided Dolphins. They hunted porpoises with harpoons thrown from specialized canoes propelled by paddles with pointed tips to reduce splashing noises that might scare their quarry. The harpoon was similar to that used to spear salmon, but larger and heavier. It consisted of a wooden shaft about four metres in length that had a double prong, each with a barbed harpoon point fashioned from elk or deer antler. The harpoon points were lashed to a line about 25 metres in length that was made from the twisted and braided intestines of Steller Sea Lions or American Black Bears and fixed to a float made from the stomach of a large seal. Harbour Seals and Steller Sea Lions were hunted using harpoons from canoes but also by quietly approaching them on their haulouts and clubbing them. Kwakwa̱ka̱'wakw hunters also used nets to catch seals.

Marine mammals were less important than terrestrial mammals as a food resource to Coast Salish, but the opposite was the case for First Nations along Vancouver Island's west coast. Archaeological studies have shown than the Nuu-chah-nulth made extensive use of pinnipeds – Harbour Seals, Steller Sea Lions and especially

Northern Fur Seals – as well as Sea Otters over the past 8000 years. They also hunted, although far less frequently, Northern Elephant Seals and California Sea Lions. To obtain Northern Fur Seals, hunters had to paddle far offshore, often out of sight of land, then quietly approach seals sleeping at the surface to harpoon them. They also used canoes and harpoons to hunt Harbour Seals and sea lions, but sometimes snuck up on them by foot on haulouts and clubbed them. First Peoples along the coast hunted Sea Otters using a bow and arrow from a canoe. They also hunted Harbour Porpoises from canoes, and in some traditional Nuu-chah-nulth reports, hunters attracted the porpoises within harpooning range by throwing handfuls of sand or pebbles over the water to simulate the sound of small schooling fish at the surface.

Nuu-chah-nulth people had highly specialized hunting traditions focused on Humpback and Grey whales. Using DNA analysis of bone remains at village sites around Barkley Sound, archaeologists have determined that Humpback Whales were the primary species targeted by whalers, with about 80 per cent of identified samples, and Grey Whales were a distant second in importance at 13 per cent. Humpback Whale bones were common in the excavations dating back 5000 years. Bones of Blue, Fin and North Pacific Right whales have also been identified at excavations of Barkley Sound village sites, but it is not clear whether these species were actively hunted or salvaged as beached or "drift" whales, or both. Ethnographic accounts suggest that North Pacific Right Whales were pursued but that Fin Whales and Blue Whales were too large and fast to hunt. There has been debate in the anthropological literature about the importance of whaling to the Nuu-chah-nulth subsistence economy. In the past some have argued that whaling was primarily a prestige activity and that successful whale hunts were relatively uncommon events. But the weight of evidence now strongly suggests that whales were a vital food source for the Nuu-chah-nulth, especially around Clayoquot Sound, Barkley Sound and Nootka Sound.

Hunting large whales was a very difficult and dangerous undertaking, and the Nuu-chah-nulth developed specialized tools and techniques for the task. Whalers pursued their prey in 10-metre-long canoes with a crew of eight: a harpooner at the bow, six paddlers, and a helmsman at the stern. The harpoon consisted of a heavy six-metre long shaft made of yew wood, upon which a harpoon head was lashed. The harpoon head was made with two elk-antler barbs bound together by sea lion sinew covered with cedar and cherry bark, each barb tipped with a sharpened mussel shell glued on with spruce gum. The

Figure 15. Humpback Whale on Echachis Island, Clayoquot Sound, killed by Nuu-chah-nulth whalers around 1907. This whale is believed to be the last killed using traditional techniques in the area.
Photo: C. Moser; BC Archives AA-00038.

harpoon head was attached to a strong 8-metre-long lanyard made of twisted sinews, then to about 200 metres of three-strand twisted cedar-bark line. Four or more floats made from inflated Harbour Seal skins were attached at intervals along the line. When the whalers got within striking distance of a whale, the harpooner thrust the harpoon into its back, at which point the shaft fell away leaving the head embedded in the whale. The line and floats were pulled out of the canoe as the whale dove to escape. When the wounded whale surfaced again, the whalers tried to embed additional harpoons in its back, with lines and floats, and to tie more floats to the original line to further slow and weaken the whale. They concluded a successful hunt by using lances with tips of sharpened elk antler or mussel shell to kill the whale. Once the whale was dead, a crew member jumped into the water to tie its mouth shut to keep it from sinking, then the crew towed it to shore. Nuu-chah-nulth traditional whaling ended early in the 20th century (figure 15).

The extent to which other First Nations along BC's coast may have hunted whales is not clear. Bones of whales were more numerous than Harbour Seal and Sea Otter bones at some old village sites in southern Haida Gwaii, and it is plausible that the Haida actively pursued whales as well as salvaging beached carcasses. Accounts by early European explorers suggest that the Haida and the Tsimshian on the northern mainland coast did occasionally hunt whales, although the tradition was likely not as well developed and important as it was in Nuu-chah-nulth culture.

The hunting of marine mammals by First Nations was probably sustainable and remained so for millennia, but this was hardly the case for the hunting undertaken by Europeans. The first explorers to arrive in this region – Juan Pérez in 1774 and James Cook in 1778 – acquired Sea Otter pelts through trade with the Haida and Nuu-chah-nulth. Cook's crew later sold their furs in Canton, China, at a profit of 1800 per cent, and on their return to Britain word quickly spread about a potentially lucrative trade in Sea Otter fur on the west coast of what would later become British Columbia. In 1785, the British ship *Sea Otter* arrived at Nootka Sound expressly to trade for Sea Otter pelts with local First Nations. They acquired 560 pelts during their brief visit and sold them in Canton at a profit of $20,000, a significant fortune in that period. This triggered a rush of entrepreneurial voyages, and by 1792 there were 21 ships trading for Sea Otter pelts on the BC coast. First Nations hunters, who had long hunted Sea Otters to obtain materials for their cloaks and regalia, met the greatly increased demand for pelts, but Sea Otter numbers quickly declined as a result. By the end of the 1700s, otters were scarce off Vancouver Island and the focus of the trade had shifted northward to Haida Gwaii and southeastern Alaska. By 1809 an estimated 55,000 Sea Otter pelts had been traded in BC and by 1850 the species was considered to be commercially extinct in the region. As the abundance of Sea Otters decreased, the value of their pelts increased substantially and they continued to be killed whenever found. The last record of a Sea Otter being killed in BC was in 1931, and despite a couple of sighting reports in the late 1930s and 1950s, the species was essentially extirpated in BC until being reintroduced from Alaska beginning in 1969.

The next marine mammal to become the focus of rampant exploitation by Europeans off the BC coast was the Northern Fur Seal. Although not as prized as Sea Otter furs, Fur Seal skins were highly valuable commodities and a large sealing industry developed, first on the seals' breeding rookeries in the Bering Sea in the late 1700s and early 1800s and then on their oceanic wintering grounds to the south. Pelagic sealing began in 1868 when trader James Christiansen took several Nuu-chah-nulth hunters and their canoes aboard the schooner *Surprise* and headed for the banks off the west coast of Vancouver Island. The hunters deployed in canoes and pursued the Fur Seals in their traditional manner using harpoons, returning to the mother ship at the end of each day. This proved to be highly effective, and a lucrative pelagic sealing industry developed. With increasing restrictions on shore sealing at the Pribilof Islands breeding rookeries off mainland Alaska, pelagic sealing expanded, with many ships joining the trade

Figure 16. Sealing schooners at anchor, Victoria harbour, 1902. Photo: BC Archives B-03355.

and the range of sealing operations gradually extending along the entire west coast of North America. By 1892 the pelagic sealing fleet had grown to 124 schooners, many of which were based out of Victoria, BC. The sealers targeted the high densities of Fur Seals wintering in coastal areas, then followed the migrating animals back to their rookeries in the Bering Sea and Sea of Okhotsk in Russia. In 1897 American citizens were prohibited from hunting seals in the Bering Sea, and many US schooners changed registry to Victoria, which became the sealing capital of the North Pacific (figure 16). The Victoria Sealing Company obtained and processed the products of an estimated 255,000 seals from 1886 to 1911, accounting for the vast majority of the pelagic take after 1894. A combination of uncontrolled sealing on the high seas and large kills on the breeding rookeries led to severe depletion of the Northern Fur Seal population and so the industry waned. The fleet shrank to 17 ships operating out of Victoria in 1906, then to only five in 1909. The North Pacific Fur Seal Convention of 1911 finally gave international legal protection to the dwindling population, and pelagic sealing came to an end.

Predator control programs and hunting had major impacts on the abundance of Harbour Seals and Steller Sea Lions in British Columbia from the late 1800s to 1968. Harbour Seals were hunted commercially for their pelts during two periods: 1879 to 1914 and 1962 to 1968, during which time 172,649 pelts were sold. Between these hunting periods a bounty was paid to encourage culling of the population to reduce competition with salmon fisheries, depredation of catches and damage to fishing nets. A total of 114,903 bounties (worth $1–5 each, depending on the period) were paid out, but likely far more seals were actually killed because there were high loss rates of animals shot in the water. In some years the bounty was insufficient to induce would-be bounty hunters to shoot seals due to the high cost of ammunition (especially at the end of World War I). In response, the Department of Fisheries sought other creative means of reducing Harbour Seal numbers. For example, in 1917 fisheries officials conducted an experiment to determine if seals could be efficiently exterminated in the mouth of the Fraser River by means of land mines. A sand bar in the North Arm of the Fraser that was routinely used as a haulout for 200–300 Harbour Seals was selected for the experiment. Fourteen mines, each containing more than six kilograms of dynamite overlaid with a heavy layer of broken chunks of iron, were wired together and buried in the sand bar. Once the seals returned to the haulout, the mines were detonated. In his description, the fisheries engineer in charge of the experiment noted that the seals lying immediately over the mines were "blown to atoms, not a piece larger than two inches square being found". Although effective, this novel culling technique was not pursued by the department, which instead raised the bounty paid per seal. By 1968, when hunting for pelts and culling ended, the Harbour Seal population in BC had been reduced to about 10 per cent of its historical size.

Steller Sea Lions were similarly depleted, mostly through a culling program undertaken by the Field Services Branch of the Department of Fisheries, again to reduce the impacts on fish stocks and fishing gear. Between 1912 and 1968 fisheries officers visited Steller Sea Lion breeding rookeries and shot animals from vessels or land, and often clubbed pups that were too young to escape. Culling took place at all rookeries in the province, but it was particularly intense at two locations off the central coast, Virgin Rocks and Pearl Rocks, which resulted in their abandonment as breeding sites (although Virgin Rocks has recently become a breeding rookery once again). The culling program was suspended during World War II, but during this period the Canadian navy and air force used rookeries and haulouts as bombing targets for training and may have killed large numbers of animals. There were also commercial ventures to market hides and sell sea lion meat as mink food, but these did not prove economically viable. By 1968, when the killing ended, the Steller Sea Lion population in BC had been reduced to between one-quarter and one-third of its historical size.

As pelagic sealing was winding down in British Columbia in the early 1900s, commercial whaling was entering a new era driven by demand for the oil that could be rendered from the whale carcasses. This oil was a valuable commodity used in the production of margarine, lard compounds, and soaps. Some whaling had taken place in BC during the 1800s, but this was relatively short lived. An intense period of pelagic whaling for North Pacific Right Whales by American and European whale ships occurred in the 1840s and 1850s, mostly centred well offshore to the north and west of BC – some Right Whales were taken in what are now Canadian waters west of Haida Gwaii. Several small fisheries for Humpback Whales developed in the late 1860s in the Strait of Georgia, based from a sailing schooner and shore stations in Whaling Station Bay on Hornby Island, Whaletown Bay on Cortes Island and Blubber Bay on Texada Island. These efforts ended around 1873 with fewer than 100 Humpbacks having been killed.

Then, in 1905, two former sealing captains, Captain Sprott Balcom and Captain William Grant, established the Pacific Whaling Company in Victoria, and incorporated new technologies into their operations. Steam-powered engines and winches on catcher boats together with new harpoons with exploding heads that could be fired from bow-mounted cannons allowed whalers to pursue the large and fast Blue and Fin whales. The Pacific Whaling Company opened its first whaling station at Sechart in Barkley Sound in 1905 and its second in Cachalot Inlet in Kyuquot Sound in 1907. It opened a third station that same year in Page's Lagoon (now Piper's Lagoon) near Nanaimo. Humpback Whales found in the Strait of Georgia during winter were the focus of the Page's Lagoon station, but evidently the supply of whales was insufficient and this station closed in 1910 after a catch of only 30 whales in the previous season.

Although catch records from the first few years at Sechart and Cachalot are incomplete, whaling was successful enough to prompt the Pacific Whaling Company to apply to the federal government for permission to build additional stations in northern BC. A station was built and opened in 1910 at Rose Harbour at the south end of Haida Gwaii, and another in 1911 at Naden Harbour, on the north side of Graham Island, Haida Gwaii. By 1911 the company had four whaling stations in operation with a total of 10 steam-powered catcher boats. Catches peaked in that year, with a total of 1624 whales: 1022 Humpbacks, 203 Blue Whales, 364 Fin Whales, 34 Sperm Whales and a single Grey Whale (see Appendix 3). After 1911 catches began to decline, precipitously so for Blue Whales, for which the annual catch exceeded 50 animals only once over the remainder of the whaling era in BC.

As whale stocks declined and markets for whale products fluctuated during the 1920s and 1930s, the whaling business became less profitable and, at times, quite tenuous. The Sechart station was closed permanently in 1918 and the Cachalot station in 1925. The two remaining stations each closed temporarily for varying periods in the 1930s then were finally shut down (Naden Harbour in 1941 and Rose Harbour in 1943). In total, 14,094 whales were reported landed at the BC shore stations between 1907 and 1943.

Following World War II, the newly formed Western Whaling Company opened a whaling station in 1948 at a former seaplane base in Coal Harbour on northern Vancouver Island. This was a far bigger operation than any of the earlier stations, with up to six catcher boats operating with modern equipment used to find whales, including aerial reconnaissance. The whaling season was longer and the range of operations much greater, almost 200 km offshore. The Coal Harbour station operated from 1948 to 1967, except for 1960 and 1961 when it was temporarily closed due to poor market conditions. The station closed permanently in 1967 after it had processed 10,353 whales killed in Canadian waters. About 90 per cent of this catch consisted of Fin, Sei and Sperm whales. Humpback and Blue whales had been so depleted by earlier whaling that they were rare and only comprised a small portion of the catch (figure 17). These two species were both finally given international legal protection in 1965.

Figure 17. A 23-metre female Blue Whale at the Coal Harbour whaling station, northern Vancouver Island, in the early 1960s. Photo: I. MacAskie.

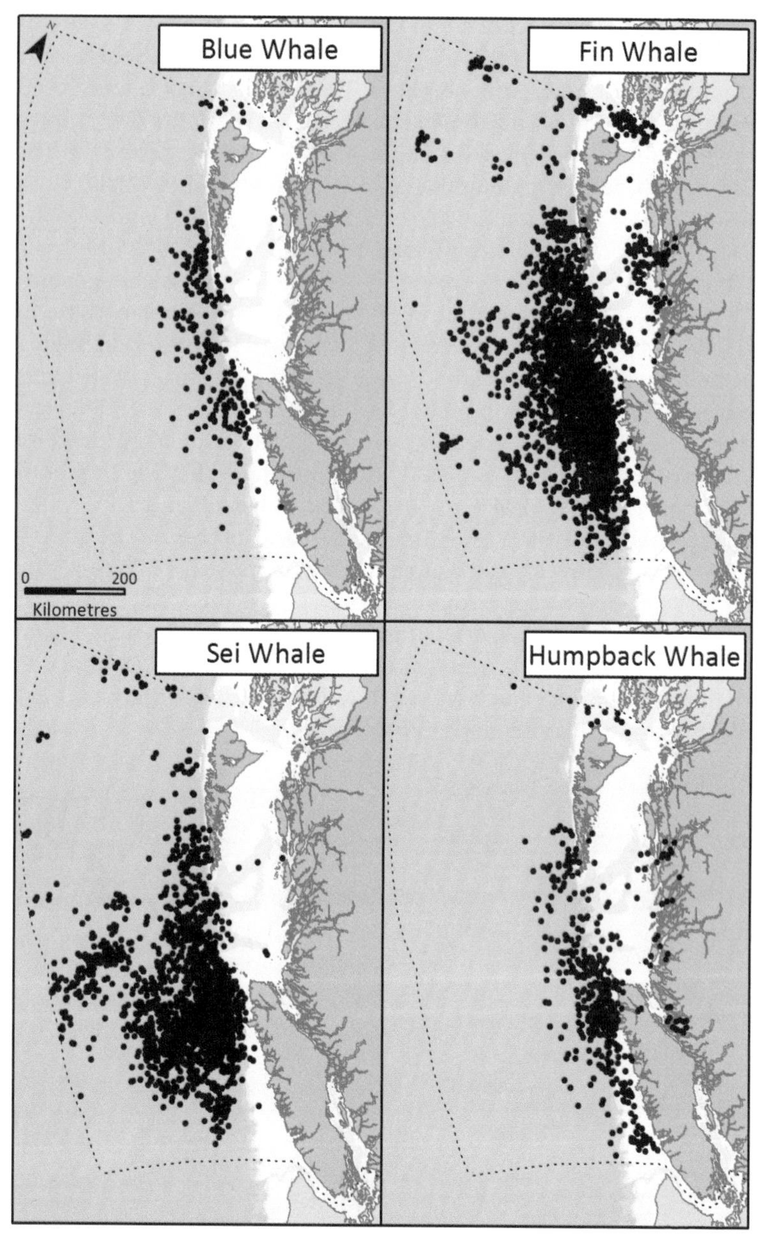

Figure 18. Locations of Blue, Fin, Sei and Humpback whales killed by shore-based and pelagic whaling operations in Canadian waters. Catches of Humpback Whales in the Strait of Georgia during the 1800s and in Barkley Sound in the early 1900s are not shown because locations were not recorded.

Although shore-based whaling operations in BC ended in 1967, whaling in Canadian waters did not. Japanese and Soviet whalers continued to hunt whales in the eastern North Pacific using factory ships with smaller catcher boats. Between 1964 and 1975, when this form of whaling ended in the region, at least 1013 whales were killed in offshore BC waters, mostly Sperm, Sei and Fin whales (see Appendix 3). This is a minimum tally, because it is now known that the Soviet Union falsified whaling records during most of this period and concealed or underreported its catches, which included protected species such as North Pacific Right Whales. All told, industrial whaling in BC waters during the 20th century killed at least 25,460 whales. Figures 18 and 19 show the locations of kills for the primary targeted species. Commercial whaling greatly depleted most populations of large whales, and only a few have made substantial recoveries since then.

Although whalers had no interest in Killer Whales due to their comparatively small size, a live-capture fishery developed during the 1960s in BC and adjacent waters of Washington to supply Killer

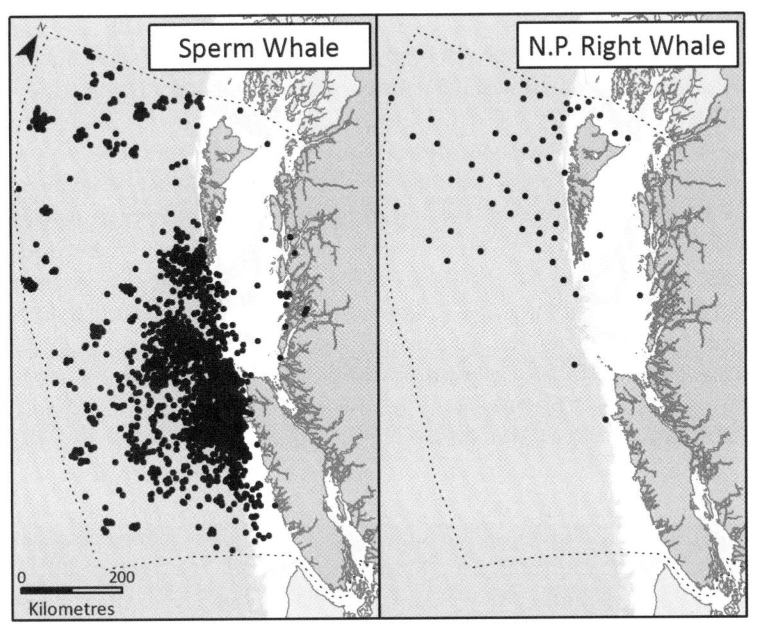

Figure 19. Locations of Sperm Whales and North Pacific Right Whales killed by shore-based and pelagic whaling operations in Canadian waters. The Right Whale map includes animals taken in the 1800s as reported by Townsend (1935). See Appendix 3 for a summary of whale-catch statistics.

Figure 20. A Killer Whale captured alive at Pender Harbour, March 1968. Photo: *Vancouver Sun*.

Whales to aquariums and marine parks for display or in one case, research. The first successful capture was of a young male known as "Moby Doll", which was harpooned at East Point, Saturna Island, in 1964. Although the intent was to collect a specimen for an accurate life-sized model at the Vancouver Aquarium, the harpoon wound was not lethal and the whale was held in a temporary enclosure in Vancouver harbour until it died three months later. The next year, an adult male Killer Whale accidentally netted near Namu on the central BC coast was purchased by the Seattle Marine Aquarium and towed there in a floating pen. "Namu", as this whale was named, was displayed in a net pen until it died a year later. The great attention and public interest attracted by these animals stimulated demand from aquariums and encouraged further captures. Fishermen made two captures in 1968 and 1969 by at Pender Harbour (figure 20), and sent 12 whales to aquariums in Vancouver, California and Europe. Sealand of the Pacific, an aquarium in Victoria, also took 8 Killer Whales in a series of captures at Pedder Bay in the early 1970s, and exported most of them. The live-capture fishery ended in 1977 due to strong public opposition; by this time, 62 whales had been removed from the waters of BC and Washington.

Today, direct exploitation of marine mammals in British Columbia is limited to a small take of Harbour Seals and Steller Sea Lions by First Nations for food and for social and ceremonial purposes. In the near future, small numbers of Sea Otters may be hunted off the west coast by the Nuu-chah-nulth for similar purposes.

Selected References: Acheson and Wigen 2002, Arima 1988, Arndt 2011, Bigg and Wolman 1975, Boas 1909, Drucker 1951, Fisheries Branch 1918, Gibson 1992, Goddard 1997, McKechnie and Wigen 2011, McMillan et al. 2008, Monks et al. 2001, Murie 1981, Nichol et al. 2002, Pike and MacAskie 1969, Suttles 1987, Webb 1988.

Conservation and Management

Globally, marine mammals have been exploited by humans for millennia, but it was only in the last 200 years or so that the magnitude of killing brought many species to the brink of extinction. In many cases, self-regulation through market feedback prevented outright extinction – as abundance declined, the animals became increasingly difficult and unprofitable to hunt and so hunting effort also declined. But this mechanism was inadequate to protect Sea Otters, whose pelts became more valuable as their numbers declined. Nor did it protect the stocks of large whales, because whaling was a multispecies endeavour. As the whale species of choice became scarce, whalers switched to more abundant species but kept taking the rare animals whenever they were encountered, further depleting them.

Actions to conserve marine mammals and manage human activities that affect them began in the early 20th century and continue to expand today. There are now a multitude of conservation efforts at many different levels, from international to national, regional and local, involving governments, non-governmental organizations, local communities and private citizens. Early conservation was focused on preventing extinction of marine mammal species, but today efforts extend to the preservation of genetic diversity by conserving regional populations and their habitat.

Although commercial exploitation of marine mammals in British Columbia dates back to the late 1700s, most regulations to manage hunting and conserve these species came far too late to avert serious depletion. Some of the first attempts to manage marine mammal exploitation in the province were initiated at the outset of modern industrial shore-based whaling. As early as 1904 the Canadian government imposed a 50-mile (80-km) minimum distance between shore stations on the west coast (later increased to 100 miles or 160 km) and a limit of one catcher boat per station (the latter regulation was soon abandoned). But rather than being motivated by concern for the conservation of the exploited populations, these regulations were intended to protect this new industry from its own demise due to overhunting. The measures were not very successful, however, as abundance of the animals began to decline soon after commercial whaling began.

Because the large whales are migratory animals, management of their exploitation required cooperation among many nations, and Canada was active in early efforts to develop international policies. International agreements and conventions on the regulation of whaling gave legal protection to Right Whales, including the North Pacific Right Whale, in 1931 (ratified in 1935) and to the Grey Whale in 1937. Unfortunately, both species had already been depleted well past the point of commercial extinction by the time this protection came into effect. The International Convention for the Regulation of Whaling of 1946 established the International Whaling Commission, which developed various regulatory measures such as restrictions on numbers, size and sex of catches for exploited species. These measures did little to slow the decline of many species, and in 1965 full protection was given to the Blue Whale and Humpback Whale due to their serious depletion on a global scale. Commercial whaling ended in BC with the closing of the Coal Harbour whaling station in 1967, and Canada withdrew from the International Whaling Commission in 1982.

The Northern Fur Seal is another migratory species that required international cooperation and agreement for protection. After being severely depleted by over-exploitation on breeding rookeries and by pelagic sealing on its winter grounds, the species was protected by the Fur Seal Treaty of 1911, which was signed by Great Britain (acting for Canada), the US, Japan and Russia. This was the first international convention for the conservation of a wildlife species, and it brought exploitation of Northern Fur Seals to an end in Canadian waters. The Fur Seal Treaty also included an article banning the killing of Sea Otters by signatory nations outside their three-mile (5-km) territorial waters. Of course, Sea Otters were virtually extinct in the North Pacific by then and, because they are a predominantly shallow-water, nearshore species, the treaty did little to protect remaining animals. The Sea Otter was finally protected in BC by the Game Act of 1931, the same year the last animal was killed and the species became extirpated in the province.

Viewed as nuisances and threats to fisheries in BC, Harbour Seals and Steller Sea Lions were the focus of many decades of efforts to reduce their populations rather than to conserve them. As a result, both species were at a small fraction of their historical abundance in the province when the hunting and culling ended in 1968. In 1970 protection was given to these species as well as other pinnipeds and the Sea Otter under Section 21 of the British Columbia Fishery (General) Regulations in the federal Fisheries Act. Prior to being amended in 1984, however, these regulations had a blanket exclusion for commercial fishermen, who were allowed to disturb or kill seals and sea lions to protect their gear and catch. In 1993 the regional regulations were superseded by the national Marine Mammal Regulations, Section 7 of which stipulates: "No person shall disturb a marine mammal except when fishing for marine mammals under the authority of these Regulations", although provisions allow limited takes by First Nations for food and for social or ceremonial purposes. Since these regulations were enacted, all Canadian waters off the Pacific coast have been closed to commercial exploitation of marine mammals.

The federal Species At Risk Act of 2002 (SARA) provides for the protection of Endangered and Threatened species and their habitat in Canada, and the conservation status of a species is determined by the Committee On the Status of Endangered Wildlife In Canada (COSEWIC). This non-governmental committee of experts assesses and designates the status of species in the following categories: Extirpated, Endangered, Threatened, Special Concern, Not at Risk or Data Deficient. Different taxonomic groups of animals and plants have species specialist subcommittees (SSCs), composed of experts in those particular species categories, that undertake assessments and propose species status to the full committee. Marine mammals have their own SSC to assess conservation status of species in Canadian waters. The conservation status assigned by COSEWIC is reviewed by Environment Canada or Fisheries and Oceans Canada (the latter for aquatic species), which then respond by listing the species – if Extirpated, Endangered, Threatened or Special Concern – under SARA or not (e.g., species may be referred back to COSEWIC for further assessment). Some species in Canada comprise two or more separate populations that are so distinct geographically, ecologically and/or genetically they are considered separate "species" for conservation ranking. These are referred to as "Designatable Units" (DUs). In BC, for example, the Killer Whale is divided into four DUs. Once a species or DU is listed as Endangered or Threatened, the federal government must develop a recovery strategy and identify and protect the species' critical habitat. Of the 31 marine mammal species known from BC waters, six have Endangered or Threatened status (Table 2). SARA makes it illegal to kill or harass a listed species or to destroy designated critical habitat

Table 2. Provincial and national conservation status of marine mammals in British Columbia as of July 2014. BC List: Red = extirpated, endangered or threatened; Blue = special concern, being sensitive or vulnerable to human activities or natural events; Yellow = secure, not at risk of extinction; Accidental = infrequent here or unpredictably outside its usual range. COSEWIC: Endangered = facing imminent extirpation or extinction; Threatened = likely to become endangered if limiting factors are not reversed; Special Concern = BC Blue List.

SPECIES or DESIGNATABLE UNIT	RANKING	
	BC List	**COSEWIC**
North Pacific Right Whale	Red	Endangered
Grey Whale	Blue	Special Concern
Common Minke Whale	Yellow	Not at Risk
Sei Whale	Red	Endangered
Blue Whale	Red	Endangered
Fin Whale	Red	Threatened
Humpback Whale	Blue	Special Concern
Sperm Whale	Blue	Not at Risk
Pygmy Sperm Whale	Accidental	Not at Risk
Dwarf Sperm Whale	Accidental	Data Deficient
Baird's Beaked Whale	Unknown	Not at Risk
Hubbs' Beaked Whale	Unknown	Not at Risk
Stejneger's Beaked Whale	Unknown	Not at Risk
Cuvier's Beaked Whale	Yellow	Not at Risk
Long-beaked Common Dolphin	None	None
Short-beaked Common Dolphin	Accidental	None
Short-finned Pilot Whale	Yellow	Not at Risk
Risso's Dolphin	Yellow	Not at Risk
Pacific White-sided Dolphin	Yellow	Not at Risk
Northern Right Whale Dolphin	Yellow	Not at Risk
Killer Whale – Southern Resident	Red	Endangered
Killer Whale – Northern Resident	Red	Threatened
Killer Whale – Transient (Bigg's)	Red	Threatened
Killer Whale – Offshore	Red	Threatened
False Killer Whale	Accidental	Not at Risk
Striped Dolphin	Yellow	Not at Risk
Harbour Porpoise	Blue	Special Concern
Dall's Porpoise	Yellow	Not at Risk
Northern Fur Seal	Red	Threatened
Steller Sea Lion	Blue	Special Concern
California Sea Lion	Yellow	Not at Risk
Northern Elephant Seal	Yellow	Not at Risk
Harbour Seal	Yellow	Not at Risk
Sea Otter	Blue	Special Concern

by diminishing its function, which for marine mammals in BC waters is most often for feeding.

Although the conservation and management of marine mammals is a federal responsibility in Canada, the BC Conservation Data Centre in Victoria actively maintains status listings for these species. The BC Red List includes species that COSEWIC considers Endangered or Threatened, the Blue List generally includes species that meet criteria for Special Concern, and the Yellow List is equivalent to COSEWIC's Not at Risk ranking.

Both the federal and provincial governments have various mechanisms to protect important and vulnerable marine mammal habitat in British Columbia. BC Parks has established several ecological reserves for marine mammals. These include the Robson Bight (Michael Bigg) Ecological Reserve, which was created in 1982 to protect sensitive habitat for Northern Resident Killer Whales in Johnstone Strait, and the Checleset Bay Ecological Reserve, created in 1981 to provide sufficient high-quality habitat for the reintroduced population of Sea Otters off the west coast of Vancouver Island. Three ecological reserves in the Scott Islands off northern Vancouver Island have also been established, mostly to protect seabird nesting colonies, but they also include key breeding rookeries of Steller Sea Lions. Other important rookeries off southern Haida Gwaii are in the Gwaii Haanas National Park Reserve and the Gwaii Haanas National Marine Conservation Area, the latter of which also protects important habitat for Humpback Whales and other marine mammals. Portions of the waters off northeastern and southern Vancouver Island have been formally designated as critical habitat for Northern and Southern Resident Killer Whales, respectively, under SARA. Work is also underway to identify and designate critical habitat for other Endangered and Threatened cetacean species off the Pacific coast.

Although commercial hunting of marine mammals in British Columbia is a thing of the past, commercialized viewing of these animals in the wild is a thriving industry in the province. Interestingly, what was perhaps the first whale-watching enterprise anywhere began in BC in the early 1900s, when Humpback Whales in Howe Sound became an attraction for tourists in Vancouver. They went on organized boat excursions to see the whales, but whaling from the station at Page's Lagoon in Nanaimo from 1907 to 1910 depleted the Howe Sound whale population, putting an end to this nascent whale-watching enterprise. Despite this early start, whale watching did not really get underway in BC until around 1980, growing quickly through the 1980s and 1990s. Today it is a major component of the nature-focused

Figure 21. Northern Resident Killer Whale A60 in Johnstone Strait showing fresh propeller wounds that eventually healed with no long-term effects. Photo: G. Ellis, August 2003.

tourism industry in British Columbia, with multiple whale-watching companies operating out of numerous coastal locations. As the size of the whale-watching fleet has grown, so too have concerns about the potential impacts that disturbance and underwater noise from the boats may have on the animals. Vessels moving in close proximity to whales also present a risk of propeller injury, although this is uncommon (figure 21). In an effort to mitigate the impacts of both commercial whale-watchers and private recreational boaters on whales, especially endangered and threatened Resident Killer Whales, Fisheries and Oceans Canada, working in partnership with whale-watching operators and conservation groups, has developed guidelines that limit how closely boats should approach whales and how fast they should be allowed to operate when around whales.

Other recent conservation and management initiatives include Fisheries and Oceans Canada's development of a Marine Mammal Response Program in the Pacific region. In collaboration with conservation groups and non-governmental organizations, Fisheries and Oceans supports a marine mammal incident response network through this program to track and respond to entanglements (figure 22), strandings (dead and live; figure 23), ship strikes, contaminated animals (oiled) and other threats. It also works with the BC Animal Health Centre, to conduct necropsies whenever possible on beach-cast marine mammal carcasses, especially cetaceans, to determine cause of death. The Vancouver Aquarium operates a Marine Mammal Rescue and Rehabilitation Centre to assist injured and distressed marine mammals throughout BC's coastal waters. The BC Cetacean

Figure 22. This Humpback Whale entangled in a gillnet in Portland Inlet (on BC's northern coast) was successfully freed. Entanglement in fishing gear is especially hazardous for nearshore species. Photo: J. Joslin, June 2003.

Figure 23. Examination of stranded cetaceans, such as this Humpback Whale on the shore of Clayoquot Sound, can yield important information about life history and causes of mortality. Photo: C. McMillan, March 2013.

Sightings Network is a program managed jointly by the Vancouver Aquarium and the Pacific Biological Station in Nanaimo to collect information on the occurrence of whales, dolphins and porpoises in BC waters and to promote marine mammal stewardship among the public.

The current conservation status of marine mammals in British Columbia waters is generally good. Harbour Seals and Steller Sea Lions have rebounded from years of culling and hunting and are now at or above historical levels of abundance. Sea Otters are thriving and steadily expanding their range over much of the BC coast. Of the whales that were depleted by commercial whaling, Humpback Whales and Grey Whales have shown strong recovery, and Fin Whales appear to be increasing as well. Sei Whales and North Pacific Right Whales are still extremely rare, but recent sightings of both species in BC waters give reason for optimism that they too are recovering. Further details on the status of these and the other marine mammals found in British Columbia are provided in the species accounts later in this book.

Selected References: Reeves 2009, Webb 1988.

Studying Marine Mammals

Marine mammals present particular challenges for researchers compared to most terrestrial mammals. Because they are found at sea, often far from shore, during all or much of their life cycle, marine mammals can be difficult to observe. They occupy a vast environment below the surface of the ocean and the range of many species is extensive. Deep-diving whales and seals spend most of their time below the surface, out of sight for surface-bound humans; some species of beaked whales have yet to be observed and identified alive. Pinnipeds regularly come ashore to breed or moult, so they are more accessible to study than the entirely aquatic cetaceans, but they too present difficulties for study when they leave their haulouts and dive into the sea to feed.

In recent decades great advances have been made in the scientific understanding of marine mammal biology. Although new technologies have helped bring this about, many discoveries are due to the increased number of researchers active in the field and the many novel, innovative approaches developed to gather information about these animals in their natural habitat. Fifty years ago most of what was known about marine mammals was from studies of dead specimens –

animals killed and collected intentionally for research, found stranded on shorelines or examined on the ramps of whaling stations and the flensing decks of whaling factory ships. These studies provided a great deal of important information about morphology, anatomy, life history and diet (from prey remains in stomach contents), but next to nothing about behaviour patterns, social interactions and communication, for example. Studies on some of the smaller marine mammals such as dolphins, seals and sea lions held in captivity have helped us understand features of their biology such as physiology, cognitive abilities, and sound production and reception. But knowledge of many important aspects of marine mammal biology, such as movement patterns, foraging ecology and social structure, has required field studies of the living animals in the wild.

Studying the long-range movement patterns of cetaceans has been particularly challenging for field biologists. Early attempts to investigate migrations of the great whales made use of "Discovery" tags, 25-cm long stainless-steel cylinders inscribed with an identification number and return address (figure 24) that were fired into the body of whales with a 12-gauge shotgun. When the whale was later killed and rendered in whaling operations, the tags were recovered and returned to the owner, together with details on capture date and location. Thousands of Discovery tags were deployed in the first half of the 20th century and provided the earliest data linking seasonal (winter-summer) migratory destinations of whales, both the large baleen whales and Sperm Whales. These tags were limited in their usefulness, however, as they gave no details about a whale's movements between the location where the animal was tagged and where it was eventually killed, sometimes years later, and the tags had a low recovery rate (<15%). Discovery tags became obsolete when industrial whaling ended, and have since been replaced by tags that use radiotelemetry to obtain information on whale movements. Once large and bulky with limited transmission range, new technologies have allowed the miniaturization of tags so that they can be attached to a whale with

Figure 24. During the whaling era, scientists used Discovery tags like this one from the Pacific Biological Station to study the movements of whales. Photo: J. Ford.

Figure 25. A small satellite-linked tag attached to the dorsal fin of a Fin Whale in Caamaño Sound will track its movements. Photo: R. Abernethy, August 2012.

minimally invasive barbs (by crossbow or air gun) and transmit to satellites overhead that relay positions to receivers on the ground (figure 25). Satellite tags can give considerable insight into both large-scale and local cetacean movements, but they are expensive and seldom stay attached for more than a few weeks (especially the minimally-invasive "limpet" tags attached with short barbs.).

A major breakthrough in field studies of cetaceans came in the early 1970s when researchers working on several whale species in different regions began to discover, at roughly the same time, that the individual animals they were studying had unique markings that could serve as natural tags. In BC, Fisheries and Oceans Canada whale scientist Michael Bigg and co-workers discovered that Killer Whales could be individually identified from nicks and scars visible in photographs of the dorsal fin and grey saddle patch at the base of the fin. Researcher Jim Darling, a graduate student at the time, found that Grey Whales off the west coast of Vancouver Island each had distinct coloration patterns and scars on their backs. Field researchers in other regions started identifying individuals using photographs of the pigment patterns on the undersides of the tail flukes of Humpback Whales, the callosity patterns on the heads of Right Whales, and nicks on the dorsal fins of Common Bottlenose Dolphins.

Photo-identification (commonly referred to as photo-ID) is now central to many field studies of cetaceans. Individuals of most cetacean species can be reliably identified with a good photograph of some part of their anatomy that is revealed when they come to the surface (figure 26). By compiling and comparing regional photo-ID catalogues, researchers have learned much about local movement patterns, fidelity of individuals to particular localities, and long-distance migrations across ocean basins. For example, Humpback Whale photo-IDs we collected in BC during the mid 2000s were used in a major international research effort known as SPLASH, which compared photo-IDs from throughout the North Pacific. It showed that the majority of Humpbacks north of Vancouver Island migrate to Hawaii for the

Figure 26. Examples of photographs used for the identification of individual whales in six species (clockwise from top left): Killer Whale, Blue Whale, Fin Whale, Humpback Whale, Grey Whale and Sperm Whale. Photos: J. Ford.

winter, whereas most off southwestern Vancouver Island migrate to Mexico. The photo-IDs have also revealed that individual Humpbacks tend to have preferred feeding areas in BC's coastal waters that they return to each year.

Photo-ID is a powerful field tool that can provide far more than just information on spatial distribution and movement patterns. Once individuals are photo-identified and assigned names, they can be followed in subsequent encounters across varying periods of time, from hour to hour or from year to year. In many cases, photo-identified animals can be tracked over their entire lifespan, yielding important information on life history such as survival and mortality rates, age at maturity, calving rates and longevity. By examining the association patterns of identified individuals – who travels with whom – researchers can describe the structure of cetacean societies. For example, annual photo-ID studies of Resident Killer Whales in British Columbia over four decades have revealed the intricacies of their social organization, from the matriline – an unusually stable social unit consisting of a mother and her living descendants – through to the community, which includes all associating matrilines in a population. Currently over 90 per cent of all Resident Killer Whales in BC have been photo-identified since they were newborn calves, and that proportion increases annually as older individuals die and new whales are born and are added to the photo-ID catalogue. Each year we learn more about the details of their population dynamics and social lives.

Some of the key indicators of a population's conservation status are overall abundance and trend – whether the number of animals in a population is increasing or declining. For cetaceans, photo-ID can play an important role in tracking these indicators. In rare cases, such as in the closed and relatively small populations of Resident Killer Whales, all individuals can be photo-identified to provide an accurate census. More often, however, it is only possible to photo-ID part of a population, so statistical approaches must be used to estimate the numbers missed and, hence, overall abundance. A variety of complex mathematical models are used to estimate abundance from photo-ID data, but all rely on a technique known as "mark-recapture". This technique is based on the idea of marking a number of animals in a population and then using the proportion of marked individuals "recaptured" in a subsequent sample of animals as an estimate of the marked proportion in the population at large. The resulting estimate can then be used to infer the total abundance. In photo-ID studies, the equivalent of marking an animal is photographing and naming it, and recapturing means to photo-identify it again at a later date.

Although mark-recapture is a good way of estimating the abundance of some cetaceans, other survey techniques can be more effective for particular species or circumstances. A few migratory species, such as Grey Whales, can be counted as they swim past fixed points on land, and these numbers can be corrected to account for various factors (e.g., periods of poor visibility) in order to arrive at an abundance estimate. Another technique used for certain species is line-transect sampling in shipboard or aerial surveys. This process involves estimating the density of the target species in strips of ocean sampled by surveying along a series of pre-determined transect lines and then extrapolating the density of animals observed in the sampled strips to the entire survey area. Also known as "Distance Sampling", this technique provides an estimate of the number of animals in a defined area at a particular time or over a period.

Another field technique that has provided a great amount of useful information about cetacean populations is biopsy sampling. This involves shooting a small dart (from an air gun or crossbow) that hits the animal's back or side and collects a small sample of skin and underlying blubber in its tip before bouncing off the animal and onto the water, where it floats until it can be retrieved by the researchers. By analysing DNA from biopsies, researchers can determine the genetic structures of cetacean populations and compare them across regions, which can help to define discrete populations within a species' total range that are the appropriate units for conservation and management. DNA fingerprinting can also identify individuals, and this can be used in making mark-recapture abundance estimates or in studying mating patterns through paternity tests. Some insight into foraging ecology can also be gained from biopsies. The ratio of stable isotopes (elements with different numbers of electrons) of carbon and nitrogen in a skin sample reflects the trophic position of the animal's prey (i.e., whether the whale has been feeding on low-level consumers like zooplankton or higher-level predators like salmon). Similarly, fatty acids in the blubber portion of the sample contain signatures of prey that have been consumed in the past, although it is generally difficult to identify the prey to species using this technique.

For a species-level understanding of diet, researchers often collect samples of prey or prey fragments near feeding cetaceans. For example, we tow nets to collect zooplankton near feeding baleen whales, or use dip nets to retrieve bits of tissue or fish scales from the surface water where toothed whales and dolphins have been consuming prey. Tissue can be identified to species using DNA analysis, and scales can be identified to species by examining them under a microscope. DNA

Figure 27. Prey remains from the stomachs of a stranded Transient (Bigg's) Killer Whale (left) and a stranded Resident Killer Whale (right). The Transient's stomach contained the teeth, vibrissae and trachea of pinnipeds, while the Resident's stomach held bones of Chinook Salmon, Lingcod and other smaller fish species. Photos: J. Ford (left) and G. Ellis.

from fecal samples collected near the surface is also used to identify prey species that have been recently consumed. In the past, stomach contents of whales killed in whaling operations were used to study feeding habits, but nowadays prey remains found in the stomachs of stranded cetaceans can provide much information on diet (figure 27).

A field technique that is being used with increasing frequency is passive acoustic monitoring of cetacean vocalizations. The calls of the large baleen whales tend to be loud and very low in pitch, which allows them to propagate for tens or even hundreds of kilometres under water and makes them particularly well suited to remote monitoring. Cetacean calls are distinctive for each species and, in many cases, differences in the structure or timing of calls can identify different populations of a species. In Killer Whales, calls encode the identity of the particular population or ecotype – Resident, Transient (Bigg's) and Offshore Killer Whales in BC waters can be easily distinguished acoustically. Distinctive dialects in the calls of Resident Killer Whales even allow different communities and pods to be identified by sound. Underwater acoustic monitoring can be accomplished using a single portable hydrophone (underwater microphone) suspended from a stationary boat or an array of hydrophones towed behind a ship. Fixed acoustic stations are useful for long-term monitoring to investigate the seasonal occurrence and relative abundance of cetaceans in particular locations.

Sounds can be recorded in all seasons, day and night, and in all weather conditions. Over the past decade the Cetacean Research

Program at the Pacific Biological Station has been deploying submersible autonomous recording instruments at various locations off the BC coast (figure 28). These make recording samples on a duty cycle (typically about 5 out of every 15 minutes) over a period of up to a year, then they are retrieved and the data uploaded for analysis. Cabled undersea observatories such as Ocean Network Canada's VENUS network in the Strait of Georgia and NEPTUNE network off the west coast of Vancouver Island have bottom-mounted hydrophones that archive recordings around the clock and can also be monitored in close to real time. All these types of acoustic monitoring are helping to provide a better picture of cetaceans' year-round use of BC coastal waters.

Figure 28. An autonomous acoustic recording instrument being retrieved from the ocean floor off Cape St James following a year-long deployment to document whale vocalizations off BC's coast. Photo: B. Koot, July 2010.

Many of the field techniques used to study cetaceans are also applied in studies of pinnipeds and Sea Otters. Photo-ID using natural markings is possible in some species, but it is typically used only for short-term research efforts in small study areas rather than in range-wide assessments of abundance and distribution. Because pinnipeds can be readily captured and temporarily restrained on their haulouts, researchers can apply marks or tags to them for recognition over the course of the study. Coloured and numbered plastic tags attached to flippers can be read at a moderate distance using binoculars; these tags may last for several years. Freeze-branding or hot-iron branding of pinniped pups is often used to create a permanent, bald, unpigmented mark that can be recognized at long distances. Although this technique has not been used in British Columbia, some Steller and California sea lions have been observed here with numbered brand marks that have been applied on US haulouts and rookeries (figure 29). Resightings of brands provide information on dispersal and movement patterns of sea lions from their natal rookeries.

Figure 29. A branded female Steller Sea Lion at a haulout on Norris Rocks, Denman Island. The code "101Y" indicates that this sea lion was born in 2002 at a rookery on St George Reef, northern California.
Photo: S. Majewski, March 2014.

Aerial surveys are the most common method of estimating the abundance of pinnipeds. Line-transect surveys can be used for some species, such as seals that haul out to pup or moult on expansive areas of pack ice, but those that haul out on rocks and small islets, like the pinnipeds found in BC, require a different approach. Harbour Seals are counted from photographs taken with a hand-held camera while the survey aircraft – generally a small float plane – flies from haulout to haulout, circling when necessary for full coverage. Researchers conduct these surveys during low tide when the most animals are on their haulouts. Corrections for the proportion of seals likely to be at sea are used to derive an abundance estimate. There are hundreds of Harbour Seal haulouts scattered along the BC coast, so conducting a coast-wide survey is a major effort.

Steller Sea Lions in BC are also surveyed photographically from aircraft, typically during the peak of the breeding season in late June or early July. Because a substantial portion of the population is at sea at any given time or spread among non-breeding haulouts, abundance estimates are often derived from the number of pups counted from photographs of the breeding rookeries. Using life table data for population demography (i.e., the proportions of males and females at various life stages, especially reproductive females, in the population), an overall abundance estimate can be obtained by extrapolation from the total pup count.

Information on pinniped diet is based mostly on identification of prey remains in scats collected on haulouts and rookeries. In the past, species were identified from inspection of hard parts found in the scats – typically vertebrae and other fish bones, especially the very hard and distinctive *otoliths*, or ear bones, as well as the beaks of squid and octopus. In recent years, DNA analysis has greatly improved not only the identification of prey species in scats, but also the proportion of those species in the diet.

Sea Otters in BC have been surveyed by helicopter and by small boat, but the latter is generally the favoured approach given their recent expansion over much of the coast and the high cost of helicopter time. Because Sea Otters are usually scattered among labyrinths of rocky reefs and kelp beds along the exposed coast, obtaining an accurate count over their entire range is a difficult and painstaking task often pursued in challenging conditions. Studies of Sea Otter diet are generally conducted through observations from shore lookouts using high-powered telescopes to identify prey items that the animals bring to the surface to consume.

Selected References: Boyd et al. 2010.

Taxonomy and Nomenclature

Taxonomy is the academic discipline of defining and naming groups of living organisms. The term is derived from the ancient Greek *taxis*, meaning arrangement, and *nomia*, meaning method. Having an objective, standardized classification and naming system for organisms based on the relationships among them is critical to the study of all aspects of biology, from evolution to ecology to behaviour. Various classification systems have existed over at least 2000 years. In the 4th-century BC the Greek philosopher Aristotle was the first to attempt a classification of living things, recognizing in the process, for example, the fundamental distinction between fish and the fish-like whales, dolphins and porpoises. He proposed a binomial (two-name) classification system, a precursor to the more comprehensive system developed by the 18th-century Swedish biologist Carl Linnaeus, upon which modern taxonomy is based. The Linnaean method of nomenclature involves a hierarchy of taxonomic categories which starts at the *kingdom* (e.g., animal, plant), each of which is then divided into *phyla, classes, orders, families, genera* (singular, *genus*) and *species*. The organism is given a binomial name, typically italicized, consisting of the genus first and species name second (the genus capitalized and the species not). When formally referring to a species, the person who first described and named the species is often given together with the year in which the description was published. For example, the formal scientific name of the Sperm Whale is *Physeter macrocephalus* Linnaeus 1758, meaning that this whale belongs to the genus *Physeter*, has the specific name *macrocephalus* and was named by C. Linnaeus in 1758. If subspecies are described, an additional name is appended to the species name (giving it a trinomial designation).

Like the taxonomy of many vertebrates, the classification of marine mammals is the subject of much scientific debate and, as a result, it is in a constant state of flux. Although many of the changes are relatively minor at the species or subspecies level, others at higher taxonomic levels are of greater significance. These changes have usually been brought about by new technologies such as DNA analysis that provide greater insight into phylogenetic ancestry (evolutionary relationships) than was possible from morphological studies alone. For example, although morphological data have long indicated that the cetaceans (whales, dolphins and porpoises) are related to the artiodactyls (even-toed hoofed mammals), recent molecular evidence from DNA analysis has confirmed that the whales are firmly embedded in the artiodactyl clade (evolutionary lineage). As a result, it has recently been proposed that the formerly recognized orders Cetacea and Artiodactyla be merged into a new order called Cetartiodactyla. This change would disrupt the established Linnaean classification such that the former order Cetacea and suborders Mysticeti (baleen whales) and Odontoceti (toothed whales, dolphins and porpoises) would be "unranked taxa" needing formal taxonomic resolution. For the purpose of this book, I use the order Cetacea and suborders Mysticeti and Odontoceti.

The scientific and common names for marine mammal species and subspecies used in this book (see Checklist of Species) are those currently accepted by the Committee on Taxonomy of the Society for Marine Mammalogy (see www.marinemammalscience.org). These in turn generally follow *Marine Mammals of the World: Systematics and Distribution* (Rice 1998), with adjustments reflecting recent taxonomic literature. There have been many changes in marine mammal nomenclature from those used in the original *Mammals of British Columbia*

(Cowan and Guiguet, 1965). Of the 19 cetacean species described in that book, 10 now have different species names. Taxonomy of pinnipeds found in BC has been less volatile, with only one change, and the species name for Sea Otter has remained unchanged. All but two of these revisions were incorporated in *The Mammals of British Columbia: A Taxonomic Catalogue* (Nagorsen 1990): *Kogia simus* is now *Kogia sima*, and *Balaena glacialis* is now *Eubalaena japonica*. Some of the primary common names of species described in this book differ from those used in these earlier references, which are instead given under "Other Common Names" in the species accounts.

There are now 31 marine mammal species known to occur in British Columbia waters – 25 cetacean, 5 pinniped, and the Sea Otter. The 6 cetaceans that have been added to the BC list since the 1965 publication of *The Mammals of British Columbia* are Hubbs' Beaked Whale, Dwarf Sperm Whale, Pygmy Sperm Whale, False Killer Whale, Long-beaked Common Dolphin and Northern Right Whale Dolphin. All but the Northern Right Whale Dolphin are rare, being recorded from only one or a few strandings or sightings. It is likely that additional marine mammal species will be found in BC waters in the coming years. For example, Bryde's Whale – a subtropical to warm-temperate baleen whale closely related to the Sei Whale – and the Common Bottlenose Dolphin have both been sighted on more than one occasion in Puget Sound, Washington, and may well have passed through Canadian waters while en route. Also, future revisions to the taxonomy of some species described in this book can be expected. As an example, three genetically and ecologically distinct lineages of Killer Whales inhabit BC waters – known as Resident Killer Whales, Transient Killer Whales (also called Bigg's Killer Whales) and Offshore Killer Whales. These are considered Designatable Units by COSEWIC and the Species at Risk Act, but they are currently all assigned to the same species. It is likely that at least Transient Killer Whales will be given status as a subspecies, and perhaps even as a separate species, in coming years.

Selected References: Cowan and Guiguet 1965, Nagorsen 1990, Rice 1998.

CHECKLIST OF SPECIES

This list contains the scientific and common names for the 31 species of marine mammals known from the waters of British Columbia. It presents the orders and families according to their generally accepted phylogenetic arrangement, and the genera and species alphabetically by their scientific names. The author of the species description follows the Latin species name; when the author's name is enclosed in parentheses, this means the species was originally described as belonging to a different genus. Classification, scientific names and common names are those accepted by the Committee on Taxonomy of the Society for Marine Mammalogy (see www.marinemammalscience.org). The only exception is that I have retained the order Cetacea in preference to the recently proposed order Cetartiodactyla (see page 78, for more details). Alternative common names are given in the species accounts.

Order Cetacea: Whales, dolphins and porpoises

Suborder Mysticeti: Baleen whales

Family Balaenidae: Right whales

Eubalaena japonica (Lacépède) — North Pacific Right Whale

Family Eschrichtiidae: Grey whale

Eschrichtius robustus (Lilljeborg) — Grey Whale

Family Balaenopteridae: Rorquals

Balaenoptera acutorostrata Lacépède	Common Minke Whale
Balaenoptera borealis Lesson	Sei Whale
Balaenoptera musculus (Linnaeus)	Blue Whale
Balaenoptera physalus (Linnaeus)	Fin Whale
Megaptera novaeangliae (Borowski)	Humpback Whale

Suborder Odontoceti: Toothed whales

Family Physeteridae: Sperm whale

Physeter macrocephalus Linnaeus	Sperm Whale

Family Kogiidae: Pygmy and dwarf sperm whales

Kogia breviceps (Blainville)	Pygmy Sperm Whale
Kogia sima (Owen)	Dwarf Sperm Whale

Family Ziphiidae: Beaked whales

Berardius bairdii Stejneger	Baird's Beaked Whale
Mesoplodon carlhubbsi Moore	Hubbs' Beaked Whale
Mesoplodon stejnegeri True	Stejneger's Beaked Whale
Ziphius cavirostris G. Cuvier	Cuvier's Beaked Whale

Family Delphinidae: Ocean dolphins

Delphinus capensis Gray	Long-beaked Common Dolphin
Delphinus delphis Linnaeus	Short-beaked Common Dolphin
Globicephala macrorhynchus Gray	Short-finned Pilot Whale
Grampus griseus (G. Cuvier)	Risso's Dolphin
Lagenorhynchus obliquidens Gill	Pacific White-sided Dolphin
Lissodelphis borealis (Peale)	Northern Right Whale Dolphin
Orcinus orca (Linnaeus)	Killer Whale
Pseudorca crassidens (Owen)	False Killer Whale
Stenella coeruleoalba (Meyen)	Striped Dolphin

Family Phocoenidae: Porpoises

Phocoena phocoena (Linnaeus)	Harbour Porpoise
Phocoenoides dalli (True)	Dall's Porpoise

Order Carnivora: Carnivores

Family Otariidae: Fur seals and sea lions

Callorhinus ursinus (Linnaeus)	Northern Fur Seal
Eumetopias jubatus (Schreber)	Steller Sea Lion
Zalophus californianus (Lesson)	California Sea Lion

Family Phocidae: True seals

Mirounga angustirostris (Gill) — Northern Elephant Seal
Phoca vitulina Linnaeus — Harbour Seal

Family Mustelidae: Mustelids (weasels and allies)

Enhydra lutris (Linnaeus) — Sea Otter

IDENTIFICATION KEYS

The identification keys in this section facilitate identification of the 31 species of marine mammals known to occur in British Columbia. Separate keys distinguish the two orders of marine mammals found here – Cetacea (whales, dolphins and porpoises) and Carnivora (seals, sea lions and the Sea Otter) – and identify the species in each order. Two types of keys are provided for each order: one for whole animals and the other for skulls. The keys are intended primarily for identifying dead animals found on the beach or skulls discovered on the shore or prepared from carcasses. Although the whole animal key may also help identify species from sightings, the Identification section in the species accounts should be consulted for details on distinguishing features of morphology and behaviour that can be used to identify live animals at sea or on shore (in the case of seals, sea lions and sea otters). For identification of marine mammal species outside BC waters, I recommend referring to the excellent keys in *Marine Mammals of the World* by Jefferson, Webber and Pitman (2008).

All keys are dichotomous, with diagnostic features arranged into paired statements, or couplets, that offer two mutually exclusive choices. To identify a marine mammal from the whole animal or skull, start with the first couplet in the appropriate key and select the statement (1a or 1b) that best describes the specimen at hand. The selection statement will direct you to the next applicable couplet. Repeat this process until you make the final species identification, then go the account for the species in question and confirm the initial identification with the additional information provided in the species description and identification sections. The keys are written to be as clear as possible. They use discrete features such as presence/absence or absolute differences in size instead of variable or imprecise features. It is important when working through these keys to use multiple features whenever possible, rather than relying on a single characteristic. It is quite acceptable to jump in at any point in the key rather than start at the beginning. For example, if you know it is a toothed whale, find the appropriate couplet in the cetacean key and start there. Definitions of technical terms are in the Glossary.

When examining stranded carcasses of cetaceans it is important to keep in mind that skin colorations change soon after death, often darkening to the point that some of the more subtle colour patterns are obscured; in such cases, it is better to use other features for identification. The beaked whales of the genus *Mesoplodon* are poorly known, and identifying them to species from beach-cast carcasses is almost impossible, particularly for females and juveniles; in these cases, close examination and measurements of skull morphology are needed for a positive identification, and taking a skin sample for DNA analysis is highly recommended.

Keys to Orders

Whole Animals

1a. Two limbs on the front half of the body; the posterior end of the body modified into horizontal flukes; a fin usually present on the back; no fur
.... Order Cetacea (whales, dolphins and porpoises)

1b. Four limbs, two forward and two at rear; the posterior limbs modified into flippers; no dorsal fin; fur (hair) present
.... Order Carnivora (seals, sea lions and sea otters)

Skulls

1a. Skull strongly telescoped (elongated), with bony nares at the apex of the skull; elongated rostrum; auditory bullae situated ventrally and not fused to the cranium; teeth absent (baleen whales), greatly reduced in number (beaked whales) or numerous and homodont (most other toothed whales)
.... Order Cetacea (whales, dolphins and porpoises)

1b. Rostrum not elongated, with bony nares at the anterior of the skull; heterodont dentition with large, conical canines as well as incisors, premolars and molars; auditory bullae fully ossified and fused to the cranium
.... Order Carnivora (seals, sea lions and sea otters)

Keys to Whole Animals

Cetaceans

1a. Baleen plates suspended from upper jaw; no teeth present; double blowhole .... (baleen whale) 2
1b. Teeth present (though sometimes not protruding from the gums); no baleen; single blowhole .... (toothed whale) 8

2a. Long ventral pleats absent (though two to five short creases or furrows may be found on throat); dorsal fin present or absent; upper jaw arched when viewed from the side 3
2b. Long ventral pleats present; dorsal fin present; upper jaw nearly straight when viewed from the side........................ (rorqual) 4

3a. No creases or grooves on the chin or throat; no dorsal fin or hump; upper jaw and mouthline strongly arched when viewed from the side and very narrow when viewed from the top; long (≤3 m), narrow black baleen plates with fine black fringes, 220–270 per side............North Pacific Right Whale, page 103
3b. Two to five short grooves on the throat; no dorsal fin, but a small dorsal hump followed by 6–12 bumps; mouthline slightly arched; body mottled grey and usually patched with yellowish to orange whale lice and grey to white barnacles on the skin; short (<40 cm), white to pale-yellow baleen plates with coarse bristles, 130–180 per side .... Grey Whale, page 115

4a. Ventral pleats end before the navel 5
4b. Ventral pleats extend to or beyond (posterior to) the navel ... 6

5a. Ventral pleats number 50–70, the longest often ending between the flippers; white to yellowish baleen, 230–275 plates per side, <20 cm long with coarse bristles; flipper has a large white patch on the dorsal surface .... Common Minke Whale, page 130
5b. Ventral pleats number 40–65, with the longest ending behind the flippers though well before the navel; baleen <80 cm long, greyish-black with 35–60 fine bristles per cm (some anterior plates may be partly white), 219–402 plates per side .... Sei Whale, page 140

6a. Flippers 25–30% of the body length, with knobs on the leading edge; flukes with irregular trailing edge; deep, broad ventral pleats, totalling <40, with the longest extending at least to the navel; low, thick dorsal fin; numerous knobs on top of the head, one prominent cluster of knobs at the tip of the lower jaw; baleen short, <85 cm long, black to olive brown with 10–35 greyish-white bristles per cm, 270–400 plates per side .... Humpback Whale, page 171

6b. Flippers <20% of the body length, lacking knobs; 40–100 fine (narrow) ventral pleats; head lacking knobs 7

7a. Head broad at the corners of the mouthline but with a sharply pointed rostrum; dorsal fin ≤60 cm high, curved slightly to moderately, and located slightly more than 30% forward from tail to rostrum; colour of lower jaw asymmetrical: right side white, left side dark grey; 20–30% of baleen on the right front ivory to yellowish-white, the remainder dark grey to black streaked with yellowish-white; plates have 10–15 grey or white bristles per cm, 260–480 plates per side .... Fin Whale, page 160

7b. Head broad and U-shaped, as viewed from the top; dorsal fin <35 cm tall, usually triangular, and set very far back toward the tail; head colour symmetrical; body mottled grey with white under flippers; baleen all black with 10–30 black bristles per cm, plates extremely broad relative to length, 260–400 plates per side.. Blue Whale, page 149

8a. Head extending well beyond the tip of the lower jaw; concavity of the blowhole facing to the side or slightly backward; teeth normally present only in the lower jaw, which is much narrower than the upper jaw .... (sperm whales) 9

8b. Lower jaw extending at least as far as the tip of the rostrum; concavity of the blowhole facing forward 11

9a. Body blackish-brown to charcoal-grey, with white lips and inside of mouth; rostrum blunt, head large, 20–30% of the body length, almost rectangular when viewed from the side; S-shaped blowhole at the front left side of the head; dorsal fin low, shaped as a hump or series of humps on the posterior half of the body; 18–26 large conical teeth on each side of the lower jaw, fitting into sockets in the upper jaw; body length 4–18 m .... Sperm Whale, page 187

9b. Head <20% of the body length; blowhole set back from the front of the head; prominent dorsal fin; 8-16 thin, sharply-pointed teeth in each side of the lower jaw, fitting into upper jaw sockets; body length <4 m 10

10a. No creases on the throat; dorsal fin small (<5% of body length) and located well posterior to the midpoint of the body; 12–16 teeth (rarely 10–11) in each side of the lower jaw; maximum body length 3.8 m .... Pygmy Sperm Whale, page 200

10b. Inconspicuous creases on the throat; dorsal fin tall (>5% of the body length) and slender, located near the middle of the back; 7–12 (rarely up to 13) teeth in each side of the lower jaw, rarely 1–3 teeth in each side of the upper jaw; maximum body length 2.7 m.................................... Dwarf Sperm Whale, page 206

11a. Two conspicuous creases on the throat, converging anteriorly to form a forward-pointing V; notch between the tail flukes usually absent or indistinct; dorsal fin short and set far back on the body .... (beaked whales) 12

11b. No conspicuous creases on the throat; prominent median notch between the tail flukes....... (dolphins and porpoises) 15

12a. Rostrum straight, long and slender, with a transverse width at its mid-length ≤20% of the total length; one pair of teeth in the lower jaw, erupted or hidden in the gum, far posterior to the tip of the lower jaw, usually behind the tip of the rostrum with the mouth closed; teeth laterally compressed, their anterior-posterior axis ≥2 times the transverse axis
.... (*Mesoplodon* beaked whales) 13

12b. Rostrum either indistinct or distinct with steep forehead; 1–2 pairs of teeth in the lower jaw (erupted or hidden in the gum), usually at the tip and in front of the rostrum when the mouth is closed; teeth nearly round in cross section, if flat, then the anterior-posterior axis ≤1.5 times the transverse axis.......... 14

13a. Moderate-sized rostrum not sharply demarcated from the forehead; males have a white rostrum and "cap" or "toque" in front of the blowhole; adult males have a pair of large, flat teeth in the middle of the lower jaw, protruding above the upper jaw when the mouth is closed (females and subadults require museum preparation for identification)
.... Hubbs' Beaked Whale, page 220

13b. Males have white areas on the head but lack a white "cap"; a dark band ("mask") often present over the eyes and top of the head; forehead flatter and the mouth line has a stronger arch (females and subadults require museum preparation for identification) .... Stejneger's Beaked Whale, page 227

14a. Distinct long rostrum; rounded forehead rises from the rostrum at a steep angle; concavity of crescent-shaped blowhole faces posterior; two pairs of teeth in the lower jaw, anterior teeth exposed outside the closed mouth and larger than the posterior teeth; colour blackish dorsally and whitish ventrally; body length ≤13 m (females slightly larger) ....Baird's Beaked Whale, page 212

14b. Short, indistinct rostrum; head small relative to body size; forehead slightly concave in front of the blowhole; one tooth in the tip of each side of the lower jaw, exposed only in males; coloration highly variable, though the head is often whitish; body length ≤9.8 m ....Cuvier's Beaked Whale, page 234

15a. Rostrum extremely short; teeth have expanded crowns, laterally compressed and relatively small ....(porpoises) 16

15b. Rostrum blunt or prominent; teeth conical and sharply pointed (unless heavily worn), circular or oval in cross-section ....(dolphins) 17

16a. Blunt rostrum with short thick body; dark grey above, light grey to white below; dark stripe between the mouth gape and flipper; short, triangular, wide-based dorsal fin; 19–28 pairs of teeth in each jaw, ≥ 1 mm in diameter at the gumline; body length ≤2 m .... Harbour Porpoise, page 329

16b. Body robust, with small head and appendages; distinct humped ridge runs from the flukes up the middle of the back; black with a striking large, white patch on the sides and belly; dorsal fin triangular with recurved tip; white or light grey trim on the dorsal fin and flukes; 23–28 pairs of slender (≤1 mm in diameter at gumline), chisel-shaped teeth in each jaw; body length ≤2.4 m................................ Dall's Porpoise, page 338

17a. Dorsal fin present... 18

17b. Dorsal fin absent; body gracile and black with sharply contrasting white chin, breast and line along belly; 37–56 pairs of slender pointed teeth in each jaw; body length ≤3.1 m .... Northern Right Whale Dolphin, page 279

18a. Head blunt with no prominent rostrum.............................. 19
18b. Head with prominent rostrum 22

19a. Forehead blunt with a vertical crease; 2–7 pairs of teeth in the anterior part of the lower jaw only (rarely 1–2 pairs in upper jaw), but the teeth may be absent or severely worn; body grey to white, with numerous scratches on adults; flippers long and sickle-shaped; body length ≤3.8 m
.... Risso's Dolphin, page 262
19b. Forehead without a vertical crease; ≥7 pairs of teeth in both upper and lower jaws 20

20a. Flippers large and paddle-shaped with rounded tips; dorsal fin tall and erect (≤0.9 metres in females, 1.8 metres in males); striking black and white coloration with ovate white post-ocular patches, white lower jaw, white ventrolateral field, and light grey saddle patch behind dorsal fin; 10–14 pairs of large (≤2.5 cm diameter) oval teeth in each jaw; body length ≤9 m
.... Killer Whale, page 287
20b. Flippers long and slender with pointed tips; body mostly black
.... 21

21a. Dorsal fin low and thick, situated at anterior third of back; dorsal fin ≥2 times as long as its height; flipper length 14–19% of the body length; 7–9 pairs of relatively slender teeth (≤1.3 cm in diameter) in the anterior portion of each jaw; mouth directed downward at an angle of 30–45° to the horizontal body axis; body length ≤7.2 m
.... Short-finned Pilot Whale, page 255
21b. Dorsal fin thin, situated midway along the back; fin <2 times as long as its height; flipper length ≤14% of the body length; 7–12 pairs of large teeth extending along the length of both jaws; distinct hump on the leading edge of flippers; body length ≤6 m... False Killer Whale, page 314

22a. Rostrum long, >5% of the body length; distance from the rostrum tip to the eye's centre <2.6 times the rostrum length
.... 23
22b. Rostrum moderate to short, <5% of the body length; distance from the rostrum tip to eye's centre >2.7 times the rostrum length; 23–36 pairs of teeth in each jaw; dorsal fin bicoloured and falcate (sometimes extremely so); back dark grey with light

"suspender stripes" from the forehead to the tail stock, white belly, light grey flank patches (black lines separate the belly from the sides); body length ≤2.5 m
.... Pacific White-sided Dolphin, page 269

23a. Black to dark grey on the back, white on the belly, prominent black stripes from the eyes to the anus and the eyes to the flippers; grooves in the palate shallow if present; 40–55 pairs of teeth in each jaw; body length ≤2.5 m
....Striped Dolphin, page 323

23b. Back dark and belly white; tan to buff thoracic patch and light grey streaked caudal peduncle form an hourglass pattern that crosses below the dorsal fin; cape forms a distinctive "V" below the dorsal fin; chin to flipper stripe; 40–67 pairs of teeth in each jaw; palate has two deep longitudinal grooves; body length ≤2.5 m... (common dolphins) 24

24a. Body relatively stocky; beak moderately short; flipper stripe narrow and not approaching the gape; flippers and dorsal fin often have light patches; anus stripe faint or absent
.... Short-beaked Common Dolphin, page 248

24b. Body relatively slender; beak long and slender; flipper stripe wide and often contacting the gape; light patches on flippers and dorsal fin generally absent; anus stripe may be distinct
............................Long-beaked Common Dolphin, page 241

Carnivores

1a. Foot-pads present (as on dog's foot); forelimbs not modified into flippers; tail extending beyond outstretched hind foot; fur brown to blackish, sometimes white around the muzzle; dense underfur ....Sea Otter, page 397

1b. All feet modified into flippers; tail very short or absent; fur grey, tan, brown, black, yellow, white, or combination of these; underfur absent, except in fur seals .... (seals and sea lions) 2

2a. No external pinnae; claws set near the ends of the flippers or extend beyond them; five claws on the hind flipper; both surfaces of the flippers covered with fur; hind flippers cannot be rotated under the body (thus the animal cannot walk on land); testes internal .... (Phocids or "true seals") 3

2b. External pinnae present; flippers incompletely furred, with only sparse growth of short hair on the top; claws on the fore flippers vestigial or absent; three claws on each hind flipper on the central digits; hind flippers rotate under the body (permitting walking); testes scrotal.... (sea lions and fur seal) 4

3a. Pelage pattern mostly small round to oval spots, usually with smaller numbers of ring-like markings and blotches; colour highly variable, from mostly dark to very light; body small, 1.6–1.7 m long and 60–150 kg........... Harbour Seal, page 386

3b. Large, robust body and thick neck; pelage coloration varies from uniformly dark charcoal grey or dark brown to light grey or tan; adult males very large (to 4.2 m long and 2700 kg in weight), with a heavily scarred and cornified chest and neck, and a large, inflatable fleshy proboscis; females reach 2.8 metres and 710 kg............. Northern Elephant Seal, page 376

4a. Dark-brown pelage with dense underfur and long, pale-tipped guard hairs that give a thick woolly appearance; terminal flaps on hind flipper digits, all approximately equal in length and shape; relatively long prominent pinnae; males to 2.1 metres long and weighing 270 kg, females to 1.5 metres and 50 kg .... Northern Fur Seal, page 347

4b. Fur short and stiff, except for the mane of males when present; hind flipper digits unequal in length, with the innermost and fifth digits longer (the innermost digit is also wider) than the second to fourth digits; pinnae relatively short and lying close alongside the head.. (sea lions) 5

5a. No distinct mane on adult males; head of moderate size, with a long dog-like muzzle; adult males have a bulging sagittal crest; male coloration dark brown with lighter areas on the head, females lighter brown to tan; males to 2.5 metres long and weighing up to 390 kg, females 1.7 metres and 110 kg .... California Sea Lion, page 367

5b. Heavy mane on adult males; head massive with a blunt, relatively short muzzle; sagittal crest less pronounced in adult males; adult coloration is pale yellow to light tan above, juveniles dark tan to light brown; males to 3.1 metres long and weighing up to 1100 kg, females 2.4 metres and 300 kg .... Steller Sea Lion, page 357

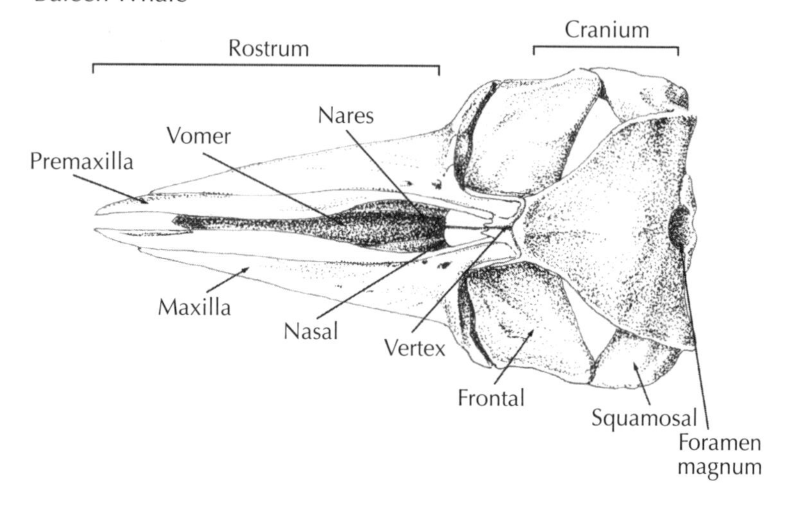

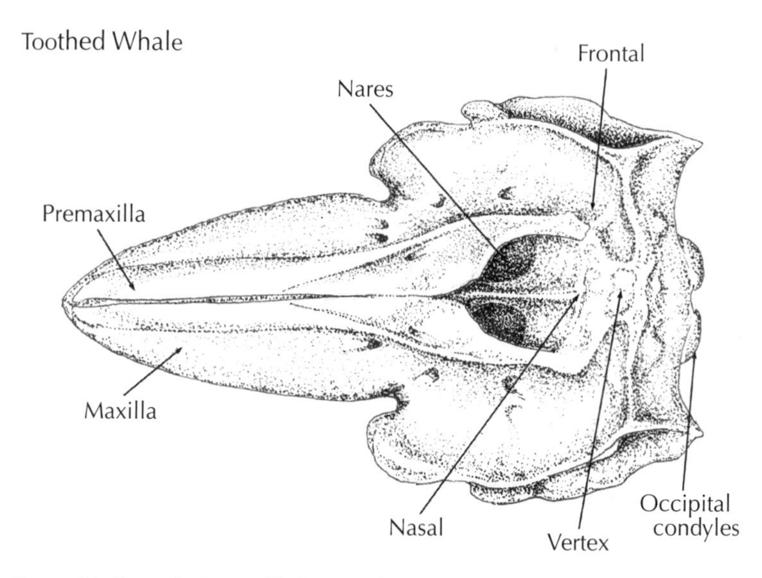

Figure 30. Dorsal views of baleen and toothed whale skulls showing bones referred to in the Key to Skulls.

Keys to Skulls

Cetaceans

1a. Large (adults >1 m long) and bilaterally symmetrical; teeth absent; mandible lacking bony symphysis ......(baleen whale) 2

1b. Generally <1.5 m long (longer in Sperm Whales) and generally asymmetrical; teeth present (though may be fully embedded in mandible in some beaked whales); lower jaw has a bony symphysis ....(toothed whale) 8

2a. Rostrum strongly arched in side view (>20° between basicranium and base of rostrum); occipital condyles do not extend to or past the posterior portion of the cranium in dorsal or ventral view; base of rostrum narrow (<30% cranial width); mandible strongly bowed laterally ....North Pacific Right Whale, page 103

2b. Rostrum moderately arched or flat in side view (<18° between basicranium and base of rostrum); occipital condyles extend to or past posterior portion of the cranium in dorsal or ventral view; mandible not strongly bowed laterally......................... 3

3a. Rostrum slightly arched in side view; nasals large; frontals exposed on vertex................................ Grey Whale, page 115

3b. Rostrum fairly flat in side view (may be slightly arched in Sei Whale); nasals reduced; frontals inconspicuous or absent on vertex ....(rorquals) 4

4a. Base of rostrum about 50% of the cranial width; anterior margin of the squamosal rounded or U-shaped (see figure 31) .... Humpback Whale, page 171

4b. Base of rostrum >65% of the cranial width; anterior margin of the squamosal pointed or V-shaped (*Balaenoptera* whales) 5

5a. Rostrum U-shaped and rounded at the tip; lateral edges of the rostrum parallel along the posterior half .... Blue Whale, page 149

5b. Rostrum V-shaped and pointed; lateral edges of the rostrum not parallel but diverge toward the posterior........................ 6

6a. Nasal bones small, <50% of the nasofrontal process, and narrow toward the anterior ....Fin Whale, page 160

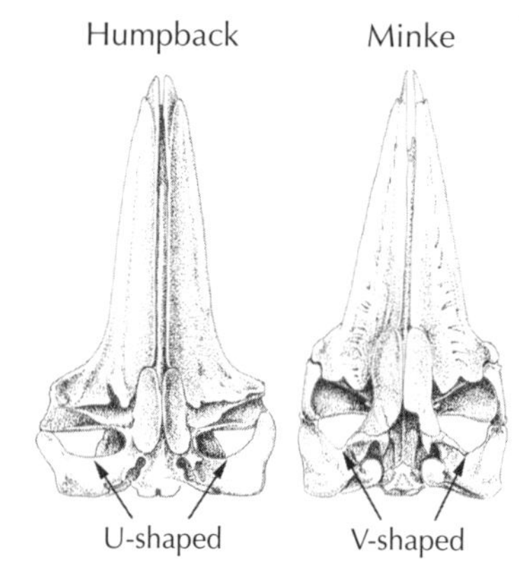

Figure 31. Ventral views of the skulls of the Humpback Whale and the Common Minke Whale, a representative species in the genus *Balaenoptera*, showing different shapes of the anterior edge of the squamosal bone.

6b. Nasal bones large, >50% of the nasofrontal process, and widen toward the anterior 7

7a. Nasal length approximately equal to the foramen magnum diameter .... Common Minke Whale, page 130
7b. Nasal length approximately twice the foramen magnum diameter .... Sei Whale, page 140

8a. Anterior cranial region dish- or bowl-shaped or with elevated maxillary ridges; rostrum deeper than its width in some species; teeth restricted to the lower jaw or few in number if present in the upper jaw ....(sperm and beaked whales) 9
8b. Anterior cranial region neither dish- or bowl-shaped nor with elevated maxillary ridges; rostrum wider than deep; teeth in both upper and lower jaws (except in Risso's Dolphin) .... (dolphins and porpoises) 14

9a. Nares obviously asymmetrical (left naris much larger than right); rostrum much wider than its depth 10
9b. Nares similar in size; rostrum nearly as deep as its width.... 12

10a. Rostrum long (>50% of the condylobasal length (CBL); zygomatic arches complete; >17 pairs of teeth; mandibular symphysis long (>30% of mandibular length) .... Sperm Whale, page 187
10b. Rostrum short (<50 % of the CBL); zygomatic arches incomplete; <17 pairs of teeth; mandibular symphysis short (<30 % of mandibular length) 11

11a. Adult skull has CBL >39 cm; rostrum relatively long; typically 12–16 pairs of teeth (sometimes 10–11) only in the lower jaw; teeth only slightly hooked .... Pygmy Sperm Whale, page 200
11b. Adult skull relatively small (CBL <31 cm); rostrum relatively short; typically 8–11 pairs of teeth (sometimes 12–13) in the lower jaw and occasionally up to 3 vestigial pairs in the upper jaw; teeth strongly hooked .... Dwarf Sperm Whale, page 206

12a. Two pairs of mandibular teeth .... Baird's Beaked Whale, page 212
12b. No more than one pair of mandibular teeth (occasionally extra rudimentary teeth present) .... (other beaked whales) 13

13a. Teeth (one pair) oval to round in cross-section and located at the tip of the mandible .... Cuvier's Beaked Whale, page 234
13b. Teeth (one pair, if present) laterally flat and located well back from the tip of the mandible .... Hubbs' or Stejneger's beaked whale*, page 220 or 227

14a. Teeth spade-shaped or peg-like; rounded bony bosses (bumps) on the premaxillae anterior to the nares; premaxillae do not contact nasals... (Phocoenidae) 15
14b. Teeth generally conical; no bosses anterior to the nares; premaxilla in contact with the right nasal ... (Delphinoidae) 16

15a. Face of the cranium high and nearly vertical, with strong development of the supraoccipital crest; frontal not visible where it meets the supraoccipital (in dorsal view) .... Dall's Porpoise, page 338
15b. Face of the cranium low and strongly diagonal, with weak development of the supraoccipital crest; frontal on at least one side visible where it meets the supraoccipital (in dorsal view) .... Harbour Porpoise, page 329

16a. Less than 27 teeth per row... 17
16b. More than 27 teeth per row 20

17a. Two to seven pairs of teeth near the tip of the lower jaw only (uncommonly one or two pairs in the upper jaw); lateral

* It is extremely difficult to reliably distinguish between the skulls of Mesoplodont whales; consult an expert for identification.

margins of the rostrum concave along the middle part of their length .... Risso's Dolphin, page 262

17b. Up to seven teeth per row on both the upper and the lower jaws; lateral margins of the rostrum generally convex 18

18a. Seven to nine teeth only in the anterior half of the rostrum; rostrum wide (length/breadth ratio generally <1.3) .... Short-finned Pilot Whale, page 255

18b. Teeth present in both the anterior and the posterior halves of the rostrum; rostrum relatively narrow (length/breadth ratio generally >1.3)... 19

19a. Teeth round in cross-section (greatest diameter of largest teeth generally <23 mm); adult CBL <78 cm; 7–10 teeth per row; premaxillae width >50% of the rostral basal width .... False Killer Whale, page 314

19b. Teeth oval in cross-section (greatest diameter of largest teeth generally >23 mm); adult CBL >78 cm; 10-14 teeth per row; premaxillae width <50% of the rostral basal width .... Killer Whale, page 287

20a. Palatal grooves deep (>3 mm at the midlength of the rostrum) 21

20b. Palatal grooves shallow (<3 mm at the midlength of the rostrum) or non-existent.. 22

21a. Rostrum relatively short and wide (<275 mm long, length/breadth ratio <3.2, rostrum length/zygomatic width ratio 1.25-1.62); 41–54 teeth per row; trapezoid-shaped platal ridge .... Short-beaked Common Dolphin, page 248

21b. Rostrum relatively long and slender (>275 mm, length/breadth ratio >3.2, rostrum length/zygomatic width ratio 1.46-2.06); 47–67 teeth per row; lance-shaped palatal ridge ....Long-beaked Common Dolphin, page 241

22a. Rostrum relatively wide (width at the base generally >25% CBL, length/breadth ratio <2.2); <40 teeth per row .... Pacific White-sided Dolphin, page 269

22b. Rostrum relatively narrow (width at the base generally <25% CBL, length/breadth ratio >2.1); >30 teeth per row........... 23

23a. Premaxillaries converge and meet or nearly meet along the dorsal aspect of the rostrum; orbits relatively deep; anterior portions of the mandible more robust; rostrum relatively long and narrow (length generally >2.3 times its width at the base) ....Striped Dolphin, page 323

23b. Premaxillaries remain widely separated dorsally on the rostrum (starting at the base and moving toward the tip); orbits relatively shallow; anterior portions of the mandible very narrow; rostrum relatively short and wide (length about 2.2 times its width at the base) .... Northern Right Whale Dolphin, page 279

Carnivores

1a. Post-canine teeth flat for crushing ....Sea Otter, page 397

1b. Post-canine teeth have sharp crowns .....(seals and sea lions) 2

2a. Auditory bullae inflated and round; supraorbital processes absent; teeth usually have multiple cusps; nasals extend far back between frontals on midline ....(true seals) 3

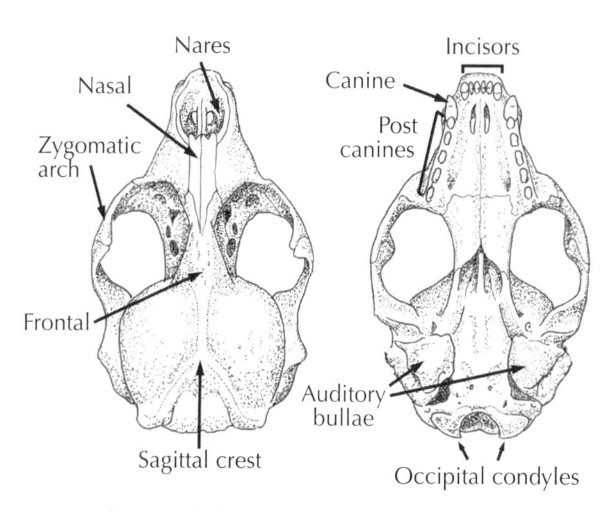

Figure 32. A Harbour Seal skull in dorsal (left) and ventral views showing bones used in the Key to Skulls for carnivores.

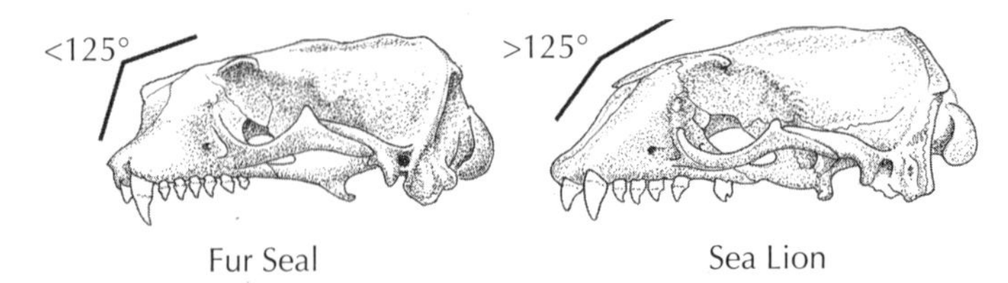

Figure 33. The angle of the facial bones of the skull in lateral view is steeper in Northern Fur Seals (<125°) and flatter in Steller and California sea lions (>125°).

2b. Auditory bullae flat, small and angular; supraorbital processes present; teeth usually unicuspid; nasals do not extend posteriorly between frontals along midline between nasals on the midline ....(sea lions and fur seals) 4

3a. Three upper incisors in each upper tooth row; muzzle relatively short and low; naso-frontal area less elevated ....Harbour Seal, page 386

3b. Two upper incisors in each upper tooth row; muzzle low and long, generally with a concave upper margin .... Northern Elephant Seal, page 376

4a. Facial angle <125° (figure 33) .... Northern Fur Seal, page 347

4b. Facial angle >125° .... (sea lions) 5

5a. Large gap between the fourth and fifth post-canines (width of about two teeth)........................... Steller Sea Lion, page 357

5b. No (or very small) gap between teeth ....California Sea Lion, page 367

Selected References: Ainley et al. 1994, Jefferson et al. 1993 and 2008, Scheffer and Slipp 1948, Würsig et al. 2000.

SPECIES ACCOUNTS

The accounts for the 31 species of marine mammals known to occur in British Columbia waters follow the same order as in the Checklist of Species. Each account includes an illustration of the whole animal and its skull to depict general features. For sexually dimorphic species – those in which the body shape or coloration varies between the sexes – the account has illustrations of both the male and the female. Features that are important for species identification are described in the preceding identification keys to whole animals and skulls. Information in the species accounts is based on observations and research in BC waters as well as on the broader scientific literature and is divided into nine sections, described below:

Other Common Names lists alternative English common names currently in use. Obsolete common names are not included unless they appear in relatively recent literature.

Description provides details on the species' relative size and shape, notable anatomical features, coloration of skin or pelage, and other distinctive traits. This is followed by actual measurements of the total body length and weight, and either the count and description of baleen (for baleen whales), the tooth count (for toothed cetaceans) or the dental formula (for pinnipeds and the Sea Otter). These measurements are provided for both adults and neonates (calves or pups of the year). Whenever data were available, measurements for specimens from BC waters were used – these are given as average, range (in parentheses) and sample size (indicated by "n"). Sources for these measurements were records from BC whaling stations, stranded animals, specimens collected (killed) for research and from aerial photogrammetry (for Resident Killer Whales). Measurements are given for males and females whenever these differ (through sexual dimorphism). When lengths were not available for BC specimens, the typical range of lengths from the literature are provided, using North Pacific measurements whenever possible. In these cases, no average length or sample size are shown. Measured body weights are seldom available for marine mammals, particularly those from BC, so these are generally based on measurements or estimates in the literature.

The description for baleen whales includes the number of baleen plates that hang from each side of the upper jaw, usually as a range, along with the typical length and coloration of the plates. In toothed cetaceans the dentition is homodont – the teeth are all very similar in shape (unlike the teeth of most mammals, which are heterodont and can be differentiated, for example, into incisors, canines and molars). The description lists the number of pairs of teeth found in each of the upper and lower jaws, along with a description of the tooth shape. For the carnivores (pinnipeds and the Sea Otter), the dental formula denotes the number of teeth in one side of the head, the first value for those in the upper jaw and the second for the teeth in the lower jaw. For example, the Harbour Seal's dental formula is "incisors 3/2, canines 1/1, post-canines 5/5", meaning three upper and two lower incisors, one upper and one lower canine, and five upper and five lower teeth posterior to the canines (post-canines are differentiated into premolars and molars in the Sea Otter).

An Identification sub-section is intended to help distinguish the species from similar animals in its habitat.

Distribution and Habitat summarizes the distribution of the species globally and in the North Pacific, and it provides details of its range in British Columbia waters. This section also gives an overview of the most commonly used species habitat and, for those that migrate, the months of the year that the animal is most often found in BC. A map is provided for each species showing where it is known to occur in BC, from the coast to as far offshore as the boundary of Canada's 200 nautical mile (370 km) Exclusive Economic Zone. Relative water depth is depicted in colours: white shows relatively shallow water typically found over much of the continental shelf (<200 metres deep), and three shades of blue show progressively deeper water going offshore. For most species, all available recorded sightings in BC waters are illustrated, using black dots. Sighting records of marine mammals were compiled mostly from vessel and aerial surveys undertaken by the Cetacean Research Program at the Pacific Biological Station and vessel surveys by the Raincoast Conservation Society. Also included are incidental observations made during seabird surveys by the Canadian Wildlife Service, and sightings compiled by two public networks – the BC Cetacean Sightings Network, a program operated jointly by the Vancouver Aquarium and the Pacific Biological Station (www.wildwhales.org), and the Platforms of Opportunity program managed by NOAA Fisheries in the US. Both solitary individuals and groups of animals are recorded as a single sighting.

It is important to note that sightings shown on the maps are not corrected for observer effort – dots on maps reflect to some extent the numbers and distribution of observers on the water. Low densities of sightings in certain areas – particularly for widely distributed species – may be more a result of a lack of observers than a lack of animals using those waters. An example of the effect of observer effort can be seen in several maps (e.g., Dall's Porpoise) where there is an obvious line with numerous sightings that extends from southwestern Vancouver Island toward the west. This is the route taken by research ships transiting to and from Station "P", a location well offshore that has long been visited several times a year for oceanographic research, and there are often observers on board recording marine mammal sightings. For rare or uncommon cetacean species, maps also include locations of strandings (shown as red squares) or of kills due to bycatch in fishing gear or whaling (shown as blue triangles). Summaries of sighting, stranding and bycatch records for rare and uncommon cetaceans are also tabulated in Appendix 2.

Maps for pinnipeds and the Sea Otter differ somewhat from the cetacean maps. For the Harbour Seal, Steller Sea Lion and California Sea Lion, the maps show the locations of sites that they use to haul out or breed, instead of sightings in the water. For the Northern Elephant Seal, they depict both sightings at sea and locations where animals have been seen to haul out for resting and moulting. Maps for Northern Fur Seals, which do not normally haul out in BC waters, show only ocean sightings. The Sea Otter map shows its current year-round range in red and sightings of single animals, usually males, as black dots.

Natural History describes aspects of the species' biology under three sub-headings: Feeding Ecology discusses preferred prey species and hunting and capture tactics. Behaviour and Social Organization covers how deep the animals dive and for how long, the size and kinds of groups they form, their mating system and behaviour associated with reproduction, and their vocalizations. Life History and Population Dynamics examines vital rates and parameters such as ages at sexual and physical maturity, duration of gestation and lactation, birth rates and longevity.

Exploitation describes how humans have made use of BC's marine mammals over the years. Many First Nations traditionally hunted a variety of marine mammals for subsistence and used body parts for ceremonies and other purposes. A number of species were later exploited in BC waters by non-aboriginal people for commercial purposes or for predator control. (Appendix 3 tabulates catch statistics by species for whales killed in BC waters by shore-based and pelagic whaling operations in the 20th century.)

Taxonomy and Population Structure contains descriptions of subspecies or discrete populations that may exist in BC or over the species' range. Species and subspecies classifications are those currently accepted by the Committee on Taxonomy of the international Society for Marine Mammalogy.

Conservation Status and Management describes the historical (where known) and current abundance for the species in British Columbia as well as population trends. It lists the species' current conservation status, according to the International Union for Conservation of Nature (IUCN) Red List, the Committee on the Status of Endangered Wildlife in Canada (COSEWIC) and the BC Provincial List. It outlines the main threats to species at risk, along with actions being taken by government and conservation groups to promote recovery and protect habitat.

Remarks includes interesting anecdotes and facts that do not fit readily into other sections, such as the derivation of the common or scientific name for the species.

Selected References lists important publications for the species generally as well as in British Columbia. It is not meant to be comprehensive but should provide the reader with useful sources of additional information.

North Pacific Right Whale

Eubalaena japonica

Other Common Names: None.

Description

The North Pacific Right Whale is a large, very robust baleen whale with a broad back and massive head. Its maximum girth may reach up to three-quarters of its overall length. Right Whales are easily distinguished from other great whales by the absence of even the vestiges of a dorsal fin or ridge on the back. The flukes are triangular in shape, smooth along the edges, and very wide (i.e., long across, perpendicular to the body axis) – their combined width can be up to a third of the total body length. The flippers are large and paddle-shaped with a somewhat pointed tip. The large mouth has a strongly arched jawline and a narrow rostrum. The fleshy margin of the lower jaw has several crenulations or scallop-shaped indentations.

The North Pacific Right Whale is mostly black, although there is usually a white patch of irregular size and shape on the belly and occasionally others elsewhere on the body. The head and rostrum have numerous *callosities*, which are raised areas of roughened skin that are lightly pigmented due to large infestations of whale lice, or cyamids. The largest callosity, known as the *bonnet*, is at the front end of the rostrum. Other callosities are found along the sides of the lower jaw, along the rostrum, around the blowholes and over the eyes. Callosity patterns are variable and provide a reliable means of identifying individuals from photographs.

Measurements

length (metres):

male: 16.5
female: 17
neonate: 4–4.5

weight (kg):

adult: ~ 90,000
neonate: ~ 1000

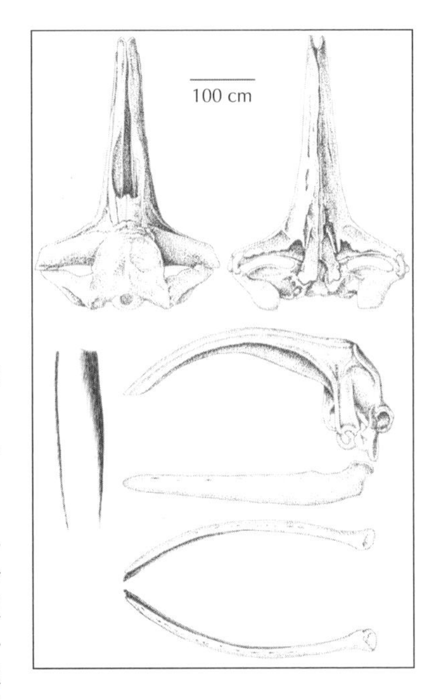

Baleen

Plates on each side: 220–270
Up to 2.8 m long, narrow (~ 25 cm), fine bristles, black.

Identification

Right Whales are among the easiest of the great whales to identify at sea. Their bulky size, broad back lacking any sign of a dorsal fin or ridge, and large head with a narrow rostrum, an arched mouthline and callosities make them immediately recognizable (figure 34). At a distance they could be confused with Humpback Whales or Grey Whales. All three species often raise their flukes above the surface when diving and, in certain conditions, the blow of Grey Whales can resemble the characteristic V-shaped blow of the Right Whale (figure 35). Humpbacks and Greys also have patches of barnacles on the head that could be mistaken for callosities. But closer inspection would

Figure 34. A rarely seen North Pacific Right Whale off the northwestern coast of Haida Gwaii. Note the highly arched lower jaw. Photo: J. Ford, June 2013.

Figure 35. A North Pacific Right Whale near Haida Gwaii displaying the distinctive V-shaped blow. Photo: J. Pilkington, June 2013.

reveal the stubby dorsal fin of the Humpback Whale or the hump and "knuckles" running along the dorsal ridge of the Grey Whale, both of which contrast with the broad, smooth back of the Right Whale.

Distribution and Habitat

Due to the current rarity of North Pacific Right Whales sightings, most information on their distribution comes from catches by 19th-century whalers, but this may not represent their actual range today. Right Whales once ranged across the North Pacific, although densities in the central part of the ocean basin – between longitudes 170°E and 160°W – were far lower than in the east and the west. Populations in the western and eastern North Pacific are considered to be discrete. In summer Right Whales were found mostly north of 40°N latitude, in the Gulf of Alaska and Bering Sea, along the eastern coast of Kamchatka, in the Sea of Okhotsk, around the Kurile Islands and in the Sea of Japan. In late fall and winter there was a southward shift in distribution to at least 30°N and possibly as far south as 20–25°N, including the Sea of Japan, Taiwan Strait and waters around Ogasawara (Bonin Islands) in the west, to coastal waters of Mexico and California in the east. Movements to lower latitudes in winter likely represented seasonal migrations for calving, as we see in other species of Right Whales and in many other baleen whales. But locations of calving and nursery grounds for North Pacific Right Whales remain a mystery. In waters off western Canada, North Pacific Right Whales were taken by 19th-century whalers mostly between May and September to the north and west of Haida Gwaii. These waters were likely used during those months primarily for feeding.

North Pacific Right Whales currently occupy only a fraction of their former range. In recent years most sightings in the eastern North Pacific have been concentrated in two areas – one in the southeastern Bering Sea and the other over the continental shelf and slope south of Kodiak Island, Alaska. There are only eight confirmed records of North Pacific Right Whales in Canadian waters since 1900, six of which were catches by whalers. Five of these catches were to the west of Haida Gwaii, and all were in the month of June between 1914 and 1929. The last kill in BC was a 12.5-metre male taken off Brooks Peninsula, Vancouver Island, on July 18, 1951, by whalers from the Coal Harbour whaling station in Quatsino Sound (see figure 7). Right Whales were legally protected from whaling after 1931, and this whale was reportedly taken in error.

Since 1951 there have been two dubious sighting reports and two confirmed records in Canadian waters. One report was of two Right Whales sighted in 1970 by a Japanese whaling scout vessel well offshore, somewhere between 50–55°N and 130–140°W. Although there is no reason to doubt the species identification, the reported sighting location is vague, so it may not have been in Canadian waters. Another uncertain report was of two Right Whales in Juan de Fuca Strait in 1983, but the species identification is highly questionable. The only confirmed records of North Pacific Right Whales in BC since the 1950s were sightings of two separate individuals made during Cetacean Research Program surveys in 2013. The first sighting was made by James Pilkington and crew aboard the CCGS *Arrow Post*, on June 9, 2013, about 15 km off the west coast of Graham Island, Haida Gwaii (figures 34 and 35). Four days later, my colleague Graeme Ellis and I joined the *Arrow Post* and we found the whale again and spent several hours observing it as it fed on zooplankton at the surface. The sighting of the second individual was made by colleague Brian Gisborne on Swiftsure Bank off the entrance to Juan de Fuca Strait on October 25, 2013. The next day, we were were able to join Gisborne and found the whale mixed with a small group of Humpback Whales in the same area. This was a large Right Whale with deep healed cuts on its rostrum, likely resulting from a past entanglement in fishing gear or some other flotsam (figure 36).

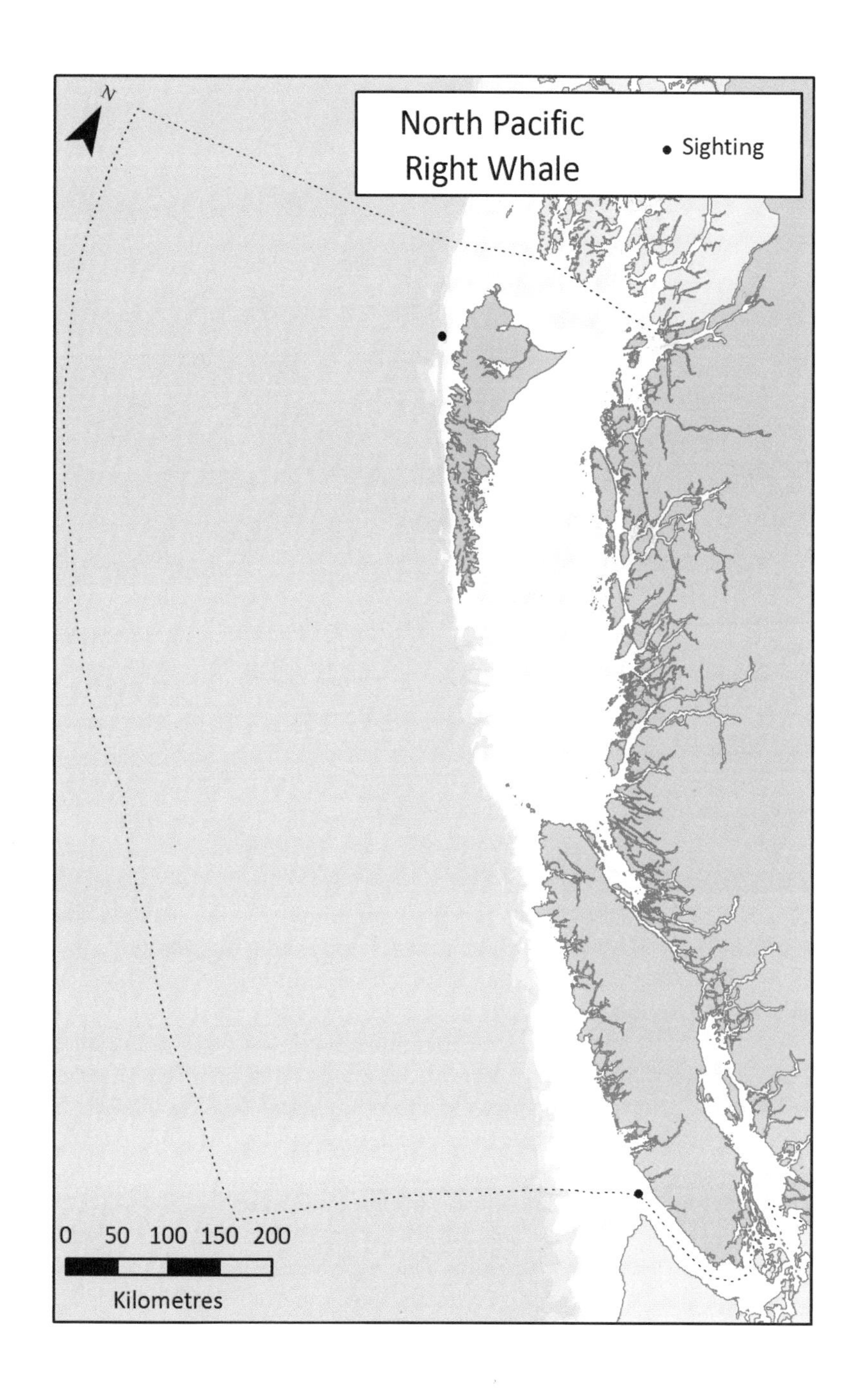
N
North Pacific
Right Whale
Sighting
0 50 100 150 200
Kilometres

Figure 36. This head-on view of a North Pacific Right Whale at Swiftsure Bank shows the broad finless back, callosities ahead of the blowhole and on the rostrum, and a healed injury near the tip of the rostrum.
Photo: J. Ford, October 2013.

Natural History

Feeding Ecology

North Pacific Right Whales are low trophic-level filter feeders that consume only zooplankton. Their diet is dominated by calanoid copepods, tiny crustaceans that reach only 5–10 mm in size, but they also take larger zooplankton such as the shrimp-like euphausiids, which can be over 20 mm in length. Most information on the diet of North Pacific Right Whales comes from the stomach contents of those killed by whalers and from recent field observations of feeding whales. In the early 1960s, Japanese whalers operating under research permits took three whales off Kodiak Island and six in the Bering Sea. Their stomachs mostly contained two species of calanoid copepod, *Neocalanus plumchrus* and *N. cristatus*. Net tows conducted near a Right Whale observed south of Kodiak Island in September 2006 revealed a dense layer of zooplankton near the bottom, at around 175 metres deep, composed of a mixture of four species of copepods and two species of euphausiids. In BC waters, the stomach of the male North Pacific Right Whale killed off Vancouver Island in 1951 contained euphausiids, otherwise known as krill. The North Pacific Right Whale observed in June 2013 off the west coast of Haida Gwaii was feeding at the surface on the copepod *Neocalanus plumchrus*.

Right Whales are large animals with prodigious metabolic requirements, so they must feed on vast numbers of their tiny prey. UBC graduate student Sarah Fortune estimated from bioenergetics studies that adult North Atlantic Right Whales require 270,000–290,000 calories per day, and that estimate would more than double for lactating females. This means adult whales must consume roughly 170–180 million copepods (*Calanus finmarchicus*, their primary prey) each day. To meet this daily requirement, the whales must find prey concentrations well in excess of 3000 copepods per cubic metre of seawater. Since the copepods consumed by North Pacific Right Whales tend to be larger than those in the North Atlantic, more modest densities of 1000–1500 copepods per cubic metre may be sufficient for this species. But even these densities are not common, which is why the whales must focus their foraging on nutrient-rich waters with high zooplankton productivity and where oceanic processes such as eddies, fronts, upwellings and convergence zones concentrate their prey into exploitable patches. Researcher Ed Gregr developed a model to predict important feeding habitat of North Pacific Right Whales using locations of historical whaling kills and ocean features that contribute to high copepod densities. This model indicates that the best habitat for Right Whales is likely to be found to the north and west of the Canadian EEZ. But it also shows that suitable feeding habitat can be found in our waters north and west of Vancouver Island, especially off the west coast of Haida Gwaii.

Behaviour and Social Organization

Very little is known about the behaviour and social organization of North Pacific Right Whales, but similarities can be expected with the closely related and better-studied North Atlantic and Southern Right Whales. When surface feeding, Right Whales swim steadily at speeds of about five kilometres per hour with their mouths agape at or near the surface to skim and strain zooplankton from the water. They also feed at depth, descending rapidly to dense layers of zooplankton up to 200 metres below the surface, foraging for periods of 5 to 14 minutes, then ascending rapidly back to the surface.

Right Whales are not swift swimmers, but they are powerful animals and can display impressive aerial behaviour such as breaching, tail slapping (also known as lobtailing) and flipper slapping. Because they are slow but powerful, Right Whales defend themselves, rather than attempting to flee, when attacked by Killer Whales. They group together, roll and thrash at the surface and slash at the predators with their tail flukes. The roughened callosities on their heads may serve as armour when the Right Whales ram attacking Killer Whales.

Right Whale courtship takes place in aggregations referred to as *surface-active groups*, which may include 10 or more males focused on a single female. Although there may be much rolling and splashing as males manoeuvre to position themselves closest to the female, these groupings lack the violent interactions often observed in Humpback Whale surface-active groups. Unlike Humpback Whales and other baleen whales, the mating system of Right Whales involves sperm competition, where females mate with multiple males and the males that produce the most sperm have a reproductive advantage by swamping the sperm of rival males. To maximize sperm production, Right Whales have the largest testes in the animal kingdom – two metres long and up to 500 kg each. Sexual behaviour is seen throughout the year, but since calving takes place primarily in winter following a 12–13 month gestation, non-winter sexual behaviour is likely social in function.

North Pacific Right Whales communicate with a variety of distinctive low-pitched vocalizations. Most common are "up calls", which start at around 90 Hz and sweep up to 150 Hz over about 0.7 seconds, and a brief but intense pulse known as a "gunshot call" made by males. Less commonly heard are "down calls", which start with a brief downsweep before sweeping upward in frequency like an "up call".

Life History and Population Dynamics

Right Whales were depleted early in the history of whaling, long before data on life history parameters were routinely collected from whale catches. As a result, most of what is known of the Right Whale's life history and population dynamics has been determined from photo-identification studies of Southern Right Whales and North Atlantic Right Whales. It's likely these parameters are similar for North Pacific Right Whales.

Right Whales reach sexual maturity at a body length of 13–16 metres. Females typically give birth to their first calf at 9–10 years of age, although sometimes as early as 5 years. Calves are born in winter and are nursed for about a year. Calf growth is extremely rapid – they double in size to reach about three-quarters of the length of their mother by weaning. While being nursed, calves grow at an average of 1.7 cm and 34 kg per day.

The typical reproductive cycle of mature females is three years. Gestation lasts about one year, lactation another year, and then the female conceives once again. Average longevity is not known. But one distinctively marked female North Atlantic Right Whale was first photographed in 1935, then again at various times over the years until 1995, when she is thought to have died from a vessel strike. She was likely older than 70 years when killed.

Exploitation

Although Humpback Whales and Grey Whales were the species most commonly taken by First Nations whalers on the west coast of Vancouver Island, there is evidence that North Pacific Right Whales were also caught on occasion. A type of whale that was like a Grey Whale "but bigger, with something growing on the back of its head" was evidently pursued by Nuu-chah-nulth people whenever it was seen. This most likely was the Right Whale, and the growth on the head refers to the bonnet and other callosities on the rostrum and around the blowhole. One Nuu-chah-nulth informant recognized this whale as *kw'utski*, meaning "mussels on head". North Pacific Right Whales have been identified using ancient DNA in whale bones recovered from archaeological excavations at the old villages of Ts'ishaa and Huu7ii in Barkley Sound, as well as in Clayoquot Sound.

Although North Pacific Right Whales were hunted by shore-based whalers in Japan as early as the 1500s, large-scale whaling for the species on the high seas by European and American whalers did not begin until the late 1830s. This whaling expanded very rapidly in the eastern North Pacific, and within a decade the species was seriously depleted there. The offshore, ship-based whalers first targeted Right Whales in the Gulf of Alaska, then within a few years shifted their effort to include the western North Pacific near Kamchatka, then the Bering Sea. In 1842, 29 American whaling vessels were operating; by 1843, there were 108, and by 1845, 263. Encounter rates with North Pacific Right Whales in the Gulf of Alaska declined by an order of magnitude within 10 years of the beginning of the hunt, yet whales continued to be taken for the next 50 years or so, by which time they were nearly exterminated. An estimated 26,500–37,000 North Pacific Right Whales had been killed by 1900. The whales were so scarce that few were caught after 1900, and the whales were given legal protection in 1931 by the international Convention for the Regulation of Whaling in Geneva (though this was not ratified and implemented until 1935). It is likely that numbers began to increase slowly through the first half of the 20th century, but beginning around 1950 the former Soviet Union began large-scale illegal whaling and at least 529

North Pacific Right Whales were killed during the 1960s alone. This likely represented the majority of the remaining eastern North Pacific population.

Taxonomy and Population Structure

There has long been uncertainty and disagreement about the taxonomy of Right Whales. For many years Right Whales in the northern and southern hemispheres were considered to be distinct species: *Eubalaena australis* in the south and *E. glacialis* in both the North Atlantic and North Pacific (the latter was referred to as the Northern Right Whale). At least one authority, however, argued that there is no evidence of consistent morphological differences between the northern and southern forms and that they should all be considered a single global species. More recently, analyses of mitochondrial and nuclear DNA markers have revealed clear and significant genetic differences among Right Whale lineages in the southern hemisphere, North Atlantic and North Pacific. Three separate species are now widely recognized. Interestingly, the North Pacific species, *E. japonica*, is more closely related to the southern species than it is to *E. glacialis* in the North Atlantic.

Conservation Status and Management

The North Pacific Right Whale is one of the most critically endangered of all the great whales, and its future is by no means secure. There are thought to be at least a few hundred Right Whales in the western North Pacific, but numbers in the eastern North Pacific and Bering Sea are dangerously low. Extensive surveys by NOAA in the southeastern Bering Sea between 1997 and 2008 resulted in abundance estimates of only 31 animals by photo-identification, 28 by genotyping from skin biopsy samples and 25 through acoustic detections. Two additional individuals were photo-identified south of Kodiak Island.

Without any confirmed sightings of North Pacific Right Whales off the BC coast in over six decades, it was uncertain, until recently, whether the species continued to exist in Canadian waters. But our observations of two different individuals off Haida Gwaii and Vancouver Island in 2013, described above, prove that the species has not been extirpated from BC. Yet there is no question that this species is extremely rare in our region. The Cetacean Research Program of the Pacific Biological Station in Nanaimo, BC, has conducted over 50,000 km of vessel-based cetacean surveys (using ships and small boats) since 2002, with only this single sighting of a Right Whale. It is important to realize, however, that even if North Pacific Right Whales occurred regularly in Canadian waters, they would be at such low densities that sighting one during short-term surveys would be highly unlikely. There is a much greater probability of detecting Right Whales with long-term passive acoustic monitoring for their distinctive vocalizations. We recently developed a network of autonomous underwater recorders that are moored in strategic locations off BC's outer coast for periods of up to a year. These – together with recordings from hydrophones deployed along the University of Victoria's NEPTUNE Canada cabled observatory network off Vancouver Island – offer a good chance that North Pacific Right Whale vocalizations will be detected if the species occurs regularly in our waters.

Although small population size alone is probably the most serious challenge facing North Pacific Right Whales, they are also vulnerable to the impacts from a range of human activities, especially ship strikes and fishing gear entanglements. Noise from seismic surveying and ship traffic is also a serious concern.

North Pacific Right Whales are listed as Endangered by IUCN and by COSEWIC, and they are on the BC Conservation Red List. The eastern North Pacific subpopulation has been assessed by IUCN as Critically Endangered.

Remarks

Of all the species of great whales, North Pacific Right Whales suffered the most dramatic and rapid depletion at the hands of whalers. When the "Northwest Ground" – their main concentration area west of British Columbia and in the Gulf of Alaska to the north – was first discovered, they were abundant. The crew of the American whaler *Ganges*, one of the first ships to work in the area, reported seeing "millions" of Right Whales in 1835–36. Although this was certainly a wild exaggeration, the usually sober and reliable whaling captain and naturalist Charles Scammon remarked that in that area they were "scattered about as far as the eye can reach from the masthead". Intensive whaling began in 1840, and within 10 years the North Pacific Right Whale was rare; 20 years later it was nearly extinct. By the early 20th century, the species was described in the American press as "one of nature's rarities", despite the fact that it was commercial whaling rather than natural circumstances that had caused its near demise. The current population in the eastern North Pacific is so small that the potential for recovery is uncertain – one can only hope that the observations of two North Pacific Right Whales off the BC coast in 2013, after decades without any reliable sightings, are signs that the species is returning to our waters.

Selected References: Arndt 2011, Brownell et al. 2001, Clapham et al. 2004, Fisheries and Oceans Canada 2011, Ford and Reeves 2008, Fortune et al. 2012, Gregr 2011, Gregr and Coyle 2009, Ivashchenko and Clapham 2012, Josephson et al. 2008a and 2008b, Kenney 2009, Monks et al. 2001, Nichol et al. 2002, Scammon 1869, Wade et al. 2011, Webb 1988.

Grey Whale *Eschrichtius robustus*

Other Common Names: California Gray Whale.

Description

The Grey Whale looks quite different from any other large whale. It is a medium-sized baleen whale with a stockier body than most of the rorqual whales. Like most baleen whales, females are slightly larger than males; they reach maximum lengths of 15.3 metres and 14.6 metres, respectively. The Grey Whale's head is moderately arched when viewed from the side and rather narrow and triangular when viewed from above. Instead of a dorsal fin, there is a hump about two-thirds of the way back from the rostrum, followed by 6–12 smaller bumps, or "knuckles", along the dorsal ridge of the tail stock. There are typically 2–7 short, deep creases along the throat, equivalent to but much less developed than the ventral pleats of rorquals. The flippers are broad with fairly pointed tips, and the flukes are relatively broad, reaching over three metres wide in adults. It has the fewest baleen plates of any baleen whale

The Grey Whale is mottled grey, as the name implies. Calves are darker than adults. The density of the irregular, light blotches that give the whale its mottled appearance varies among individuals, and many animals also have white scars on their bodies. There can also be extensive patches of white to yellow or orange coloration on the body, especially along the top of the head, which result from infestations of barnacles and whale lice, or cyamids. The baleen is creamy white to pale yellow.

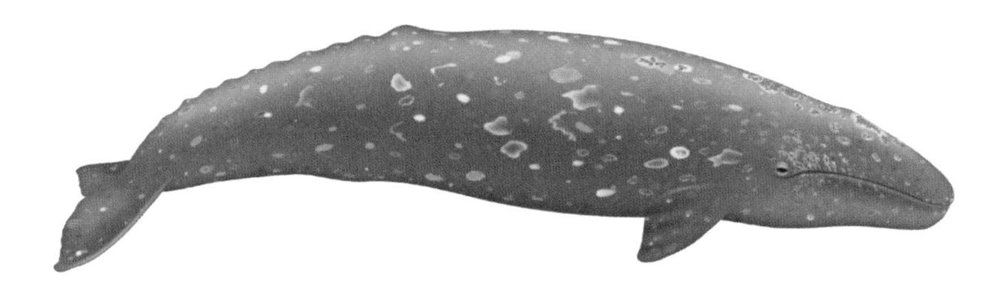

Measurements

length (metres):

male: 11.5 (11.1–12.2) n=6
female: 12.2 (11.8–12.7) n=4
neonate: 4.6–4.9

weight (kg):

adult: ~ 35,000
neonate: 700–9000

Baleen

Plates on each side: 130–180
Short (<40 cm), with coarse bristles, cream colour to pale yellow.

Identification

Identifying Grey Whales at sea is generally easy, especially at close range. In British Columbia they are usually found within a few kilometres of shore, except when crossing open bodies of water such as Queen Charlotte Sound and Hecate Strait, and are sometimes seen in very shallow areas just beyond the surfline along the outer coast. Their mottled grey colour, narrow arched head and lack of a well-defined dorsal fin are their most distinctive features. Individual Grey Whales can be reliably distinguished by unique colour patterns, scars and barnacle patches. Their blow is variable in shape, from low, bushy and sometimes V-shaped to tall and narrow. They frequently raise their flukes when diving.

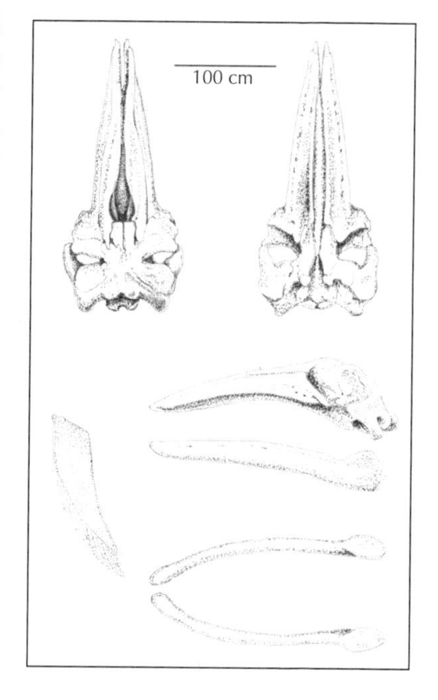

The most common large whale species that may be encountered in the same nearshore habitat typical of Grey Whales is the Humpback Whale, which has a black rather than grey back and usually has a distinct – though highly variable – dorsal fin. Humpbacks also have much longer flippers and a broader head with numerous knobs. Like Grey Whales, Sperm Whales have a dorsal hump rather than a shapely fin and a series of "knuckles"

along the tail stock, but they lack the mottled coloration and barnacle patches seen on Grey Whales and are generally found in deep water far from shore. North Pacific Right Whales have callosities on their head that could be mistaken for barnacle patches, but Right Whales have a very broad, smooth back with no fin or hump and are mostly black in colour (they are also extremely rare in British Columbia).

Distribution and Habitat

Grey Whales are found only in the North Pacific Ocean. A population once existed in the North Atlantic, but it was extirpated in the 17th century, possibly by hunting. Two populations are found in the North Pacific: an eastern population associated with the west coast of North America and a western population that occurs along the coast of Asia. The western population once found along the coasts of Korea, China, Japan and Russia has been seriously reduced in size and now comprises fewer than 150 animals. The eastern Pacific population has recovered from 19th-century whaling and is found throughout its historical range, which stretches along the coast from Baja California and the adjacent northwestern coast of mainland Mexico in the south to the Chukchi and Beaufort seas in the Arctic. The species is strongly migratory, with the majority of the eastern population travelling 15,000–20,000 km (round trip) between winter breeding lagoons on the west coast of Baja California and summer feeding grounds in the Bering, Chukchi and Beaufort seas. A subpopulation of a few hundred animals – known as "summer residents", the "southern feeding group" or the "Pacific Coast Feeding Aggregation" (PCFA) – does not migrate to the main northern feeding grounds but stays through summer and fall at scattered feeding locations along the coast from central California to Kodiak Island, Alaska.

The western Pacific population of Grey Whales formerly migrated between summer grounds in the Sea of Okhotsk to winter grounds off Korea, Japan and China. Currently, most of the small remnant of this population can be found during summer and fall feeding in the waters around Sakhalin Island and the Kamchatka Peninsula, Russia, but researchers have not yet established its winter range. In an attempt to solve the mystery of where these whales go for the winter and how they get there, Bruce Mate of Oregon State University and his Russian colleagues affixed satellite tags to several whales at Sakhalin Island prior to the winter migrations in 2010 and 2011. The tags on two whales – a male called Flex in 2010 and a female called Varvara in 2011 – transmitted long enough to reveal their migratory destinations, which were quite unexpected. Rather than heading south toward Korea, Japan or China, both whales migrated eastward across the Bering Sea, through the eastern Aleutian Islands, then across the Gulf of Alaska – the same "great circle route" taken by ships travelling between Japan and Vancouver. Flex remained 20 or more kilometres offshore as he passed Vancouver Island, and he finally came close to shore off Oregon, where the tag stopped transmitting. Varvara, on the other hand, reached the coast off Haida Gwaii, crossed to northern Vancouver Island and continued southward within a few kilometres of shore, finally reaching the breeding lagoons of Baja California in January 2012. At the end of February, she migrated back north to the Bering Sea, using the same nearshore route as the rest of the eastern population migrants, and arrived back at Sakhalin Island in May. Flex's surprising migration in 2010 prompted scientists to compare the photo-identification catalogue for Sakhalin Island whales with catalogues for the Pacific Northwest (Oregon to southeastern Alaska) and San Ignacio lagoon, Mexico. To everyone's continued surprise, six western Grey Whales – one of which was Flex – had been previously photographed by researchers Wendy Szaniszlo and Brian Gisborne in Barkley Sound and off the West Coast Trail. An additional four whales were matched to the much larger San Ignacio lagoon catalogue. This suggests that a substantial proportion of the Sakhalin Island whales mixes with the eastern population on the breeding grounds in Baja California, which calls into question the discreteness of the current western population.

For the majority of Grey Whales, British Columbia waters serve solely as a migration corridor between southern breeding and northern feeding grounds. Southbound Grey Whales en route to Baja California transit the BC coast between November and January, with a peak off the west coast of Vancouver Island during the latter half of December. Northbound migrating whales begin appearing off Vancouver Island in late February, increase to a peak in the last two weeks of March, then decline during April and May. Both southward and northward migrations are segregated by age, sex and reproductive condition. Near-term pregnant females lead the southbound procession with estrous females, adult males and immature whales of both sexes following behind, in that sequence. Pregnant females give birth to their calves in or while approaching the shallow lagoons along the west coast of Baja California. Newly pregnant females that conceived on the breeding grounds are the first to head back north, followed by adult males and juveniles. Females with newborn calves migrate

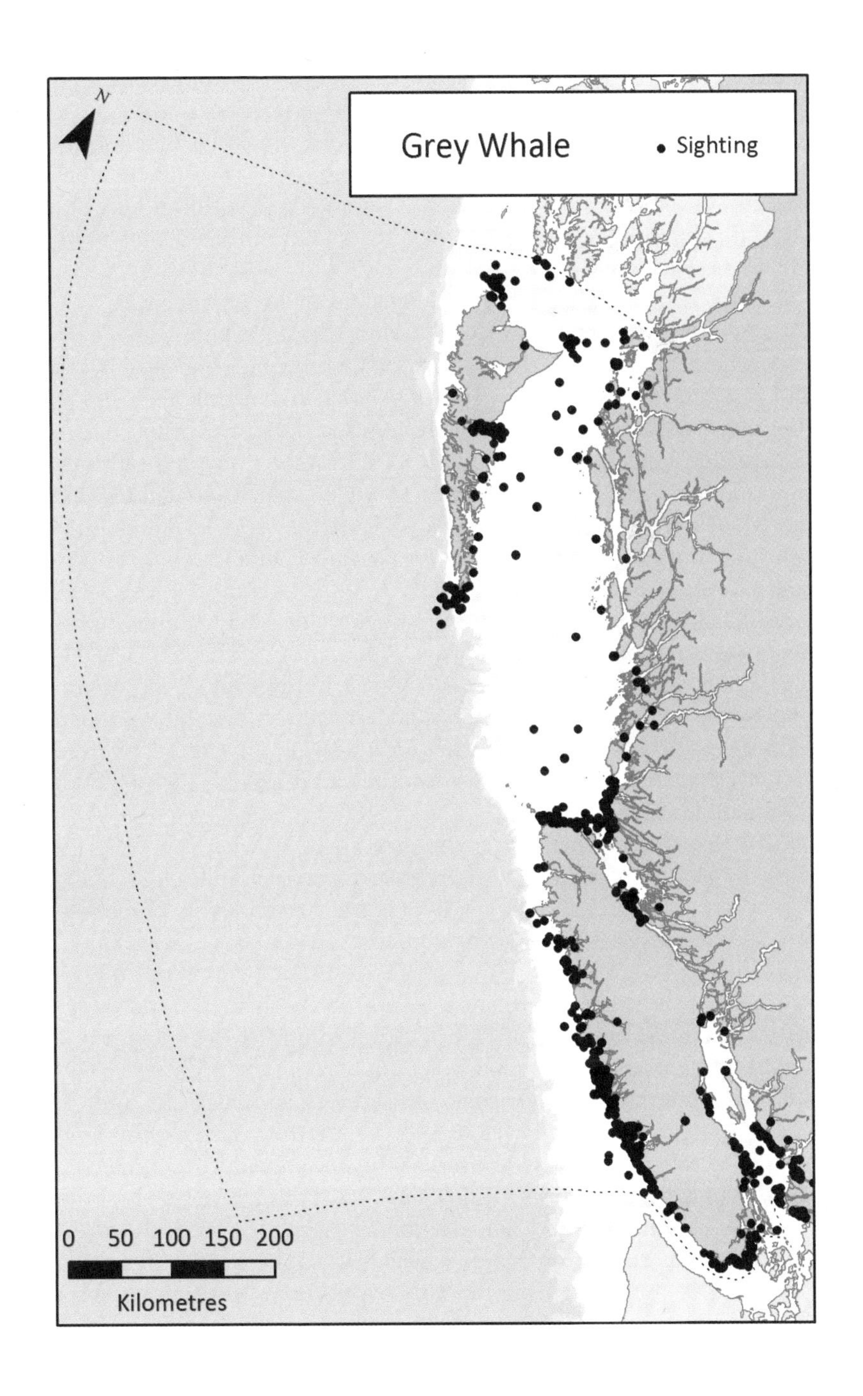
Grey Whale
Sighting
N
0 50 100 150 200
Kilometres

northward up to four weeks later than the rest, passing Vancouver Island mainly during May. The southward migration is difficult to observe off the BC coast due to typically poor winter weather and the tendency of migrants to pass several kilometres or more from shore. Northbound migrants swim closer to shore (most within 5 km) at a more moderate pace and are readily observed from headlands along the west coast of Vancouver Island.

Until recently, the migration route taken by northbound Grey Whales once they leave Vancouver Island was uncertain. Some early researchers suggested that they turn offshore from Vancouver Island and cut directly across the Gulf of Alaska to Unimak Pass in the Aleutian Islands, then enter the Bering Sea there. But whale scientist Gordon Pike at the Pacific Biological Station compiled sightings from mariners and lightkeepers and, in a 1962 article, concluded that the whales cross directly from Cape Scott to Cape St James, then continue migrating close to the outer coast of Haida Gwaii. This proposed route was actually based on very little evidence – just a few Grey Whale sighting reports from Langara Island light station on the northwestern corner of Haida Gwaii. Pike attributed the lack of sightings along the west coast of Haida Gwaii during the spring migration to the remoteness of the area and suggested that the scattered sightings of Grey Whales in Hecate Strait during this period involved animals that were "temporarily confused" after leaving the north tip of Vancouver Island.

To determine whether northbound Grey Whales use the west coast of Haida Gwaii or an alternative route, my colleagues and I deployed small satellite tags on five migrating whales in Clayoquot Sound, Vancouver Island, during March of 2009–2011. Upon reaching Cape Scott, none of the five whales continued westward to Haida Gwaii, but instead all turned to the north-northeast and swam up the centre or east side of Hecate Strait, rounded Rose Spit at the northeastern corner of Haida Gwaii, cut across Dixon Entrance and then continued up the outer coast of southeastern Alaska. To confirm that this unexpected route was indeed preferred by most northbound Grey Whales, we dispatched observers to simultaneously count passing whales during the spring migration of 2011 at two strategic shore locations – Bonilla Island light station in Hecate Strait and Langara Island light station on Haida Gwaii. Of a total of 306 migrating Grey Whales counted during the three-week survey, 97 per cent were sighted at Bonilla and only 3 per cent at Langara. This indicates that the primary northward migratory corridor for Grey Whales is Hecate Strait and Dixon Entrance rather than the west coast of Haida

Gwaii, which has important conservation implications given proposed industrial developments and other planned uses for these inside waters. The route taken by southbound Grey Whales in BC waters is poorly known, but it seems likely to be off the outer coast.

The regular occurrence of "summer resident" Grey Whales in British Columbia waters was described in 1974 by biologists David Hatler and Jim Darling working in Clayoquot Sound. They were the first to use individual photo-identification of Grey Whales from natural markings, and they found that a distinctive animal occurred regularly over three summers in the early 1970s. By 1981 Darling had identified 93 different whales along the west coast of Vancouver Island, many of which were seen repeatedly over several summers. It is now known that there are several hotspots along the BC coast where summer resident whales can be found – most important are the southwestern shore of Vancouver Island where it parallels the West Coast Trail, Barkley Sound, Clayoquot Sound and near Cape Caution on the mainland coast. Other summer residents can be found in scattered locations along the coast, including the outer shores of Porcher, Aristazabal and Dundas islands on the north coast. Some can also be seen, most years, in Boundary Bay, south of Vancouver, especially early in the season. A total of about 200 summer resident Grey Whales have been identified between northern California and southeastern Alaska during the past 40 years, over half of which have been in BC waters. There is considerable movement of these animals among feeding locations in and outside British Columbia, and so, typically, fewer than 100 summer resident Grey Whales are likely to occupy BC waters in any given year.

Natural History

Feeding Ecology

Grey Whales are unlike other baleen whales in that they are primarily, though not exclusively, bottom-feeders. They feed on a wide variety of benthic and epibenthic invertebrates, especially gammaridean amphipods – small crustaceans like the sand-hoppers seen on beaches – including the tube-dwelling ampeliscid amphipods. Other important bottom or near-bottom prey include polychaete worms, Bay Ghost Shrimp, and mysids (also known as opossum shrimp), which swarm close to the bottom over shallow rocky reefs (figure 37). The whales also feed by skimming through swarms of free-swimming zooplankton, such as crab larvae, at or near the sea surface.

Figure 37. A Grey Whale foraging for mysiid shrimp in a kelp reef offshore of the West Coast Trail, southwestern Vancouver Island. Photo: B. Gisborne, September 2008.

Long-term studies by Jim Darling and colleagues as well as research by University of Victoria's David Duffus and his graduate students in Clayoquot Sound have provided insight into patterns of habitat use and diet of summer resident Grey Whales. The most prevalent foraging activity during summer is feeding on swarms of mysids, most commonly *Holmesimysis scultpa*, which are very small (<11 mm) but occur in high densities just above the sea floor in and near kelp beds. One study near Cape Caution north of Vancouver Island found average densities of 440,000 mysids per cubic metre in swarms targeted by feeding Grey Whales. During summer months, Grey Whales seen close to the shoreline along the West Coast Trail between Port Renfrew and Pachena Bay on Vancouver Island's west coast are feeding mostly on mysids.

Another important foraging activity in parts of Clayoquot Sound, especially later in the season, is benthic feeding in shallow, soft-bottom bays for amphipods, mainly *Ampelisca* spp. Amphipods are tiny – only about 3–8 mm in length – but occur in or on the sandy seabed at densities averaging over 25,000 per square metre. Benthic feeding is often seen in Cow Bay on Flores Island and Ahous Bay on Vargas Island. Grey Whales typically roll on their right side and swim slowly forward while sucking sediment into the side of their mouth and filtering out the prey with their baleen. The whales forage from the intertidal zone to depths of around 30 metres, and this feeding behaviour creates long trails of sediment called "mud plumes" that are visible in the water column and on the surface. It leaves pits in the soft bottom that are about 1–3 metres long, 0.5–1.5 metres wide and perhaps 0.5 metres deep. Back in the early 1970s benthic feeding could be observed on a daily basis during summer in Wickaninnish Bay, which is bordered by Long Beach in Pacific Rim National Park. But the Grey Whales' use of this area declined sharply in the late 1970s for unknown reasons, and they have only used this bay sporadically in recent years. Benthic feeding has also been observed in the mud bottom of Grice Bay, a shallow cul-de-sac on the inshore side of Esowista Peninsula, close to the Tofino Airport. Mostly it is subadult whales that seem to feed at this location, and they are primarily targeting Bay Ghost Shrimp.

Grey Whales have also been seen in Clayoquot Sound skim-feeding on free-swimming zooplankton near the surface. Primary prey species of this type are crab larvae – mostly of porcelain crabs and occasionally Dungeness Crabs. They also occasionally target dense patches of free-swimming amphipods, *Atylus borealis*, which are found above sandy seabeds. Finally, Grey Whales are regularly observed to feed on Pacific Herring eggs, which are laid in great quantities on eelgrass or algae in certain nearshore shallow waters during the spawning period of March to April. Key sites for such feeding near Vancouver Island are Hesquiat Harbour at the western end of Clayoquot Sound, Toquart Bay in Barkley Sound and along the east side of Moresby Island in Haida Gwaii. Herring eggs are probably targeted by summer residents and some northbound migrants.

Behaviour and Social Organization

Grey Whales are generally found alone or in small groups of two or three. Group composition is dynamic, and there do not appear to be any stable, long-term bonds among individuals. But some well-known summer residents have been observed to return annually to Clayoquot Sound for many years, and some of these individuals frequently travel together. Large aggregations occur in breeding lagoons and on northern feeding grounds in the Bering and Chukchi seas.

During migration, Grey Whales swim steadily day and night. On their spring northward migration, they average 4–5 km per hour and may cover 88–127 km per day. Southbound migrants in winter travel faster, averaging 7–9 km per hour and 144–185 km per day. While migrating or on the breeding grounds, they are frequently active at the surface – breaching, headstanding or spyhopping. Aerial displays are seen less commonly on the feeding grounds (figure 38). A study of the Grey Whales' activity budget during summer in the Cape Caution area found that about 77 per cent of their time was spent foraging, 15 per cent travelling between feeding sites and 8 per cent socializing. While feeding in kelp beds for mysids, the whales typically dove for an average of 2.2 minutes, surfaced for three breaths about 15 seconds apart, then dove again. Individuals averaged about 26 dives per hour, spending 83 per cent of their time under water. Grey Whales are not fast swimmers but can reach a top speed of about 16 km per hour for brief periods.

Nineteenth-century whalers hunting in the breeding lagoons of Baja California called the whales Devil Fish because of their ferocity when harpooned or when their calves were threatened. Grey Whale mothers also vigorously defend their calves from mammal-hunting Killer Whales, which target the vulnerable neonates during their northward migration. Lacking the speed to escape from Killer Whales, Grey Whales avoid predation by seeking the cover of kelp beds or shallow water, if close by, or by physically defending themselves and their calves if caught in the open. During the northward migration, mothers and calves hug the shoreline whenever possible, following the contours of bays and indentations rather than swimming from point to point as do the older cohorts. When crossing open stretches of water, the mothers and calves adopt a low-profile swimming pattern to escape detection by Killer Whales, surfacing to breathe in a manner known as *snorkelling* – raising just the blowholes above the surface and breathing very quietly. If attacked by Killer Whales, the mother will roll and thrash at the surface and strike out at the attackers with her tail flukes. Hotspots for Killer Whale attacks on migrating mothers and calves are Monterey Bay, California, and near False Pass, Alaska, but attacks have also been observed several times along the west coast of Vancouver Island.

Grey Whales are not known for their vocal prowess, but they do produce a variety of low-frequency sounds with energy occasionally extending as high as 4 kHz. The most notable is a rapid series of sharp knocking sounds that may have a hollow or metallic quality. Sounds are mostly heard during social activities in or near the breeding lagoons. Grey Whales are relatively quiet when migrating or on their feeding grounds.

Despite their reputation for combative defence when hunted during the whaling era, Grey Whales have become increasingly habituated to humans thanks to frequent non-threatening interactions with whale-watchers, particularly in the breeding lagoons in Baja Califor-

Figure 38. Grey Whales are often seen in shallow water close to shore along the west coast of Vancouver Island. This whale is spyhopping, raising its eyes above the water, and revealing patches of barnacles and whale lice on its rostrum. Photo: B. Gisborne, August 2009.

nia. There, certain whales – including mothers with calves – will seek out boats full of people and seem to enjoy being patted and scratched on the head or back. These "friendly" whales are becoming increasingly common and are seeking out interactions with boaters in other areas, such as the west coast of Vancouver Island. While thrilling for the whale-watchers, it carries risks to the whales, which may be inadvertently injured by boat propellers or entangled in fishing gear.

Life History and Population Dynamics

Grey Whales attain sexual maturity at about 8 years of age (range 6–12 years) at mean lengths of 11.7 metres in females and 11.1 metres in males. Mating and calving is strongly seasonal and synchronized with migration. Females typically have only one estrus cycle every two years, which means that, at most, only half the females are available each year for mating. Grey Whales appear to have a promiscuous mating system, with both sexes copulating with several partners during the mating season, which is during the southward migration or early in their stay in the breeding lagoons. A single calf is born after an 11- to 13-month gestation. The peak of calving is mid January, with many births either during the southward migration or on or near the breeding grounds. Calves are nursed with a very rich milk, which has a higher fat composition (53%) than that of most other cetaceans. Calves are weaned at 7–9 months, typically during the summer feeding period and prior to the southward migration. Longevity is not well known, but it is likely at least 40 years and may reach 80 years.

Exploitation

First Nations in various parts of the Grey Whales' range hunted them for thousands of years. The Chukchi and Yu'pik people of the Chukchi and Bering seas, respectively, hunted Grey Whales, as did the Aleuts living along the Aleutian chain of islands in Alaska. The Aleuts hunted whales from skin-covered baidarkas (kayaks), using lances or spears tipped with a poison, aconite (it is likely that the poison was ceremonial and did not actually debilitate and kill the whales). Grey Whales were also hunted by the Nuu-chah-nulth people along Vancouver Island's west coast and the by Makah of western Washington using open canoes with harpoons and floats. Although the Grey Whale is the species most commonly associated with Nuu-chah-nulth whaling, recent evidence from archaeological studies suggests that the Humpback Whale was of greater importance. At one site in Barkley Sound, Humpback Whales made up 77 per cent of bone fragments identified by DNA and Grey Whales only 11 per cent. The Nuu-chah-nulth name for Grey Whale is *ma·ʔak*.

Commercial exploitation of Grey Whales in the northeastern Pacific began in the mid 1800s. American "Yankee" whalers used sailing ships that launched small oar-powered boats to pursue Grey Whales in the breeding lagoons of Baja California as well as along their coastal migration routes. This "fishery" expanded rapidly, peaking in 1855–65 with annual catches averaging 480 whales, then declining around 1875. An estimated 6000–8000 whales were taken during this period. Because the whalers often targeted females with calves in the lagoons, the population lost much of its reproductive capacity, which further accelerated the decline. By the turn of the century, the population was commercially extinct but some whaling continued in various parts of the range. Grey Whales received partial protection under the International Agreement on the Regulation of Whaling in 1937, which was ratified by Canada, the US and Mexico but not Russia. Full protection came with the International Convention for the Regulation of Whaling in 1946, which included the Soviet Union. The International Whaling Commission (IWC, established by the 1946 convention) currently allows a subsistence hunt by Chukchi people off the Chukotka Peninsula, Russia, which has averaged about 150 whales annually over the past 50 years. The Makah in Washington resumed subsistence hunting of Grey Whales in 1999. To date, they have taken only two animals.

Taxonomy and Population Structure

The Grey Whale is the only living species in the family Eschrichtiidae (formerly Rhachianectidae, as was used in the original *Mammals of BC* by Cowan and Guiget). The extinct North Atlantic Grey Whale is also considered to belong to this species. The recent discovery that Grey Whales identified around Russia's Sakhalin Island travel to our shores and mix with eastern Grey Whales raises questions about population structure in the North Pacific. Recent studies by geneticist Tim Frasier, Jim Darling and others have shown that the whales belonging to the "summer resident" group or Pacific Coast Feeding Aggregation (PCFA) are genetically differentiated from the larger migratory population and therefore may warrant separate management for conservation.

Conservation Status and Management

The eastern North Pacific population of Grey Whales has recovered well from 19th-century whaling. Various estimates put the abundance of this population prior to the start of commercial whaling at 12,000–24,000 whales, and this was reduced to only a few thousand by 1900. The population likely began to recover early in the 20th century, especially after it was legally protected in 1937. Annual shore-based counts in California suggested that the population was increasing at an annual average rate of 3.2 per cent during 1967–88 and reached a peak abundance of 21,000 in 1998. But very high mortality and low calf production in 1999 and 2000 reduced the population to about 16,000 in 2001. This decline may have been caused by poor feeding conditions in the Bering and Chukchi seas due either to unusually extensive ice cover, which may have limited the area over which the whales could forage, or a decline in benthic productivity. The whales migrated south in poor nutritional condition and experienced high mortality due to starvation, especially during the return northward migration. Many of the stranded whales were emaciated, and aerial measurements using photographs of live whales showed that they too were skinnier than normal. Migration is energetically demanding for Grey Whales, which feed little if at all while migrating and lose as much as two per cent of their body weight per week while away from their feeding grounds. The population began to recover after 2001 and the most recent estimate is again around 20,000 animals, which is thought to be close to the current carrying capacity of their habitat.

Today the greatest human threat to the eastern population of Grey Whales is increasing activity and development in and around the breeding lagoons in Mexico. Fortunately, the Mexican government has taken substantial steps toward protecting the whales and limiting human activities in these critical areas.

Other threats to Grey Whales include fishing gear entanglement, ship strikes, and noise and disturbance from shipping and industrial development. Grey Whales are particularly vulnerable to entanglement in crab trap lines since they swim in relatively shallow areas that are often fished for Dungeness Crab. Like other large whales, Grey Whales are vulnerable to vessel strikes, particularly from ships and small boats moving at speeds of over 15 knots. Their tendency to migrate in nearshore coastal waters generally keeps Grey Whales away from major shipping lanes, but our recent finding that the northbound migration corridor for the population is Hecate Strait and Dixon Entrance raises concerns about more extensive interactions with ships. Much of this route currently overlaps with moderate levels of vessel activity, and this is projected to grow sharply in the coming years with the proposed expansion of ports at Kitimat and Prince Rupert. Hecate Strait is also the site of potential future oil and gas exploration and wind farm development, which could also affect Grey Whale migration. Seismic testing and wind farm construction both produce noise at levels that have been shown to cause avoidance behaviour in migrating Grey Whales.

The eastern Pacific population of Grey Whales is listed by the IUCN as Least Concern, and the western population is listed as

Critically Endangered. In Canada Grey Whales have been designated as Special Concern by COSEWIC and are on the BC Blue List.

Remarks

The genus name for the Grey Whale, *Eschrichtius*, and that of its family, Eschrichtiidae, were given by British zoologist John Edward Gray in honour of 19th-century Danish cetologist Daniel Eschricht. Although Eschricht never studied Grey Whales, he wrote some of the earliest descriptions of the natural history of Killer Whales and Bowhead Whales. The common name of the species refers to its coloration, not the person who named it – hence the different spellings in Canada (Grey) and the USA (Gray). The species name *robustus* means "strong" or "oaken" in Latin.

Selected References: Allen and Angliss 2011, Darling 1984, Darling et al. 1998, Dunham and Duffus 2002, Fisheries and Oceans Canada 2010, Ford et al. 2013, Frasier et al. 2011, Hatler and Darling 1974, Jones and Swartz 2009, Mate et al. 2011, McMillan et al. 2008, Monks et al. 2001, Pike 1962, Pike and MacAskie 1969, Reeves et al. 2010, Rice and Wolman 1971, Stelle 2001, Stelle et al. 2008, Swartz et al. 2006, Weller et al. 2012.

Common Minke Whale

Balaenoptera acutorostrata

Other Common Names: Minke Whale.

Description

The Common Minke Whale is a small baleen whale with a sleek, streamlined body. The rostrum is narrow and sharply pointed, and there is a single dorsal ridge that runs along the midline of the head. The dorsal fin is relatively tall and falcate, and it is positioned about two-thirds of the way along the body. The flippers are narrow with pointed tips. There are 50–70 fairly short throat pleats extending from the tip of the lower jaw to just behind the flippers.

Minke Whales are dark grey to black dorsally and white ventrally, with swaths of lighter coloration on the sides, including a "shoulder streak" behind the flippers which can form a chevron pattern over the back. These swaths may be indistinct on some individuals. Oval white scars are often present on the sides and back, likely as a result of Cookiecutter Shark bites. A distinct white band extending laterally across both the outer and inner surfaces of the flippers is unique to the species (figure 39). The baleen is generally white to cream-coloured.

Measurements

length (metres):

- male: ~ 7.9
- female: ~ 8.5
- neonate: ~ 2.5–2.8

weight (kg):

- adult: ~9200
- neonate: ~320

Figure 39. This Minke Whale breaching in Cormorant Channel displays the white band over its flipper. Photo: J. Towers, July 2011.

Baleen

Plates on each side: 230–275 (North Pacific form)
Coarse bristles, up to 21 cm long, white or cream-coloured.

Identification

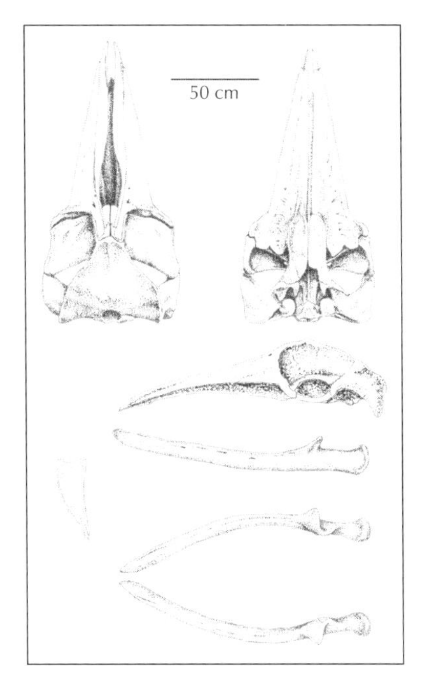

Common Minke Whales can be difficult to spot at a distance because of their inconspicuous blows and brief, erratic surfacing behaviour, but once sighted they are fairly easy to identify. They share a similar body form with Blue, Fin and Sei whales, but they are much smaller than these species, which typically have a highly visible blow. Minkes usually begin surfacing with their rostrum pointed up, then quickly roll forward as they breathe, with the dorsal fin appearing very soon after the blowholes have submerged. In contrast, Sei Whales usually surface with their blowholes and dorsal fin above

Figure 40. Common Minke Whales can be curious and will sometimes approach boats or kayaks for closer inspection.
Photo: B. Collen, Barkley Sound, May 2006.

the surface at the same time, while the dorsal fin of Fin and Blue whales usually appears long after the blowholes have submerged. At a distance, the dorsal fin of a Minke could be confused with that of a small Killer Whale or perhaps a Pacific White-sided Dolphin, but it is positioned further back on the animal's body. The dorsal fin of beaked whales is also positioned fairly far back on the body, but the fin of these deepwater species tends not to be as falcate. At close range, the Minke Whale's sharply pointed rostrum and diagnostic white flipper band should be visible. Minkes are often difficult to approach, but they occasionally come close to slowly moving or drifting boats, as if curious (figure 40).

Distribution and Habitat

Common Minke Whales are distributed widely in tropical to cold waters in both the northern and southern hemispheres. Although they can be found offshore, they are most often observed in shallow, coastal waters. Like most baleen whales, Common Minke Whales are considered migratory, moving seasonally from winter breeding areas to summer feeding areas in higher latitudes. But their migratory patterns are poorly known, and the species can also be found in temperate waters throughout the year.

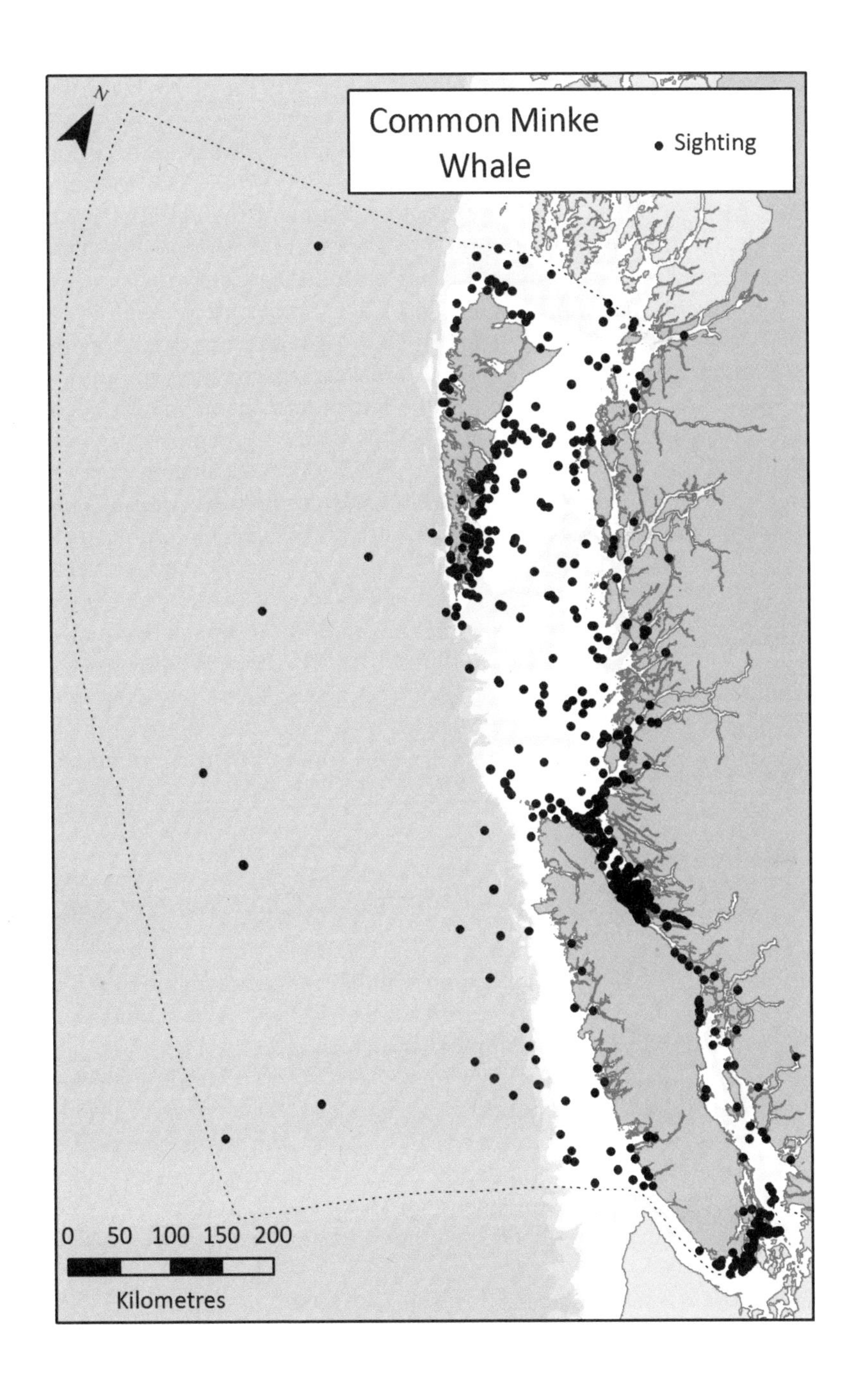
Common Minke Whale
Sighting
N
0 50 100 150 200
Kilometres

In the North Pacific, Common Minke Whales occur from the Bering and Chukchi seas to the equator. In both the western and eastern North Pacific, Minkes are thought to migrate northward in spring and summer and to move southward in fall and winter. During summer, Minke Whales occur in nearshore waters of the eastern North Pacific, from the Aleutian Islands to Baja California, but they have been sighted in all seasons off California and Baja California.

In British Columbia, Common Minke Whales are present along the entire coast, from Dixon Entrance to Juan de Fuca Strait. Sightings are most frequent around the Gulf Islands, particularly in Boundary Pass and Haro Strait, and in the straits and channels near Alert Bay off northeastern Vancouver Island. They are also regularly sighted along the east side of Moresby Island, Haida Gwaii, in shallow parts of Hecate Strait, Queen Charlotte Sound and off the west coast of Vancouver Island. Sightings have been reported in British Columbia in all months of the year, with the highest rates in July and August and relatively few reports for December through February. Most or all Minke Whales in BC waters probably migrate south for at least a portion of winter, but the timing may be diffuse. Of 44 Minke Whales photo-identified in BC by Jared Towers and colleagues, all but one exhibited white oval scars that were likely caused by Cookiecutter Sharks, which are found only in warm waters well to the south of British Columbia.

Natural History

Feeding Ecology

Common Minke Whales feed on a wide variety of small fish and zooplankton. In the North Pacific, important prey species include Pacific Herring, Pacific Saury, Japanese Anchovy, Walleye Pollock and krill (euphausiids). In British Columbia and the adjacent waters of the US San Juan Islands, Minke Whales have been observed feeding on Pacific Herring and Pacific Sand Lance (figure 41). Most observed feeding activity is concentrated over shallow banks and submarine slopes with a moderate incline and depths of 20–100 metres. Of two Minkes taken by Japanese whalers offshore of Haida Gwaii in July 1971, one had been feeding on krill and the other on fish (species unrecorded).

Behaviour and Social Organization

In British Columbia, Common Minke Whales are typically seen alone – on rare occasions, pairs or trios are seen. Loose aggregations can sometimes be observed in good feeding areas, but these groupings do not persist. Pioneering studies by researcher Eleanor Dorsey and colleagues in the San Juan Islands showed that individual Minke Whales have high site fidelity, returning each year to feed in three particular locations around the San Juan Islands. From 1980 to 1984, thirty Minkes in that area were photo-identified from natural markings on their dorsal fins and backs, and eighteen of these were observed repeatedly feeding in the same areas in multiple years. Individuals specialized in one of two feeding tactics: some simply lunged at patches of schooling fish, while others targeted "bait balls" (schools of herring or sand lance driven to the surface and densely concentrated by diving seabirds or predatory fish). Researcher Jared Towers of the Marine Education and Research Society, Alert Bay, has been studying Minke Whales in BC, especially in the channels and straits off northeastern Vancouver Island. Like Eleanor Dorsey in the San Juan Islands, he has found strong site fidelity, with six individuals feeding repeatedly each summer in the channels between the islands near Alert Bay. A few of these individuals have also been photographed off Victoria and near Tofino in spring and fall, perhaps returning from or heading to southern wintering areas.

Until recently, very little was known about the vocal behaviour of Common Minke Whales. Low-frequency downswept calls about half a second long had been described for Minke Whales feeding in

Figure 41. A Common Minke Whale lunge feeding on Pacific Sand Lance. Photo: J. Towers, Alert Bay, June 2014.

the St Lawrence estuary, Quebec, but no vocalizations had been recorded from North Pacific Minkes. Finally in 2005 cetacean scientists Shannon Rankin and Jay Barlow with the Southwest Fisheries Science Center in California solved a long-standing mystery by linking the presence of Common Minke Whales observed off Hawaii during ship surveys with a distinctive underwater sound known as a "boing". This strange signal – a brief pulse followed by a two- to four-second pulsed tone with energy reaching over 4 kHz – was first described in the 1960s from US Navy submarine recordings off California and Hawaii, but its source was uncertain. Recordings from a seafloor hydrophone array off Hawaii have detected boing sounds from November through March, which corresponds to the probable period of occurrence of breeding Minkes. It is possible that boing sounds are produced only by males and play a role in courtship or mating, as is true of the songs of Fin and Humpback whales. Minkes feeding along the coast of BC are very quiet, but Jared Towers has recorded a few low-pitched sounds from Minkes using an autonomous recording device moored underwater near Alert Bay.

One reason Minke Whales may be reluctant to vocalize in local waters is that this would increase the likelihood of their being detected by mammal-hunting Killer Whales, which are common in coastal waters of BC. Being smaller than other baleen whales in the region, Common Minke Whales are more vulnerable to predation by Transient (Bigg's) Killer Whales. Between 1994 and 2012, my colleagues and I documented 13 chases or attacks on Minke Whales in coastal BC and southeastern Alaska, 8 of which ended with the Minke being killed. Like other species in the genus *Balaenoptera*, Minke Whales use a "flight" strategy to escape predators, and can often outdistance Killer Whales if they have sufficient open water to maintain a prolonged high speed. Killer Whales can reach a similar speed to that of Minkes – about 30 km per hour – but cannot sustain it for as long. If Killer Whales manage to chase the Minke into a narrow or confined bay, then it is invariably cornered against the shore and killed. In October 2002 four Transient Killer Whales chased a Minke Whale at high speed into Ganges Harbour, Salt Spring Island, where it partially stranded in the shallows along the breakwater in front of town. Hundreds of onlookers watched as the Killer Whales took turns ramming the whale for several hours before it attempted to escape into deeper water and was killed and eaten. As is typical of *Balaenoptera* whales that use flight to avoid predation, the Minke made no effort to defend itself from the attacking Killer Whales.

Life History and Population Dynamics

Breeding in the northern hemisphere is diffusely seasonal, with calves born mostly during winter following a gestation of about 10 months. Calves are then nursed for five or six months. Pregnancy rates for Common Minke Whales are high – approaching 90 per cent in some populations – suggesting that reproduction is annual. Sexual maturity has been estimated at 7 years in males and 6 years in females. Longevity is poorly known, but is probably at least 50 years and may be up to 60 years.

Migration of Common Minke Whales appears to be staged by age and sex, and there is some evidence that this may also be the case in British Columbia feeding areas. All stranded Minke Whales for which sex was reported in BC (six) and Washington (two) were female. In contrast, two Minkes taken by whalers in deep water off the BC coast were both males. These samples are obviously small and more data are needed to confirm segregation. There are no confirmed sightings of whales accompanied by calves in BC, suggesting that weaning takes place before the whales return to these waters.

Exploitation

Common Minke Whales were too small to be important to commercial whalers until the larger species were depleted in the first half of the 20th century. Focus then shifted to smaller species, and substantial numbers of both Common and Antarctic minke whales were taken in the southern hemisphere and in the North Atlantic and North Pacific. Today, Common Minke Whales are still hunted in the North Atlantic by Norway and in the western North Pacific by Japan.

In British Columbia, Minke Whales appear not to have been targeted by First Nations whalers. There is no reference to the species in ethnographic accounts of Nuu-chah-nulth whaling, nor have remains been identified from archaeological excavations along the west coast of Vancouver Island. But Minkes were evidently hunted occasionally by Makah whalers off Cape Flattery and in Juan de Fuca Strait. Minke Whales were of little interest to commercial whalers in British Columbia. There are records of only three Minkes being taken by shore-based whalers: two off Haida Gwaii in 1923 and one off the west coast of Vancouver Island in 1967. Japanese ship-based whalers killed an additional two Minkes west of Haida Gwaii in 1971.

Taxonomy and Population Structure

Until fairly recently, all Minke Whales were considered to belong to the same global species, *Balaenoptera acutorostrata.* But evidence for a separate species in the southern ocean began accumulating, and by the late 1990s *B. bonaerensis* became widely accepted as a distinct species, the Antarctic Minke Whale. This triggered the need for a new common (as opposed to scientific) name for *B. acutorostrata* to distinguish it from this new species, hence the Common Minke Whale. Three subspecies of Common Minke Whales are currently recognized – the North Atlantic Minke Whale (*B. a. acutorostrata*), the North Pacific Minke Whale (*B. a. scammoni*, formerly *B. a. davidsonii*) and the as-yet-unnamed southern hemisphere Dwarf Minke Whale. In the North Pacific there are at least three genetically distinct populations of Common Minke Whales: one in the Sea of Japan and East China Sea, another in the western Pacific west of 180°W, and at least one in the eastern Pacific. Significant differences in the structure of Minke Whale boing calls recorded off Hawaii compared to California suggest that there is subpopulation structure in the eastern North Pacific as well.

Conservation Status and Management

Despite continued whaling for Common Minke Whales in some regions, their populations are thought to be in better condition than those of many other baleen whales. Abundance estimates for some areas of the North Pacific show about 20,000 in the western Pacific, about 2000 in the Bering Sea and 1000 off the US west coast. No estimate is available for the whole of British Columbia waters, and although the species is commonly seen in some areas along the coast, it does not seem to be particularly abundant. Minke Whales were sighted on only 45 occasions during almost 30,000 km of shipboard survey effort by DFO's Cetacean Research Program from 2002 to 2011. In contrast, 2807 Humpback Whale and 462 Fin Whale sightings were tallied during these same surveys. Rob Williams and colleagues with the Raincoast Conservation Society conducted surveys in continental shelf waters from Dixon Entrance to Juan de Fuca Strait (excluding the west coasts of Vancouver Island and Haida Gwaii) in 2004–05, which resulted in 14 Minke Whale sightings, mostly in the open waters of Hecate Strait and Queen Charlotte Sound. From these sightings, the researchers estimated an abundance of 388 whales in those waters. Photo-identification studies in inside waters also suggest low densities. For example, the channels between western Johnstone Strait and eastern Queen Charlotte Strait are considered excellent areas for finding Minkes, but researcher Jared Towers and colleagues found that close to 90 per cent of about 300 photo-identification encounters with Minkes from 2005 to 2012 involved repeated identifications of just 6 individuals. In total, they have identified only 44 whales in over 400 encounters from the central BC coast to Victoria. This certainly suggests that the species is not plentiful in coastal BC waters.

Aside from whaling, the main human-caused threats to Common Minke Whales are fishing-gear entanglement and ship strikes. It is not known if these are important sources of mortality for Minkes in BC.

The species is ranked by the IUCN's Red List as Least Concern, is considered Not at Risk in Canadian waters by COSEWIC and is on the BC Yellow List.

Remarks

The common name of the species is said to originate in Norway, where a neophyte whale spotter named Meincke misidentified a Minke Whale as a Blue Whale, and these small rorquals were thereafter referred to as "Meincke's Whale" (pronounced mink-ee). The specific name *acutorostrata* means "sharp rostrum", referring to the whale's pointed head.

Selected References: Carretta et al. 2011, COSEWIC 2006, Cowan 1939, Dorsey et al. 1990, Ford et al. 2005 and 2010a, Perrin and Brownell 2009, Pike and MacAskie 1969, Reeves et al. 2002, Scheffer and Slipp 1948, Stewart and Leatherwood 1985, Towers 2011, Towers et al. in press, Williams and Thomas 2007.

Sei Whale *Balaenoptera borealis*

Other Common Names: None.

Description

The Sei Whale has a large, sleek and highly streamlined body typical of the genus *Balaenoptera*. It has a sharp rostrum and a single prominent longitudinal ridge that runs forward along the centre of the rostrum ahead of the blowholes. The dorsal fin is relatively taller than that of Fin and Blue whales, and is usually falcate. Sei Whales have about 50 ventral grooves or throat pleats (range 40–65), fewer than Blue, Fin or Common Minke whales. These pleats end forward of the navel, unlike in other *Balaenoptera* species, where they extend past the navel.

Sei Whales are dark grey dorsally, as well as on the ventral surfaces of the flukes and flippers, while the ventral surface of the body is a lighter grey to creamy white. Many individuals have light greyish streaks sweeping up and backwards from the area of the eye. Numerous white oval scars scattered over the body, particularly the posterior half, are thought to be mostly from Cookiecutter Shark bites but may also be caused by the bites of Pacific Lampreys or by parasitic copepods *Pennella* sp., which burrow through the skin and into the blubber.

Measurements

length (metres):

male: 13.2 (7.9–17.1) n=2144
female: 13.6 (9.4–19.5) n=1631
neonate: ~ 4.5

weight (kg):

adult: <45,000
neonate: ~ 700

Baleen

Plates on each side: 350 (219–402)
Up to 80 cm long, black with a fine light grey or whitish fringe.

Identification

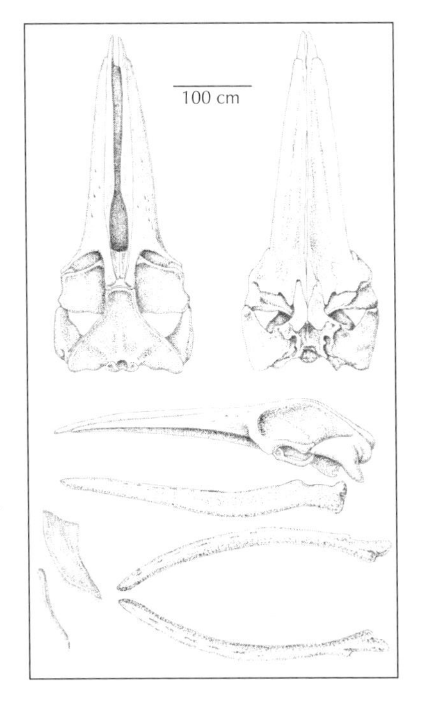

In British Columbia waters Sei Whales are most easily confused with Fin Whales and, less so, with Blue Whales and Common Minke Whales. Besides being larger, Fin Whales have a dorsal fin that is smaller relative to body size, rises at a shallower angle from the back and is positioned slightly further back on the body compared to Sei Whales. However, in the North Pacific there is so much variation in dorsal fin shapes of both species that it is not a foolproof way to tell them apart. A diagnostic feature of Fin Whales is asymmetrical coloration of the lower jaw, which is white on the right side only, whereas the lower jaw of the Sei Whale is dark on both sides. The white jaw is usually visible when a Fin Whale surfaces, but seeing it requires being positioned off the whale's right side (or above, if viewing from an aircraft). When a Sei Whale surfaces, the dorsal fin is usually visible at the same time as the blowholes, while a Fin Whale's dorsal fin appears after it has completed its blow and is beginning to dive. Fin Whales also tend to arch their back more than Sei Whales when starting a deep dive. Blue Whales are larger and have blue-grey mottling on the body as well as a much smaller dorsal fin than Sei Whales. Common Minke Whales are much smaller and rarely have a visible blow. They also have white patches on their flippers, but these can be difficult to observe at sea.

Distribution and Habitat

The Sei Whale has a cosmopolitan distribution, occurring in the North Pacific, North Atlantic and southern hemisphere. It is an oceanic species and seldom found in coastal waters. Sei Whales migrate between subtropical to warm-temperate waters in winter and cold-temperate to subpolar waters in summer. They prefer water temperatures of 8–18°C and therefore don't frequent high latitudes like most other

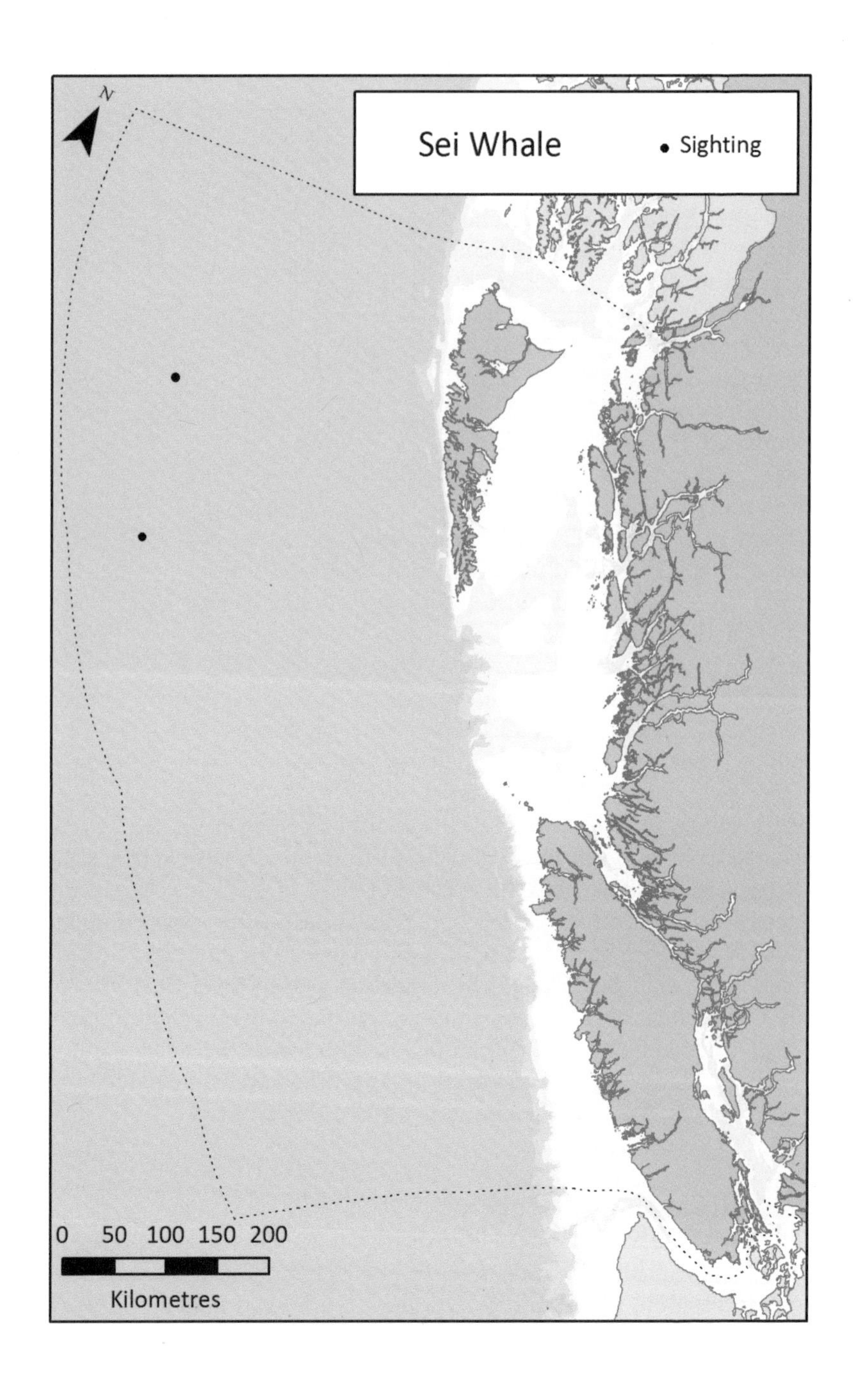
N
Sei Whale
Sighting
0 50 100 150 200
Kilometres

rorquals. Sei Whales exhibit large-scale shifts in distribution; they may abruptly disappear from areas where they had occurred regularly for years and appear in others where they had long been absent. Naturalist Roy Chapman Andrews remarked in 1916 that the Sei Whale "has a roving disposition and sometimes travels great distances in its wanderings". Such nomadic movements may be related to changes in prey abundance and ocean-scale regime shifts, such as the Pacific Decadal Oscillation.

Our understanding of the distribution of Sei Whales in British Columbia is based entirely on catches during the whaling era. The great majority of Sei Whales were taken in deep water beyond the continental shelf break (99 per cent in depths greater than 1000 metres; see figure 18). Most catches were off the west coast of Vancouver Island and relatively few off BC's north coast, but this may be due to the general lack of interest in this species prior to the early 1940s when the two whaling stations in Haida Gwaii ceased operations. Pregnant females were the first to be taken in the whaling season, which began in early to mid April, followed by non-pregnant females and males. There appeared to be a strong offshore shift in distribution as the season progressed through September. It is possible that avoidance of whaling vessels caused some movement offshore, or that whales inclined to remain closer to shore were removed first, but no similar effect was apparent in the other rorqual species being hunted. Recovery of embedded Discovery tags (see figure 24) at the Coal Harbour whaling station provided some indication of the wide-ranging movements of Sei Whales. A 12.5-metre male marked by Soviet whalers on April 22, 1964, in the open ocean about 2000 km southwest of Tofino was killed about 200 km off central Vancouver Island on August 14. That same year, a Japanese scout boat tagged a 14.3-metre male on May 29 about 270 km north-northwest of Haida Gwaii, and it was killed on July 12 about 100 km west of Estevan Point, Vancouver Island.

Natural History

Feeding Ecology

Sei Whales are unusual among baleen whales in having two modes of feeding: they are "skimmers", filtering zooplankton from the water while swimming with the mouth open, and "gulpers", lunging at prey patches and taking a mouthful at a time. The fringes on their baleen plates, fine compared to other rorquals, allow them to skim the surface for small zooplankton, especially copepods that are only a few millimetres in size. When in lunge- or gulp-feeding mode, Sei Whales mostly target small schooling fish. The throat pleats of Sei Whales are relatively shorter than those in other members of the genus, suggesting that they take smaller mouthfuls than Fin and Blue whales. The use of multiple feeding techniques appears to give Sei Whales more flexibility in the types of prey they target. Stomach contents of 463 Sei Whales killed in BC waters during 1963–67 were examined at the Coal Harbour whaling station to determine diet (1453 Sei Whales were actually taken, but most had empty stomachs), and a striking shift in prey was observed over this period. From 1963 to 1965 copepods were present in more than 90 per cent of the stomachs, with euphausiids (krill) in only 3–10 per cent and fish in 5–10 per cent. Fish species were mostly Pacific Saury, with Pacific Herring, Walleye Pollock, lanternfish (myctophids) and Ragfish found occasionally. In 1966 and again in 1967 – the last year of whaling in BC – copepods were found in only 3–4 per cent of stomachs. In 1966 euphausiids dominated the diet of Sei Whales, then in 1967 fish were predominant and zooplankton relatively scarce in the diet. Off California, Northern Anchovy and euphausiids were the most common prey of Sei Whales taken from 1956 to 1971, and copepods were rare.

Behaviour and Social Organization

The social behaviour of Sei Whales is poorly known. They typically travel alone or in small groups of two to five, although larger aggregations of individuals may be seen in feeding areas. There is no information on the stability of groupings over time. Together with Fin Whales, Sei Whales are the fastest swimmers of the baleen whales. They reportedly can exceed 40 km per hour in short sprints when chased by whalers, although their routine swimming speed is probably less than 10 km per hour. Their high-speed swimming ability enables them to outpace Killer Whales, their only natural predator.

Like other baleen whales, Sei Whales produce low-frequency calls for communication. Most common in the North Pacific is a simple pulse, lasting 1.3 seconds and repeated at irregular intervals of 5–25 seconds. The pulse sweeps down in frequency from 39 Hz to 21 Hz, too low to be audible by human ears.

Life History and Population Dynamics

Sei Whales of both sexes reach sexual maturity at 8–10 years old. Whales in the northern hemisphere mate in winter, with a peak in conceptions in December. Calves are born following a gestation of 11–12 months and are weaned at 6–8 months after travelling with their mothers to summer feeding grounds. Intervals between calves are usually 2–3 years. The Sei Whale's lifespan probably exceeds 50 years.

Exploitation

Because of their mainly offshore distribution as well as their speed and elusiveness, Sei Whales were not generally available to First Nations whalers. The species has not been identified in archaeological investigations of ancient Nuu-chah-nulth village sites on the west coast of Vancouver Island or elsewhere in the province. Because they are smaller than Blue and Fin whales, Sei Whales were also generally overlooked by modern whalers until the larger species had been depleted. Sei Whales were first taken off the BC coast in 1917 from coastal whaling stations at Kyuquot, Rose Harbour and Naden Harbour. Total annual kills were less than 200 animals until the early 1960s, when interest in Sei Whales increased at the Coal Harbour whaling station due to the demand for fresh whale meat in Japan. Between 1962 and 1967 (when the station closed), 2155 Sei Whales were killed and processed (see Appendix 3). In addition, at least 355 Sei Whales were taken by ship-based Japanese whalers, and an unknown number were killed by Soviet whalers between 1964 and 1975 in offshore waters within Canadian jurisdiction. Commercial whaling of this species ended in 1975, when Sei Whales came under IWC protection.

Taxonomy and Population Structure

Sei Whales are closely related to Bryde's Whales, found in tropical to warm-temperate waters, which overlap with Sei Whales in lower latitudes. The two species are difficult to distinguish, and whalers often mistakenly recorded Bryde's Whales as Sei Whales. (Interestingly, two strandings of Bryde's Whales occurred in Puget Sound in 2010. There is no evidence that either of these whales entered Canadian waters before stranding, but they may have.)

Two subspecies of Sei Whales are recognized, *Balaenoptera borealis borealis* in the northern hemisphere and *B. b. schlegellii* in the southern hemisphere. There is some evidence that distinct populations of Sei Whales exist in the North Pacific. Studies of blood types and morphological measurements of whales collected by Japanese whalers showed differences in whales from the western and eastern North Pacific, with a division at roughly 155°W longitude.

Conservation Status and Management

Sei Whales were seriously depleted by commercial whaling throughout the world's oceans. Much of this depletion took place relatively late in the whaling era, after the larger Blue and Fin whales had become scarce and the focus shifted to Sei Whales. Prior to whaling, there were an estimated 42,000 Sei Whales in the North Pacific, and by the time the hunt ended in 1975, only about 9000 are thought to have remained. There is no sign that Sei Whales have recovered significantly in the eastern North Pacific. Sightings of the species have been rare in recent ship surveys for cetaceans off the US mainland west coast. Only three sightings were recorded during a survey covering an area of more than 1.1 million square km off the US west coast in 2008, which resulted in an abundance estimate of 215 whales.

Despite being the most common species taken during the last few years of whaling in BC, Sei Whales appear to be scarce in the region today. No Sei Whales were seen during 28,000 km of ship surveys off the BC coast by the Pacific Biological Station's Cetacean Research Program from 2002 to 2008, but most of these surveys were likely inshore of their preferred habitat. Only 13 per cent of the surveys were in deep water beyond the continental shelf (more than 1000 metres deep), where the great majority of the Sei Whale catches were made. Another reason for Sei Whales not showing up in these surveys is the difficulty of distinguishing them from Fin Whales at a distance, even with the "Big Eye" binoculars used on our ship surveys. During these surveys, we had a total of 519 sightings of baleen-type whales that could not be identified to species due to their distance from the ship track. Most of these were probably Fin or Humpback whales, but some could have been Sei Whales. During August 2–3, 2012, two sightings of a total of four Sei Whales were made in the outer portion of Canada's EEZ during a joint IWC/Japan cetacean survey of the eastern North Pacific (figure 42).

Since Sei Whales are generally found in oceanic waters far from land, they are at relatively low risk from most human impacts. Whaling was a major threat, but this ended in 1975 with protection of the species in the North Pacific. However, about 100 Sei Whales are taken annually in the North Pacific by Japan under a self-assigned scientific permit. Some Sei Whales die from ship strikes, but the extent of this problem is not known. A dead Sei Whale was found wrapped around the bow of a ship at Port Angeles, Washington, in 2003, but it is not known where this whale was hit or whether it was dead or alive when struck. Entanglement in fishing gear is a threat to all large whales, although is likely a minor one for Sei Whales due to their offshore

Figure 42. One of four Sei Whales sighted about 250 km off of Haida Gwaii during a cetacean survey in August 2012. Photo: S. Mizroch.

distribution. Ocean noise, pollution and ecosystem changes due to global warming are all potential risks to cetaceans, including Sei Whales.

Seven per cent of the Sei Whales taken off California during the 1950s and '60s had an unusual disease of unknown origin that caused progressive shedding of baleen plates and their replacement with an abnormal papilloma-like growth. These whales only had fish in their stomachs, suggesting that their ability to filter-feed on zooplankton was inhibited. None appeared to be emaciated, however, so the disease did not appear to be life-threatening. The disease has not been observed in other Sei Whale populations, and it was not noted in records from BC coastal whaling stations.

Globally, the Sei Whale is red-listed as Endangered by IUCN. In Canada, the Pacific population was assessed as Endangered by COSEWIC in 2003 and again in 2012, and the species is on the BC Red List. Fisheries and Oceans Canada developed a recovery strategy in 2006 and prepared an action plan to implement recovery actions in 2012. These actions include continuing research efforts to assess current abundance and distribution of Sei Whales through surveys and acoustic monitoring.

Remarks

The name "Sei" Whale originates from the Norwegian word "seihval", as these whales would arrive off the Norwegian coast at the same time as a fish known as "sei", or Saithe, a species of Pollock. It is usually pronounced "say", although some use the pronunciation "sigh". The species name *borealis* is Latin for "northern". Although Sei Whales were once abundant in BC waters, there is no stranding record for the species in the province likely because of their deepwater, offshore habitat.

Selected References: Andrews 1916; COSEWIC 2003; Ford and Reeves 2008; Ford, Abernethy et al. 2010; Ford, Koot et al. 2010; Gregr et al. 2000, 2006; Gregr and Trites 2001; Horwood 2009; McMillan et al. 2008; Rice 1974 and 1977.

Blue Whale *Balaenoptera musculus*

Other Common Names: Sulphurbottom Whale.

Description

The Blue Whale is not only the largest cetacean, but also the largest animal known to have existed on Earth. The largest Blue Whales on record were killed by whalers in the first half of the 20th century in the Antarctic and reached lengths of up to 31.7–32.6 metres (104–107 ft). The maximum reported weight is 190 tonnes, but 70–150 tonnes is more common. Blue Whales off British Columbia do not attain such massive sizes – the largest recorded Blue Whale off our coast was a 29.8 metre-long (98 ft) female with a girth of 19.2 metres (63 ft), brought into the Cachalot whaling station in Kyuquot Sound, Vancouver Island, in September 1915.

The Blue Whale has a body shape similar to others in its genus – very streamlined, tapered and elongated. It has a large, broad U-shaped head (viewed from above) that comprises nearly 25 per cent of its body length. The head is flat when viewed from the side except for a single prominent ridge along the centre of the rostrum, which ends in a large, raised *splash guard* around the blowholes. The dorsal fin is proportionately smaller than those of other rorquals and is positioned about three-quarters of the way back on the body. Fin shape varies from a small nubbin to triangular and falcate. The flippers are long and pointed, reaching about 15 per cent of the body length. The tail flukes are broad and triangular with a straight or slightly curved trailing edge from each tip to the median notch. Blue Whales have 60–88 throat grooves or pleats that run along the ventral surface from the tip of the lower jaw to the navel.

Blue Whales are generally bluish-grey and somewhat lighter on the ventral surface. They have distinctive mottled patterns on their flanks, back and ventral surface. The mottling pattern is permanent and unique on each animal, and this feature is used by researchers to photo-identify individuals. Blue Whales sometimes have a yellow to orange-brown cast on their ventral surface and lower sides, caused by diatom films on their skin. This coloration led whalers to call them "sulphurbottoms".

Measurements

length (metres):

male: 21.3 (9.8–25.0) n=191
female: 22.4 (9.8–29.8) n = 252
neonate: 6–7

weight (kg):

adult: 70,000–150,000
neonate: 2700–3600

Baleen

Plates on each side: 260–400
Up to 1 m long with a coarse fringe, black.

Identification

From a distance, Blue Whales can be confused with other large rorquals, especially Fin and Sei whales. But Blue Whales are much paler than these other species, their appearance varying from slate grey-blue on cloudy days to silvery or turquoise on bright sunny days in clear water. Grey Whales can also appear very pale due to light-coloured patches of pigmentation and encrusting barnacles, but Grey Whales are much smaller and differ substantially in body shape. Blue Whales tend to raise their massive "shoulders" and blowhole region out of the water more than other rorquals when surfacing. Their

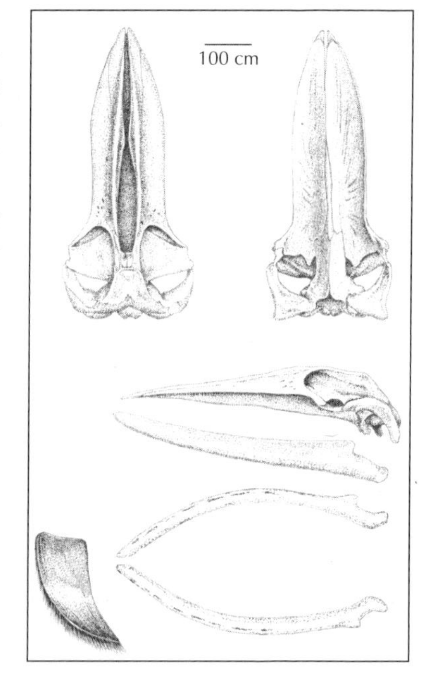

Figure 43. A Blue Whale diving near the CCGS *John P. Tully* during a cetacean survey off Haida Gwaii. Photo: L. Nichol, July 2013.

head is blunt and rounded, unlike the sharp heads of Fin and Sei whales. Their spout is taller (up to 10–12 metres), denser and broader than other whales. The dorsal fin of Blue Whales appears tiny relative to their body size and is set further back than in other rorquals, often appearing above the surface only during the last in a series of short surfacings prior to a long dive. About one in five Blue Whales has the habit of lifting its tail flukes above the surface when diving (figure 43), behaviour that is rarely seen in Fin or Sei whales but is common in Humpback, Grey and Sperm whales.

Distribution and Habitat

Blue Whales have a cosmopolitan distribution in the world's oceans, from tropical waters to the pack ice edges in both hemispheres. They are usually found in oceanic waters but also inhabit coastal and shelf waters in some regions. Their seasonal movements can be extensive but are complex and poorly understood. Many Blue Whales move between productive, high-latitude feeding areas in summer and fall and warmer, low-latitude regions in winter. In some areas, however, they can be found year-round.

In the North Pacific Blue Whales were widely distributed throughout coastal and pelagic waters before being depleted by commercial whaling. In the western North Pacific they ranged from southern Japan to Kamchatka and the western Aleutian Islands, in the central Pacific from Hawaii to the Aleutian Islands, and in the eastern North

Figure 44. A Blue Whale surfacing close to the southwestern shore of Moresby Island, Haida Gwaii. This individual has also been identified off southern California. Photo: J. Calambokidis, August 2006.

Pacific from Central America to the Gulf of Alaska. It's not known if Blue Whales currently make use of their entire historical range. Sightings in the western North Pacific are very rare, and these whales also appear to be scarce in the Gulf of Alaska. Satellite tracking of Blue Whales tagged off the California coast by whale researcher Bruce Mate and his colleagues during 1993–2007 revealed extensive migratory movements along the Central and North American continental margin. There was a general southward movement during winter to Baja California, especially the Gulf of California, and to an area west of the Costa Rica Dome, a productive offshore upwelling area. During spring and summer the whales tended to move north mostly to areas off California but occasionally as far as Vancouver Island and the Gulf of Alaska. Fine-scale movement patterns suggested that foraging takes place throughout the year and over much of their range.

In British Columbia waters, Blue Whales were taken by whalers mostly offshore of the continental shelf edge from southern Vancouver Island to the Alaska border (see figure 18), with a few also taken in Hecate Strait and in western Dixon Entrance. Sightings of Blue Whales in recent years have been rare, but this may be partly due to the lack of observer effort in oceanic waters off the BC coast. During ship surveys conducted by the Pacific Biological Station's Cetacean Research Program between 2002 and 2013, we saw Blue Whales on 16 occasions, all to the south or west of southern Haida Gwaii during June–August (figure 44). We have also logged four sightings of Blue

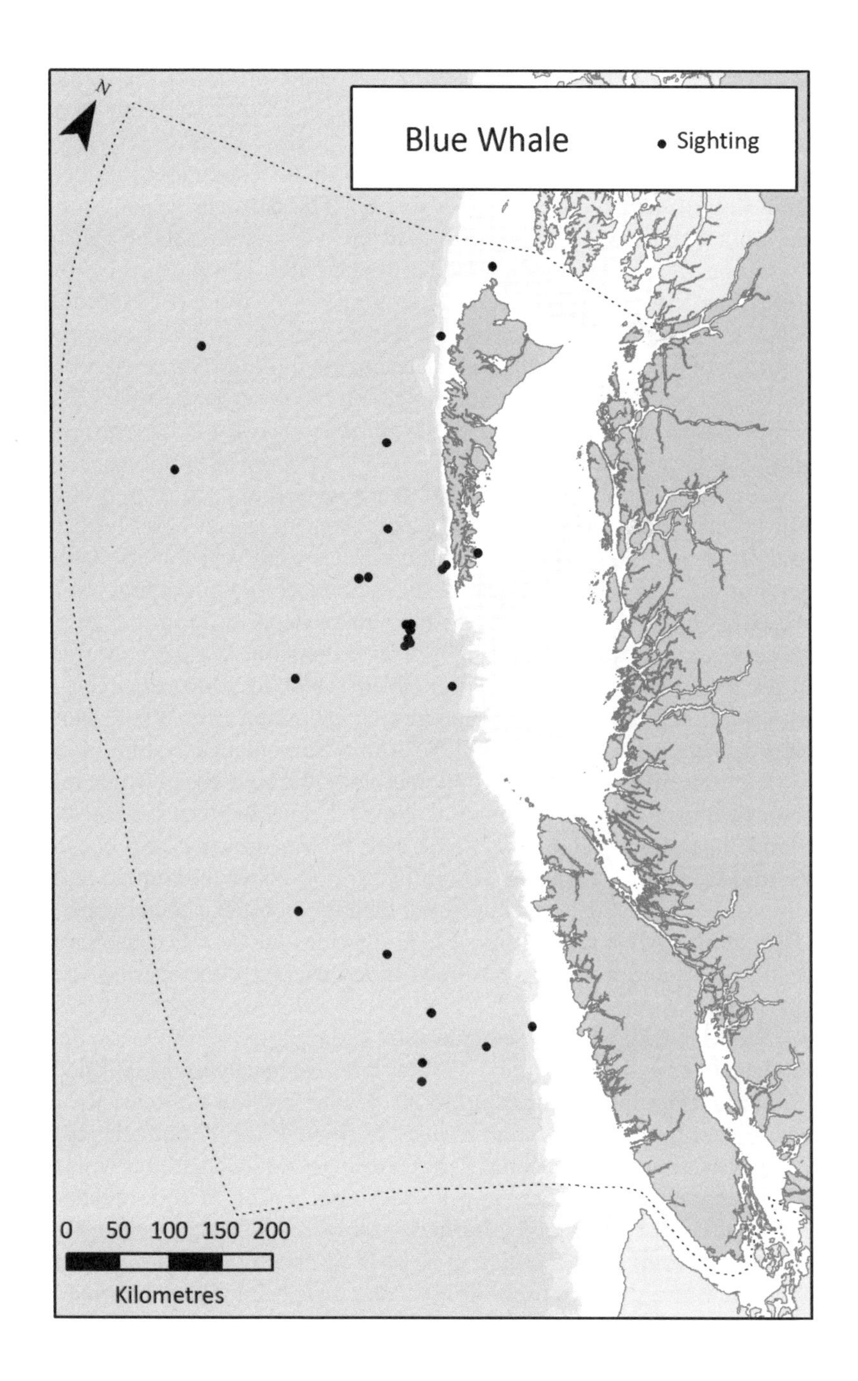
N
Blue Whale
Sighting
0
50
100
150
200
Kilometres

Whales during aerial surveys off the west coast of Vancouver Island in November and December, 2011–13. Several other sightings have been reported to the BC Cetacean Sightings Network in recent years, all in deep waters beyond the continental shelf. There are only two stranding records for Blue Whales in BC. The first was an 18.3 metre animal that was reported stranded on Wickaninnish Island near Tofino on August 12, 1959; it evidently died after being wounded by a harpoon from a whale catcher operating out of the Coal Harbour whaling station and the carcass drifted to shore. The second stranding record is of a decomposed Blue Whale carcass found in Survey Bay on the west coast of Banks Island on the north BC coast on February 22, 2014. Based on the length of the skull, this animal was between 20 and 25 metres long.

Blue Whale catches off the BC coast were lowest in March and April, when the whaling season began, then increased steadily to a peak in August and September. Total catches were very low in October, but this is more likely because the stations closed in the first week or two of the month than because the whales departed from the area. Underwater acoustic monitoring indicates that Blue Whales continue to use BC waters well into winter. From 1994 to 2000 researchers with Scripps Institution of Oceanography examined Blue Whale calls recorded on the US Navy's SOSUS (Sound Surveillance System) network of bottom-mounted hydrophones along the west coast, which included an array on the continental slope off northwestern Vancouver Island. They found that Blue Whale calls off Vancouver Island began in August, increased during September and October, continued at a fairly steady intensity from November through February, then tapered off in March. Most recent Blue Whale sightings off the BC coast have been in summer, but there has been little observer effort during late fall through winter.

Working with whale researcher John Calambokidis from Cascadia Research Collective (Olympia, Washington), we have photo-identified 17 Blue Whales in BC waters up to 2013. The first was photographed by crew on the sailboat *Island Roamer* on June 12, 1997, off the east side of Gwaii Haanas National Park Reserve, Haida Gwaii. This whale was resighted 28 days later in a feeding aggregation of Blue Whales in Santa Barbara Channel, southern California – having covered a minimum distance of 2500 km at an average speed of at least 3.7 km per hour. Since then, 16 more Blue Whales have been photo-identified in BC waters, all of them to the west or southwest of southern Haida Gwaii. Of the 17 whales, 3 have been matched to the California catalogue, which contains about 2000 individuals. This indicates that at least a portion of the Blue Whales using BC waters are part of the population found off southern California. The range of Blue Whales off the west coast may shift with changes in feeding conditions that are affected by oceanic patterns. There is some evidence that Blue Whales may move further north during cool-water years in the Pacific Decadal Oscillation, which alternates between warm and cool conditions on a 20–30-year basis.

Natural History

Feeding Ecology

Blue Whales feed almost exclusively on euphausiid crustaceans, commonly referred to as krill. They are lunge feeders, engulfing as much as 150 tonnes of seawater at a time, thanks to their expanding throat pleats. They then strain out the krill with their baleen as their throat contracts and expels the water. In BC waters Blue Whales feed predominantly on two species, *Euphausia pacifica* and *Thysanoessa spinifera*, which are only 2–3 cm long but often occur in dense patches. When foraging, the whales seek out these patches of krill, sometimes finding them near the surface but more often at depths of over 100 metres. Adult Blue Whales can consume up to 3600 kg of krill per day. Of 38 Blue Whale stomachs examined from 1949 to 1955 at the Coal Harbour whaling station, 30 (79%) contained euphausiids and 8 (21%) were empty.

Behaviour and Social Organization

Blue Whales are usually found alone, in pairs, or in loose aggregations of a few dozen individuals in productive feeding areas. Little is known about their social organization or mating behaviour. While foraging, Blue Whales normally dive for 8–15 minutes, although longer dives of 20–30 minutes have been observed. Studies of foraging Blue Whales off California by University of British Columbia graduate student Jeremy Goldbogen, working with researchers from Cascadia Research Collective, have revealed fascinating details of their underwater feeding behaviour. Using dataloggers and video cameras temporarily attached to the animals with suction cups, the research team was able to track the whales' fine-scale movements in three dimensions as they lunged through dense patches of krill at depth. Following a surface period of 2–3 minutes, a foraging whale typically dives at a rate of 9–10 km per hour to depths of 150–250 metres, which puts it directly beneath a dense layer of krill. At this point, the whale turns upwards and accelerates toward the krill layer, reaching a speed of up to 18 km per hour, then rolls upside-down and opens its mouth. The rush of water into the mouth rapidly inflates the throat, bringing the whale nearly to a halt. The whale then drifts slowly forward as it closes its mouth, expelling the engulfed water through its baleen and trapping the krill, and completes the barrel roll so that it is no longer inverted. The lunging process takes less than 10 seconds, and purging the water takes another 30 seconds on average. After completing a lunge, the whale swims beneath the krill once more and repeats the process. The whales typically make two to five lunges before heading back to the surface to breathe after a dive of 10–12 minutes. Most feeding activity takes place during daylight hours, and the whales appear to rest after dark, rarely diving below 50 metres. Starting at dusk, krill ascend to the surface to feed on phytoplankton, spreading out at densities too low for Blue Whales to feed on effectively.

While in a good foraging area, Blue Whales move slowly at speeds of 2–3 km per hour. Their speed increases to 3–5 km per hour when transiting between foraging areas or migrating. When pursued by Killer Whales – their only natural predators – Blue Whales can sprint at speeds of over 30 km per hour, almost porpoising above the surface as they flee.

Blue Whale vocalizations are loud, long and so low in frequency that they are mostly inaudible to the human ear. Their calls are the most powerful sounds made by any animal and can be heard on hydrophone arrays at distances of more than 1000 km. They vocalize regularly throughout the year with peaks in vocal activity from mid summer to late winter. Call types vary in structure in different ocean regions and can be used to distinguish Blue Whale populations. In the northeastern Pacific, Blue Whales produce four different call types. The most common are referred to as A and B calls – the A call is a burst of pulses about 17 seconds long, while the B call, which is often given about 25 seconds after an A call, is a tonal signal that lasts about 19 seconds. Both have most energy well below 100 Hz. Occasionally a shorter C call at about 11 Hz is heard between A and B calls. A and B calls are often produced by travelling males in a repetitive series lasting many hours, and such series appear to represent songs that might be part of courtship. A fourth type, the D call, is a one-second pulse that sweeps down from 60 to 45 Hz and is produced by both sexes. It seems to be exchanged among several individuals in an area, often during breaks between feeding sessions, and likely serves a contact function.

Life History and Population Dynamics

Blue Whales reach sexual maturity at 8–10 years of age (range 5–15 years). In British Columbia, whaling records indicate that sexual maturity is attained at an average length of 20.5 metres in males and 21.5 metres in females. Physical maturity is reached at lengths of 23.7–24.4 metres in males and 24.4–25.9 metres in females. Females give birth every two or three years. Mating takes place in late fall and winter, and a single calf is born following a gestation of 11–12 months. Calves weigh about three tonnes at birth and are 6–7 metres long. They are nursed for 6–8 months, gaining an impressive 90 kg per day, or almost 4 kg an hour! At weaning, Blue Whales are about 16 metres long. Lifespan is probably at least 80–90 years. Blue Whales are known to hybridize occasionally with Fin Whales.

Exploitation

Although First Nations whalers were familiar with Blue Whales, these animals were evidently considered too large and dangerous to hunt. The Nuu-chah-nulth names for Blue Whales were *yayach'im* or *kw'isasapalh* ("snow on back"). Blue Whale bone fragments have been identified from archaeological excavations at the Nuu-chah-nulth village site of Ts'ishaa in Barkley Sound. These remains may have originated from a "drift whale" (a dead whale found drifting on the ocean or washed ashore) rather than one actually hunted by whalers.

The large size and great speed of Blue Whales did not stop modern whalers from overtaking and subduing them with high-speed catcher boats armed with explosive harpoons. Because of their size, Blue Whales were highly prized and hunted relentlessly throughout the industrial whaling era in all oceans. The greatest numbers of Blue Whales were taken from the early 1900s to the late 1930s, but despite declining numbers the species did not receive IWC protection until 1966. By then the global population of Blue Whales had dropped from its pre-whaling level of about 300,000 to fewer than 10,000.

In the North Pacific about 8000 Blue Whales were killed during the whaling era. At least 1368 of these were taken in Canadian waters by whalers operating out of shore stations in British Columbia between 1908 and 1965 (see Appendix 3), but the total is likely greater than this because catches from 1905 to 1907, which included Blue Whales, were not recorded. Catches were greatest in the early years of commercial whaling, with 814 Blue Whales killed during the first 10 years of record-keeping (1908–17). The biggest season was 1911, when 203 Blue Whales were taken in British Columbia. The Blue Whale population in BC was depleted quickly, as catches declined to five or fewer animals per year during the 1930s and early 1940s despite a high demand for the species. Catches of Blue Whales increased once again with the opening of the Coal Harbour whaling station in 1948, which used larger catcher boats and operated over a wider range than the earlier stations. BC whale biologists Gordon Pike and Ian MacAskie noted in the 1960s that due to its large size, the Blue Whale was "eagerly hunted whenever the opportunity arises". Between 1948 and 1965, when the species was finally protected, 154 Blue Whales were processed at Coal Harbour.

Taxonomy and Population Structure

Three subspecies of Blue Whales are currently recognized. *Balaenoptera musculus musculus* refers to northern hemisphere Blue Whales, including those in British Columbia; the largest subspecies, *B. m. intermedia*, is found in Antarctic waters; and a smaller form, *B. m. brevicauda*, known colloquially as the Pygmy Blue Whale, is found in subantarctic waters of the southern Indian Ocean and the southwestern Pacific Ocean. Geographic variation in the structure of Blue Whale calls suggests there may be discrete populations in the North Pacific.

Conservation Status and Management

Blue Whale populations remain extremely depleted throughout their range although there are signs of recovery in some regions. In the Antarctic, Blue Whales were especially abundant but few remained after 300,000 animals were killed, mostly during the first half of the 20th century. The best recent estimate for Blue Whale abundance in the Antarctic is 2300 whales (in 1997–98), but the population is thought to be growing at about eight per cent a year. The northeastern Pacific population of Blue Whales, which ranges from Central America to the Gulf of Alaska, has recently been estimated from photo-identifications off California to number around 2500 and is increasing slowly. The number of Blue Whales currently using the waters of British Columbia appears to be very small.

The primary human-caused threats to Blue Whales are ship strikes and entanglement in fishing gear. Ship strikes are a known source of mortality in the congested coastal shipping lanes off southern California, where there are seasonally high densities of Blue Whales. Scars from past encounters with fishing nets and other gear are occasionally observed on Blue Whales, though entanglement is not considered a significant source of mortality. Increasing levels of low-frequency ambient noise from shipping may mask the communication calls of Blue Whales, and there are concerns that seismic air gun sounds from geophysical surveys and petroleum exploration, as well as mid-frequency naval sonar, may disturb Blue Whales or displace them from important habitat.

The Blue Whale is listed as Endangered globally on IUCN's Red List and in Canadian Pacific waters by COSEWIC and SARA. The species is also on the BC Red List. In 2006 Fisheries and Oceans Canada developed a recovery strategy for the Pacific population in Canada.

Remarks

The species name *musculus* means "little mouse" in Latin. It's unclear if Linnaeus was being facetious when he assigned this name to the gigantic species or if it perhaps refers to "muscular".

Selected References: Bailey et al. 2009; Bortolotti 2009; Burtenshaw et al. 2004; Calambokidis et al. 2009; COSEWIC 2011; Ford, Abernethy et al. 2010; Ford, Koot et al. 2010; Goldbogen et al. 2011 and 2012; Gregr et al. 2000 and 2006; McMillan et al. 2008; Monks et al. 2001; Oleson et al. 2007; Pike and MacAskie 1969; Sears and Calambokidis 2002; Sears and Perrin 2009; Stafford et al. 2001.

Fin Whale *Balaenoptera physalus*

Other Common Names: Finback Whale.

Description

The Fin Whale is a large, streamlined baleen whale, second in size only to the Blue Whale. Fin Whales are larger in the southern than the northern hemisphere, and females are 5–10 per cent longer than males. The rostrum is V-shaped (viewed from above) and sharply pointed with a single median ridge on the upper surface. The dorsal fin is set far back on the body and varies in shape from falcate and rounded to triangular and pointed, with the leading edge usually rising from the back at a relatively shallow angle (<40°). The caudal peduncle behind the dorsal fin has a sharp dorsal ridge, which is the origin of the whale's antiquated common name, Razorback. The flippers are long, tapered and pointed at the tips. There are 50–100 throat pleats that extend from the chin to the navel.

Fin Whales are dark grey to black on the back and flanks, grading to light grey to white on the undersides. In British Columbia waters, they often have many small round or oval whitish scars on the back and sides, thought to be caused by the bites of Cookiecutter Sharks or Pacific Lampreys, or possibly from attachment of the parasitic copepod *Pennella balaenopterae*. Fin Whales in other regions do not exhibit such extensive scarring. The undersides of the tail flukes and flippers are light grey to white. Along the back just behind the head there are usually light-coloured streaks that run upwards and back, joining at the midline of the back to form a chevron pattern. This pattern is often faint or obscured by diatoms in BC. The colour of the Fin Whale's head is distinctively asymmetrical – on the left side,

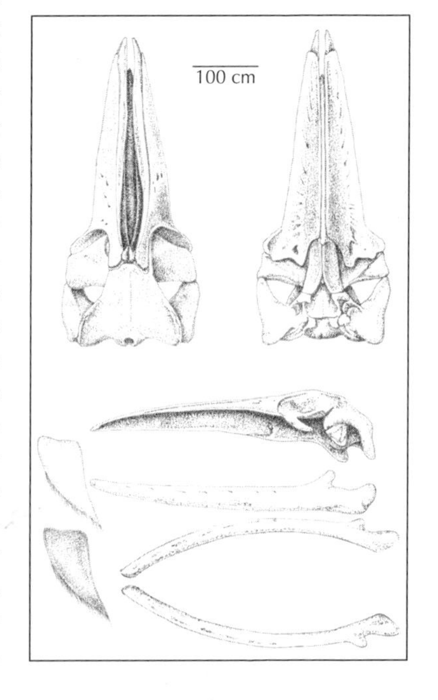

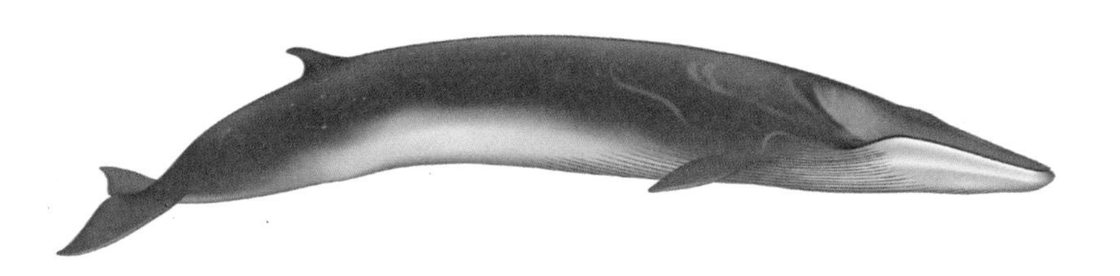

the dark grey extends further ventrally to include the lower jaw, but on the right side the lower jaw is light grey to white. In addition, on the right side the mouth cavity and the forward third of the baleen are whitish. The remaining baleen plates are dark grey to black with yellowish-grey bands, and the coarse fringes of the plates are brownish-grey to greyish-white. A band of light grey often extends from the right side of the head up and toward the rear to the midline of the back.

Measurements

length (metres):

male:17.3 (9.1–25.0) n = 2291
female: 18.0 (9.1–25.0) n = 2375
neonate: 6–6.5

weight (kg):

adult: 50,000–90,000
neonate: 1800–2700

Baleen

Plates on each side: 260–480
Up to 65 cm long and 30 cm wide with coarse fringes, dark-grey to black with yellowish-grey bands and brownish-grey to greyish-white fringes.

Identification

Fin Whales can be easily confused with Blue Whales and Sei Whales when observed from a distance. Adult Fin Whales are smaller than Blue Whales and larger than Sei Whales, but relative size can be difficult to judge on the water. Fin Whales are darker than Blue Whales and usually have oval white scars scattered along their back, which are not present on Blue Whales but are common on Sei Whales. In BC waters Fin Whales often have a brownish tinge caused by a film of diatoms on their skin. These can become blotches of brown as the diatom layer is sloughed off (figure 45). A Fin Whale's dorsal fin is larger relative to body size than that of a Blue Whale (which is tiny compared to its huge size), and it tends be somewhat more angled

Figure 45. Fin Whales in BC waters often have numerous white spots and scars as well as brownish patches caused by a film of diatoms on their skin. Photo: J. Ford, Langara Island, June 2012.

(i.e., less erect) and more falcate than that of Sei Whales. When Fin Whales surface, their dorsal fin normally doesn't appear above water until the final, arching dive as they sound. In contrast, the dorsal fin of Sei Whales is usually seen at the same time as they exhale and inhale at the surface. When diving, Sei Whales usually do not arch their back as much as Fin Whales. If in doubt, it's best to try for a clear view of the right lower jaw as the whale surfaces, which is very light grey or white in Fin Whales (figure 46).

Distribution and Habitat

Fin Whales have a cosmopolitan distribution and can be found in all major seas from the equator to the polar regions. They are mostly seen near or seaward of the continental shelf edge but occasionally occur in nearshore waters, especially where deep water comes close to the coast. Like other rorquals, Fin Whales are migratory, although their migration patterns appear to be complex and are poorly understood. There is a general shift to high latitudes in summer for feeding and back to warm waters in winter for breeding, but Fin Whales are found in some areas year-round. Fin Whales in the Gulf of California, for example, appear to be resident, and the loud low-frequency calls of Fin Whales have been heard throughout the year on hydrophones mounted on the sea floor in the Gulf of Alaska. Whether Fin Whales seen or heard in high latitudes in winter do not migrate at all or are

Figure 46. This pair of Fin Whales surfacing near Langara Island show the diagnostic white right side of the lower jaw. Photo: J. Ford, June 2012.

simply late-leaving or early-returning migrants is not yet clear. It is assumed that most Fin Whales in the North Pacific move southward to warm waters near the Tropic of Cancer in winter, but specific wintering areas where mating and calving take place remain undiscovered (assuming such areas exist).

In British Columbia, commercial whalers took Fin Whales over a wide area of outer coast waters as well as in Dixon Entrance, Hecate Strait and Queen Charlotte Sound (see figure 18). One Fin Whale that was marked with a tag in November 1962 west of Baja California was killed by whalers near the south end of Haida Gwaii in May 1963. Whalers killed Fin Whales throughout the April to October whaling season, with numbers increasing each month from April to August, then declining in September and October, likely due to reduced effort. The BC Cetacean Sightings Network database has records of Fin Whale sightings in every month of the year, although very few of them are for November through March, likely due at least in part to reduced observation effort and poor viewing conditions in winter. But Fin Whales were the second most frequently sighted large whale, after Humpbacks, during ship surveys undertaken from 2002 to 2010 by the Cetacean Research Program at the Pacific Biological Station; they were seen on surveys in all seasons, including winter. Fin Whale calls were heard on most days during the five-month deployment (mid February to mid July 2006) of an autonomous underwater acoustic recorder at Union Seamount, about 350 km west of Vancouver Island.

Fin Whales can be found near and beyond the continental shelf break off the entire British Columbia coast. Sightings are also quite common in western Dixon Entrance and in certain areas of Hecate Strait, especially along the east side of Gwaii Haanas National Park Reserve and in Moresby Trough, a tongue of deep water that extends from the shelf break south of Haida Gwaii northeastward to eastern Hecate Strait. Fin Whales are also occasionally sighted in Queen Charlotte Strait, Johnstone Strait and Juan de Fuca Strait. Although these large whales are mostly seen in deep waters, they are also found regularly in inshore waters, including protected channels of the inside passage. One of their high use areas is Caamaño Sound on the northern mainland coast, including Squally Channel and Whale Channel. The Cetacean Research Program, working with CetaceaLab based on Gil Island, photo-identified 86 individual Fin Whales in this area from 2006 to 2013. Many of the individuals have been resighted in multiple years, suggesting a high level of site fidelity, and studies are currently underway to improve understanding of Fin Whale habitat use patterns in these confined channels.

Natural History

Feeding Ecology

Fin Whales are lunge feeders that prey primarily on zooplankton – mostly euphausiids and some copepods – as well as on a variety of schooling fishes. Knowledge of their diet in British Columbia waters comes entirely from observations of stomach contents of whales caught during the latter part of the commercial whaling era. Of 959 Fin Whale stomachs with food remains that were examined at the Coal Harbour whaling station during 1955–67, 96 per cent contained euphausiids and 4 per cent contained copepods. Squid and fishes, likely Pacific Herring primarily, were each found in less than 1 per cent of the stomachs. Free-swimming pteropods (a kind of sea snail, also known as a Sea Butterfly), were found in one stomach examined in 1964.

Studies in the 1950s by cetacean scientist Gordon Pike at the Pacific Biological Station showed that two species of euphausiids were consumed by Fin Whales, *Euphausia pacifica* and *Thysanoessa spinifera*, but the proportions of these two species tended to vary over the whaling season and among years. There was a general pattern of feeding mostly on *E. pacifica* early in the season and then mostly *T. spinifera* later in the season, but this varied from year to year. The volume

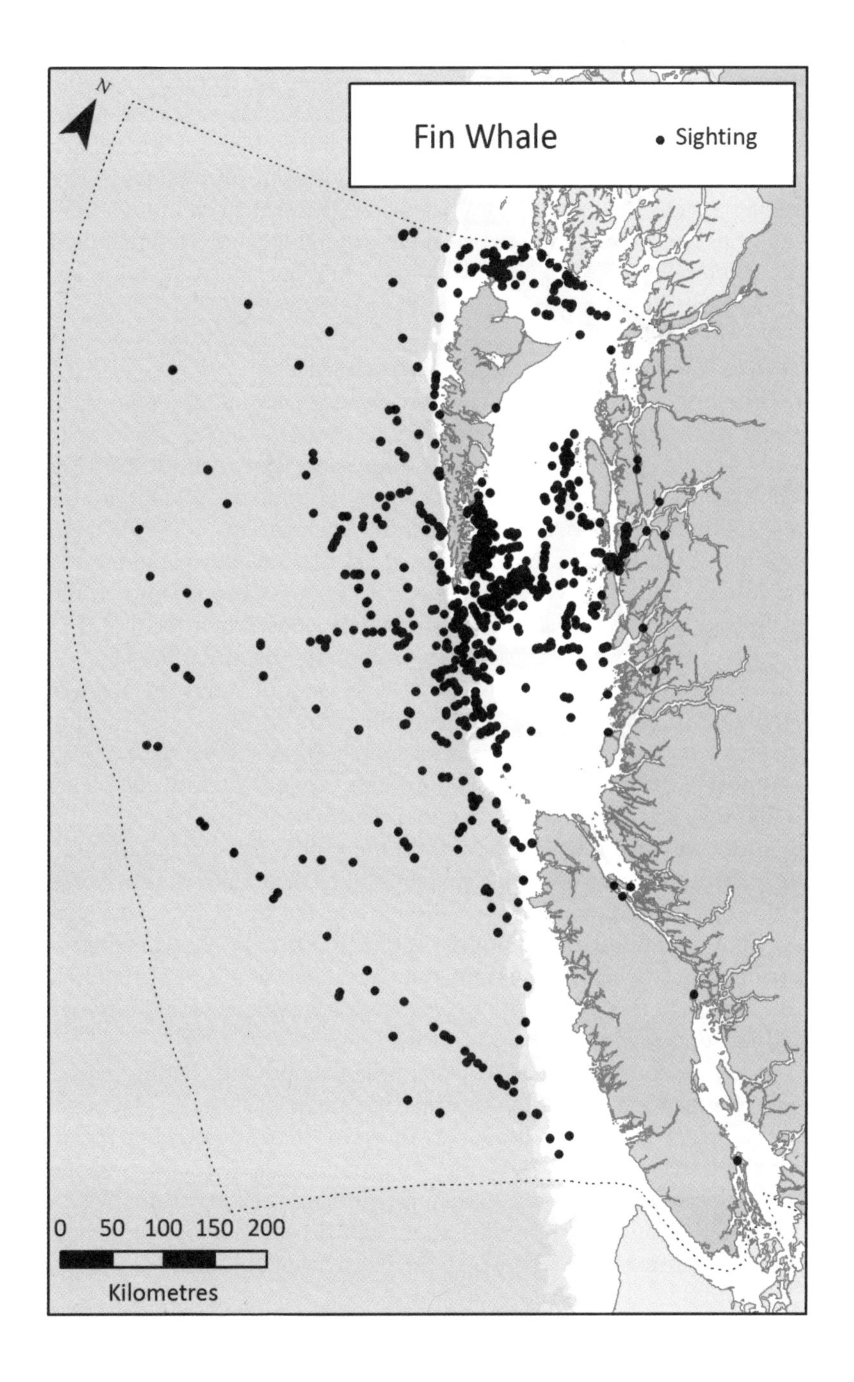
Fin Whale
Sighting
N
0 50 100 150 200
Kilometres

of prey found in the whales' stomachs tended to increase over the course of the season. Pike also made some interesting observations on what might be interpreted as selective foraging by Fin and Humpback whales in the summer of 1950 – the majority of Humpbacks were feeding on *T. spinifera* and Fin Whales on *E. pacifica*, even though both whale species were caught in the same general locality. He attributed this difference to one species (likely the Fin Whale) feeding at greater depths than the other.

Behaviour and Social Organization

In British Columbia, Fin Whales are usually seen alone or in small groups of two to four. Loose aggregations of several dozen whales may occur in areas of good feeding. In September 2013, about 100 Fin Whales were feeding together with a few Humpback Whales southwest of Bonilla Island in Hecate Strait. Aggregations of Fin Whales have also been seen near the heads of submarine canyons along the shelf break between Cape St James and Triangle Island, and to the north and west of Langara Island, Haida Gwaii. When feeding, Fin Whales often dive to a dense layer of euphausiids at depths of 100–250 metres, and lunge through this layer several times on a single foraging dive. They will also feed on krill at the sea surface by turning on their side and lunging into a dense patch. As they open their mouth to almost a 90° angle, the inrushing water expands their throat pleats so that they can engulf more than their entire body mass of water – up to 70,000 kg – from which they strain the krill as it is purged through the baleen plates. Fin Whales typically roll to their right as they lunge at the surface, exposing the left flipper and tail fluke above the surface. The asymmetrical coloration of the Fin Whale's head, which is predominantly white on the right, may be related to this right-handedness, although it is unclear how this coloration pattern would improve feeding efficiency.

Little is known about the social organization and mating behaviour of Fin Whales. Mating and calving most likely take place offshore. On feeding grounds such as those in British Columbia, group composition can be very dynamic, with individuals frequently moving between groups. As noted earlier, preliminary results from photo-identification studies show that there can be site fidelity, with individuals returning each summer to feed in particular locations, such as Caamaño Sound.

Fin Whales are fast swimmers and can move rapidly from one feeding area to another; their speed inspired American naturalist Roy Chapman Andrews to call them "greyhounds of the sea". Like other members of its genus, the Fin Whale is a "flight species", using a combination of high speed and stamina as its main escape strategy when pursued by its only predator, the Killer Whale. This is often a successful strategy – we watched a Fin Whale quickly outpace a group of nine Transient (Bigg's) Killer Whales on the chase near Langara Island, Haida Gwaii, in June 2010. But the outcome is not always favourable for the Fin Whale. In 2005 researchers in the Gulf of California observed 16 Killer Whales capture and consume a Fin Whale after chasing it for an hour at speeds reaching 40 km per hour.

Fin Whales in BC waters are quite approachable and can be curious around boats. We have observed them making repeated shallow passes under our drifting research boat, turning on their side to visually inspect the hull. Fin Whales communicate with each other using very loud low-frequency calls. Best known are repetitive series of calls with a descending pitch centred at around 20 Hz, far below the hearing range of humans. These calls, each slightly less than one second long, are repeated at intervals of a few to 45 seconds and represent a simple form of song, lasting for many hours or even days. These songs, which likely can be heard by other Fin Whales at ranges of hundreds of kilometres, are believed to be produced only by males, and may play a role in mate attraction and breeding displays. Additional call types are used in other behavioural contexts.

Life History and Population Dynamics

Fin Whales grow rapidly, reaching 95 per cent of their maximum body size by 9–13 years of age. In British Columbia, whaling data indicate they are sexually mature at about 17.6 metres in males and 18.2 metres in females, which corresponds to ages of 6–7 years and 7–8 years respectively. Physical maturity is then reached at average lengths of 19.8 metres for males and 20.8 metres for females. Mating in the northern hemisphere takes place between December and February, and a single calf is born following a gestation of 11–12 months. The calf is weaned at about 6–8 months, when it is 11–13 metres long. Intervals between calves are typically two years. Fin Whales can live to 80–90 years, and perhaps longer. They are known to hybridize occasionally with Blue Whales.

Exploitation

Fin Whales were hunted relentlessly during the 20th century by commercial whalers in all major oceans, especially after the larger Blue Whales were depleted. More Fin Whales were killed than any other whale species – about 725,000 in the southern hemisphere alone. The initial pre-whaling population of Fin Whales in the North Pacific was estimated to be 42,000–45,000, of which 25,000 to 27,000 were in the eastern portion. Whaling reduced the North Pacific population to an estimated 8000–16,000 by 1975. In British Columbia commercial whaling of Fin Whales from shore stations began in 1905 and ended with the closing of the last whaling station, at Coal Harbour, in 1967. The industry-reported kills totalled 7,497 Fin Whales over this period (see Appendix 3). An additional 201 Fin Whales were taken in offshore waters within Canada's EEZ from 1964 to 1975 by Japanese high-seas whalers, and an unknown number were killed by Soviet whalers during the same period.

Fin Whales were likely too fast and too far from shore to be of interest to First Nations whalers, which is probably why there is only vague mention of the species in Nuu-chah-nulth ethnographic accounts. But Fin Whale bones have been identified by using ancient DNA from excavations at the Nuu-chah-nulth village site of Ts'ishaa in Barkley Sound. It's not known whether these remains were from whales hunted by early whalers or from stranded or drift whales.

Taxonomy and Population Structure

The phylogeny of Fin Whales below the species level is still uncertain, but most whale biologists consider there to be at least two subspecies – *Balaenoptera physalus physalus* in the northern hemisphere and *B. p. quoyi* in the southern hemisphere. Movement data from tags as well as immunogenetic findings suggest there are distinct eastern and western populations in the North Pacific. A genetically distinct and isolated population of about 600 Fin Whales resides in the Gulf of California, Mexico.

Conservation Status and Management

Due to serious depletion, Fin Whales in the North Pacific and southern hemisphere were protected from whaling by the International Whaling Commission beginning in 1975, and commercial whaling ceased altogether in the North Atlantic by 1990. Limited Fin Whale hunting has continued in Greenland as "aboriginal subsistence" whaling, and commercial whaling of this species has recently been resumed by Iceland in the North Atlantic and by Japan in the Antarctic. No whaling for Fin Whales currently takes place in the North Pacific.

There are no reliable estimates of current Fin Whale abundance globally or in the North Pacific due to patchy survey coverage. Fin Whale numbers off the west coast of the US mainland were estimated at about 3000 from surveys in 2005 and 2008, with no indication of a trend since surveys in the mid 1990s. About 5700 Fin Whales were recently estimated for the Bering Sea, coastal Aleutian Islands and Gulf of Alaska, where they appear to be increasing. Surveys by the Raincoast Conservation Foundation in coastal BC waters in 2004–05 estimated about 500 Fin Whales. But these surveys did not include waters off the west coasts of Haida Gwaii and Vancouver Island, or beyond the shelf break, so the true abundance of Fin Whales off Canada's west coast is certainly far greater. Since 2005 we have photo-identified Fin Whales during field surveys around Haida Gwaii and off the northern mainland coast, and as of 2013 we have catalogued 512 individuals – and the rate of discovery of new whales does not seem to be slowing. Our qualitative impression from these field studies is that Fin Whale numbers in BC waters are increasing quite rapidly. This is particularly evident around Langara Island and in the Caamaño Sound area.

Aside from whaling, the most significant known human-caused threat to Fin Whales is ship strikes. Of all the great whales, Fin Whales are the most frequently reported as being struck by ships. On two occasions, in 1999 and 2009, a cruise ship arrived in Vancouver harbour from northern BC with a dead Fin Whale draped over its bulbous bow, although there was evidence in the latter case that the whale had been struck post-mortem. Two other dead Fin Whales, one observed from the air off the west coast of Vancouver Island and the other stranded on Haida Gwaii, appeared to have died from major injuries due to ship strikes. Entanglement of Fin Whales in fishing gear does occur but is not known to be a significant source of mortality. Like other baleen whales using low-frequency calls for communication, Fin Whales are potentially vulnerable to disturbance and acoustic masking by noise from ships and other sound sources, such as seismic exploration and military sonar.

Fin Whales are listed as Endangered globally on IUCN's Red List. COSEWIC and SARA list the as Threatened in Canadian Pacific waters, and they are on the BC Red List. In 2006 Fisheries and Oceans Canada developed a recovery strategy for the Pacific population of Fin Whales in Canada.

Remarks

The species name for the Fin Whale is based on the Greek *phusalis*, which means a wind instrument or something that resembles a toad that puffs itself up. This name presumably refers to the expansion of the ventral throat pleats while feeding.

Selected References: Aguilar 2009, Ford, Abernethy et al. 2010; Ford, Koot et al. 2010; Ford and Reeves 2009; Goldbogen 2010; McMillan et al. 2008; Mizroch et al. 2009; Monks et al. 2001; Pike 1953a; Pike and MacAskie 1969; Stafford et al. 2009; Williams and Thomas 2007.

Humpback Whale *Megaptera novaeangliae*

Other Common Names: None.

Description

The Humpback Whale has a large, robust body with extremely long flippers, which are about one-third of the body length and longer than those of any other species. There are numerous knobs, or *tubercles*, on the head and jaws, and along the leading edges of the flippers. These tubercles, together with prominent serrations along the trailing edges of the tail flukes, are diagnostic of the species. Tubercles are most prominent near the lips and the tip of the chin. The dorsal fin is located approximately two-thirds of the way back from the rostrum and is positioned on a raised hump of tissue. Its shape is highly variable among individuals, from a low, poorly defined knob or bump to a well-developed falcate fin. Like all rorquals, Humpback Whales have multiple parallel throat pleats. Humpbacks have 14–35 pleats reaching from near the tip of the chin back to the area of the navel.

Humpback Whales are generally black dorsally with variable amounts of white ventrally. Those in the North Pacific tend to have less white underneath than those in the North Atlantic and southern hemisphere. The flippers of North Pacific Humpback Whales are white ventrally and either black or white dorsally, but they are generally white on both sides in other oceans. The tail flukes of Humpbacks are black on the dorsal side with highly variable coloration ventrally – the underside of the tail flukes on some animals are all black, others are all white and many have variable amounts of white and black. Individuals can be identified by their distinct coloration pattern on the ventral surface of the flukes together with serration patterns on the trailing edges.

The skin of Humpback Whales is usually colonized with barnacles that concentrate on particular parts of their bodies. Two different types of barnacles attach to Humpbacks – acorn barnacles of the genus *Coronula* and stalked barnacles of the genus *Conchoderma*. Acorn barnacles are usually found along the rostrum, lips and throat, and in clusters on the flipper tubercles, at the tip of the chin (known as the *cutwater*), around the genital slit and on the tips of the tail flukes. These clusters form a substrate for the attachment of stalked barnacles, which adhere to the shells of the acorn barnacles.

Measurements

length (metres):

male: 11.9 (7.6–15.5) n=553
female: 12.3 (7.6–17.1) n=470
neonate: 4–4.6

weight (kg):

adult: ~ 40,000
neonate: 680–900

Baleen

Plates on each side: 270–400
Up to 85 cm long, black, except the front plates are brownish-white.

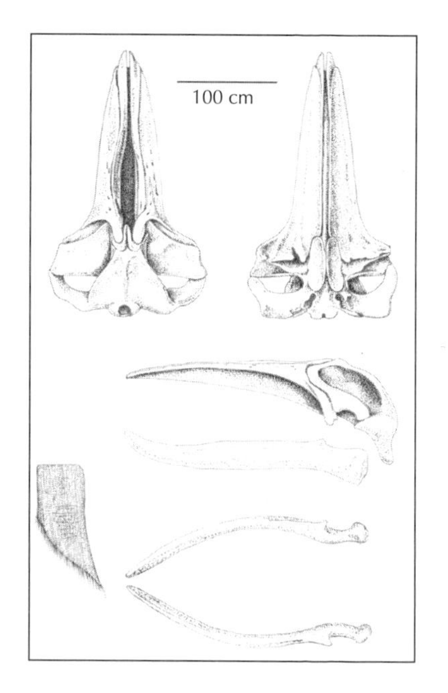

Identification

Humpback Whales are probably the most familiar and easily recognized of the great whales. At a distance, however, Humpbacks could be confused with other large whales, especially Grey Whales, Sperm Whales and North Pacific Right Whales. These species all have similarly low, bushy-shaped blows and lack a large, well-defined dorsal fin. On close inspection, Humpbacks can be easily distinguished by their large flippers, tubercles on the head, dorsal fin shape and serrated trailing edges of the tail flukes, which are usually raised above the surface as the whale sounds. Photographs of these distinctive serrations and the coloration pattern on the ventral side of the flukes are used to identify individuals. A catalogue of over 2000 individual Humpback Whales identified to date in BC waters can be viewed online at the Pacific Biological Station's Cetacean Research Program website.

Distribution and Habitat

Humpback Whales have a cosmopolitan distribution in the world's oceans. They are highly migratory, moving seasonally between summer feeding areas in productive cool to cold-temperate waters and winter breeding areas in subtropical to tropical waters. Feeding areas are mainly in coastal or shelf waters, and breeding areas are usually located near island or reef groups. In the North Pacific, Humpback Whales mostly spend spring through fall months in coastal and continental shelf waters around the Pacific Rim, from northern Japan in the west, into the Bering Sea in the north, to California in the east. There are four main breeding regions in the North Pacific: Hawaii, Mexico (southern Baja California, the central mainland west coast and the Revillagigedo Islands), Central America (southern Mexico to Costa Rica) and Asia (islands south of Japan, Taiwan, the Philippines and the Mariana Islands). Humpback Whales regularly cross oceanic waters to reach these migratory destinations, but the actual routes they use are poorly known.

An ocean-wide study of Humpback Whales called SPLASH (Structure of Populations, Levels of Abundance and Status of Humpbacks) conducted from 2004 to 2006 greatly improved our knowledge of the linkages between breeding grounds and feeding areas in the North Pacific. SPLASH is the largest whale study ever attempted, involving over 400 researchers in 10 countries: Canada, the US, Russia, Japan, Mexico, Philippines, Guatemala, Nicaragua, El Salvador and Costa Rica. The study was based primarily on two approaches: photo-identification of individual whales using natural markings on the tail flukes in order to determine population abundance and migratory connections between breeding and feeding areas, and DNA analysis of skin biopsies to learn more about genetic population structure. SPLASH revealed fascinating migration patterns in the North Pacific and complex population structure resulting from strong fidelity to breeding and feeding regions. Humpbacks wintering in the Asian area migrate primarily to feeding grounds in Russia and the Aleutian

Islands. Whales from Hawaii – the largest breeding area in the North Pacific – feed in waters from the eastern Aleutians through southeastern Alaska to southern British Columbia. Humpbacks wintering in Mexico migrate to a variety of feeding areas from Alaska to California, while those wintering in Central America feed in waters from southern BC to California.

Photo-identification matches from the SPLASH study show that most Humpback Whales that feed in British Columbia waters migrate to either Hawaii or Mexico in winter, but the proportions that use these breeding areas change with latitude. North of Vancouver Island about 85 per cent of photographically matched Humpbacks migrated to Hawaii, while the remaining 15 per cent migrated to Mexican breeding grounds. But off southwestern Vancouver Island and adjacent waters of northern Washington, 58 per cent of matches were to Mexican breeding grounds, 36 per cent to Hawaii and 6 per cent to Central America. Previous studies have also revealed that on rare occasions BC Humpbacks migrate to even more remote destinations. Whale researcher Jim Darling and colleagues photo-identified a Humpback that migrated from the Ogasawara (or Bonin) Islands south of Japan to the west coast of Vancouver Island, then back again to the Ogasawaras. This represents a minimum round-trip swim of more than 15,000 km.

The Humpback Whale has adapted to exploit different prey species in a variety of habitats. As a result, it can be found throughout inshore, outer coastal, continental shelf and offshore waters of British Columbia. As the population recovers from depletion by whaling, it is reoccupying many areas where it was formerly abundant. As recently as the early 1980s, sightings of Humpback Whales in BC were uncommon, occurring mostly in the waters off Haida Gwaii and sporadically off northern Vancouver Island and along the north mainland coast. Sightings have steadily increased since then, and Humpbacks can now be found reliably in most coastal regions, with sightings even becoming frequent in the Strait of Georgia. In fact, Humpback Whales were the most commonly sighted cetaceans during 21 ship surveys undertaken by the Cetacean Research Program, Pacific Biological Station, from 2002 to 2008. Of 2,862 cetacean sightings over 30,000 km of survey track, 1,700 (59%) were of Humpbacks. The highest densities of Humpback Whales in BC waters are usually near Haida Gwaii, particularly along the eastern side of Moresby Island in the Gwaii Haanas National Marine Conservation Area Reserve and around Langara Island. Other important areas for Humpbacks are: waters near Prince Rupert, including Chatham Sound, Work Channel and Portland Inlet;

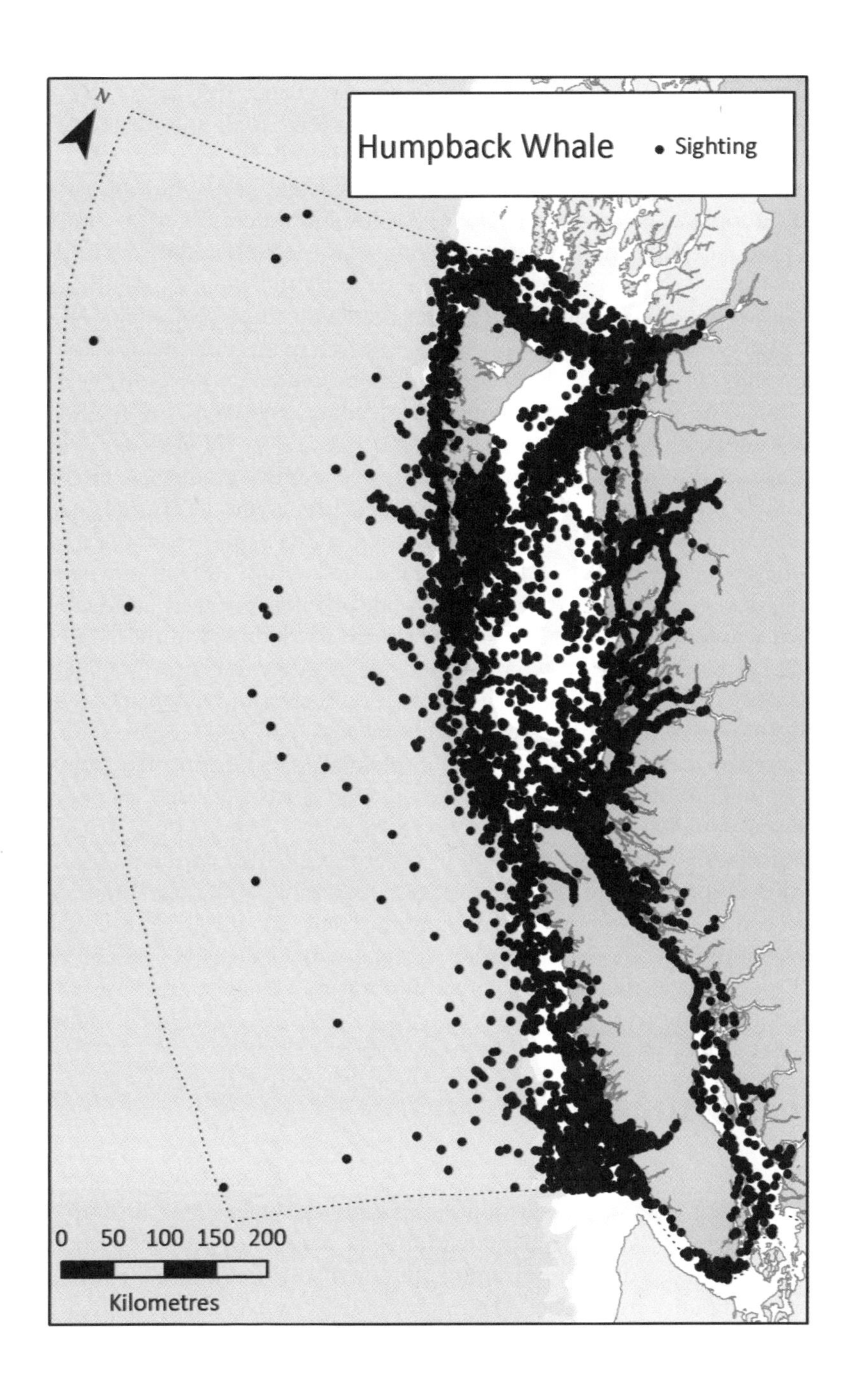

N
Humpback Whale
Sighting
0 50 100 150 200
Kilometres

along the north and central mainland coast, particularly in the channels around Gil and Gribbell islands and in Fitz Hugh Sound; and off northeastern and southwestern Vancouver Island.

Although Humpback Whales are strongly migratory, they are seen in all months of the year in British Columbia waters. They are most abundant from April to November, but some can be found throughout the winter. The migration appears to be fairly diffuse in timing, with the last animals heading south from the BC feeding grounds in late January or February, overlapping with early arrivals from the breeding grounds. Humpback Whales have a strong tendency to return each year to the same general location for feeding, known as site fidelity. Working with photo-identification data collected by DFO's Cetacean Research Program between 1992 and 2007, UBC graduate student Andrea Ahrens (nee Rambeau) examined site fidelity of Humpbacks along the BC coast. Of the 585 unique individuals photo-identified in multiple years, 57 per cent were re-identified within 100 km of where they had been seen in previous years, and 25 per cent were re-identified within 25 km. A few individuals were re-identified in essentially the same location, within half a kilometre of where they had first been identified. Despite this strong site fidelity, there can be considerable movement of whales in a feeding season as well as between years. For example, there is an influx of Humpback Whales during late summer and autumn into the channels and deep fjords between Caamaño Sound and Kitimat on the northern mainland coast. These whales presumably spend time feeding in other areas earlier in the summer prior to moving into the inlets. Another example is the late-summer occurrence of Humpbacks near Victoria. There can also be substantial inter-annual shifts in distribution, likely due to changes in availability of food – areas that support large numbers of whales in one year can be almost vacant the next.

Natural History

Feeding Ecology

Compared to most baleen whales, Humpback Whales have a diverse diet. They are "gulp" feeders that lunge at patches of prey with open mouths, taking discrete mouthfuls rather than skimming the surface or swimming through the water column with mouth agape, as Right Whales do. Humpbacks feed mostly on krill (euphausiid crustaceans) but also take a wide variety of schooling fishes. In British Columbia, information on Humpback Whale foraging ecology comes from analysis of stomach contents from whale catches and from recent field observations. Between 1949 and 1965, Pacific Biological Station whale biologist Gordon Pike examined the stomach contents of 346 Humpbacks taken off Vancouver Island by whalers based at the Coal Harbour whaling station. Of the 287 stomachs with prey remains, 263 (92%) contained only krill, 12 (4%) contained only copepods, and 2 (0.7%) contained only Pacific Herring. The remaining stomachs contained mixtures of these prey types, while one contained krill and Pacific Saury and one was full of a small (5 cm) unidentified squid. Pike identified two species of krill, – *Euphausia pacifica* and *Thysanoessa spinifera* – but did not specify the copepods.

More recently we have been collecting prey samples using dip nets and towed plankton nets in the vicinity of feeding Humpback Whales to better understand their diet in nearshore waters. Of the zooplankton prey samples collected, most were krill, but off the north shore of Graham Island, Haida Gwaii, some were copepods and megalopae (free-swimming larvae) of Dungeness Crab. In northern BC waters Humpback Whales are frequently observed feeding on Pacific Herring, both adults and juveniles, often with the use of cooperative bubble-net techniques to concentrate the fish (see Behaviour and Social Organization, below). We have also regularly observed individual whales or mothers with calves lunge-feeding on dense schools of Pacific Sand Lance close to shore in the Langara Island area. Over the past decade Humpbacks have been increasingly targeting Pacific Sardines, a formerly abundant species on the BC coast that virtually disappeared in the late 1940s due to overfishing and poor ocean survival. Sardines began reappearing off the southern BC coast in the early 1990s and their abundance increased dramatically in the 2000s. Sardines are particularly important prey off the west coast of Vancouver Island, where they have attracted large numbers of Humpbacks into Barkley Sound and Clayoquot Sound in recent years.

Behaviour and Social Organization

Humpback Whales are perhaps the best known of the great whales and have been studied extensively on both their feeding and breeding grounds. They are moderately social whales, often gathering into small groups that may consist of several short-term associates or occasionally longer-term companions, and larger aggregations may develop in breeding and feeding areas. Humpbacks are well known for their migratory behaviour, and they undertake some of the longest migrations of any mammal – the record holder is a whale that migrated from the Antarctic Peninsula to American Samoa and back, a 18,840-km round trip. The reasons why Humpbacks undertake these long migrations are the subject of scientific debate. They may migrate to high latitudes to take advantage of seasonal prey abundance associated with summer blooms, then benefit thermodynamically – use up less fat reserves – by moving to warm waters in winter even though they feed little if at all during migration or on the breeding grounds. The thermal benefit may be especially important for newborn and very young calves. Humpbacks may also migrate to low-latitude breeding areas at least in part to avoid predation on calves by Killer Whales, which are found in greater densities in high-latitude waters. How Humpbacks navigate during migration is unknown, but one possibility is that they orient to geomagnetic cues and ocean currents. Migrating whales can cover long distances very rapidly. One individual photo-identified in Sitka Sound, Alaska, in early January was observed again off the island of Hawaii 39 days later. The minimum distance between these two points is 4440 km, so the whale had to have travelled at a minimum average speed of 4.74 km per hour on migration. It seems unlikely that the whale was photographed on its last day in Alaska and its first day in Hawaii, so its actual transit time was likely shorter and thus its average speed even faster.

Humpback Whales are well known for their diverse repertoire of often-spectacular aerial displays, including flipper slaps, tail slaps and breaches (figure 47). These are seen at any time of the year, but are especially common on the breeding grounds. They occur in a wide range of social and environmental contexts, so their specific purpose is often unclear. For example, breaches may be triggered by the meeting of two individuals or the sudden arrival of a wind line, quickly changing calm seas to choppy. Aerial displays may signify excitement or annoyance and are often accompanied by distinctive loud "trumpet" or "whistle" blows.

Starting in autumn and continuing through winter on the breeding grounds, male Humpback Whales sing long, complex songs that cover a wide range of frequencies, from 20 Hz to 24 kHz. Songs are haunting medleys of whines, grunts and squeals that play important roles in mating interactions by attracting females and by sorting dominance among males. All males in a breeding area sing essentially the same 10–20 minute-long song, and although the song structure changes gradually over time, the whales coordinate the changes by mimicking each other. Males also join and compete aggressively in *surface-active* or competitive groups of 10 or more individuals that gather around a single receptive female. These competitive interactions can be quite violent, with males head-butting, tail-slashing

Figure 47. A Humpback Whale breaching in Blackfish Sound.
Photo: C. McMillan September 2011.

and ramming into each other with a force that can leave individuals scraped and bleeding. Males spend significantly longer than females on the breeding grounds, likely to take advantage of multiple mating opportunities, while newly pregnant females return quickly to feeding grounds. As a result, males typically outnumber females by more than two to one on breeding grounds.

From spring through late fall the predominant activity of Humpback Whales is foraging. They are perhaps the most inventive of all baleen whales in their techniques for corralling and concentrating their prey for capture. One method often used to feed on krill at the surface in northern BC waters is called *flick feeding*. A whale submerges head-down, raises its tail to the surface and then repeatedly sweeps it forward, each time sending a cascade of water three to five metres onto the surface ahead. After flicking its flukes in this manner up to 10–15 times, the whale, sometimes with a companion, surfaces quickly on its side and lunges through the edge of the splash zone with its mouth open to engulf the krill. How this technique improves the catchability of krill is not clear. It seems likely the fast-swimming krill flee from water landing on the surface, concentrating at the edge of the splash zone where the whales then focus their lunge. The percussion of water hitting the surface may also briefly disorient the krill, impeding their escape. We have noted krill drifting as if stunned immediately following a flick-feeding session. This feeding technique is particularly common in the waters off southeastern Moresby Island in Haida Gwaii, where we have observed dozens of Humpbacks doing it at the same time in a small area.

Another elaborate and well-known technique is called bubble-net feeding or *bubble-netting*. This involves a group of whales – usually 5–10 animals but sometimes fewer or more – working cooperatively to encircle and corral schooling fish, especially Pacific Herring (figure 48). The whales dive roughly in unison, then swim under water until they locate a school of herring. They then swim in a large circle under the school, with one whale releasing a stream of bubbles from its blowhole. As the bubbles rise to the surface, they form a cylindrical barrier blocking the fish from escaping. While the group circles the school, one whale repeatedly emits an intense, repetitive "feeding call", which may serve to acoustically herd the fish and coordinate the whales' movements. The pitch and tempo of the feeding calls increase as the bubble-net is completed, and then the calls abruptly stop. This appears to be a signal for all of the whales to simultaneously lunge upwards through the school with their mouths open, engulfing the fish. Bubble-netting is a specialized tactic used only by a subset of the Humpback population, often composed of individuals that have worked together cooperatively for many years. We have observed this tactic at several locations on the northern BC coast, particularly around Langara Island, in Work Channel and Chatham Sound near Prince Rupert, and in the Caamaño Sound area.

Life History and Population Dynamics

The breeding cycle of the Humpback Whale is strongly seasonal. Females come into estrus and conceive in winter, and after an 11.5-month gestation they give birth to a single calf the following winter. Calves are nursed for almost a year but likely begin to feed independently at about six months of age. At weaning they are almost 10 metres in length. Females give birth every one to five years, but intervals of two to three years are most common. Both sexes reach sexual maturity at five to nine years of age, at a length of around 12 metres. The maximum lifespan is not known with certainty but is likely at least eighty years. Natural predators include sharks, known to occasionally take calves on the breeding grounds, and Killer Whales. Calves are the most vulnerable to Killer Whale attacks, and pressure from these predators may be intense in some populations. For example, scars or other marks left by Killer Whale teeth are visible on the tail flukes of

Figure 48. A group of Humpback Whales bubble-net feeding on Pacific Herring in Caamaño Sound. Photo: J. Ford, July 2013.

30–40 per cent of Humpback Whales off Mexico, indicating that these individuals have survived an attack in the past. Most scarring from attacks takes place in the whale's first year of life, probably while on its first migration to high latitudes for feeding. As relatively slow swimmers, Humpback Whales are not able to outdistance pursuing Killer Whales. But large juveniles or adults can mount an effective defence by slashing at the predators with their powerful tail flukes and long flippers, which are encrusted with barnacles that may function like brass knuckles. Large Humpbacks are sufficiently intimidating that mammal-hunting Killer Whales generally ignore them as potential prey, and any predatory interest is focused on the calves. The high incidence of Killer Whale tooth scars on young Humpback tail flukes suggests that their mothers successfully defended them from attack.

Exploitation

The Humpback Whale was the most important of the large whales taken by the whale-hunting Nuu-chah-nulth people along the west coast of Vancouver Island. Their name for the species, *iiḫtuup*, is also their generic term for whales. Although ethnographic accounts suggested that Grey Whales were taken more often than Humpbacks, recent evidence obtained by researchers with Simon Fraser University's Department of Archaeology indicates that this was not the case. DNA analysis of whale bone samples collected at Ts'ishaa, a large ancient village site on Benson Island in Barkley Sound, revealed that 77 per cent were Humpbacks and that Grey Whales comprised only about 11 per cent. The proportion represented by Humpback Whales was even higher at the Ch'uumat'a and T'ukw'aa village sites on the western margin of Barkley Sound, near the entrance to Ucluelet harbour, which date back to about 2500–3000 years ago. Humpback Whales were likely available to Nuu-chah-nulth whalers during most months of the year in the relatively protected waters of Barkley Sound, whereas Grey Whales were probably abundant mostly during the spring migration along the exposed outer coast. This availability, plus the larger size and higher oil content of Humpback Whales, may have made them preferable to Grey Whales.

Humpback Whales were heavily exploited by the commercial whaling industry in many regions of the world's oceans. Because of their coastal distribution and relatively slow swimming speed compared to other rorquals, they were often the first whales to be hunted in a newly discovered area. Commercial Humpback whaling in the North Pacific began in the mid 19th century and expanded dramatically during the early 20th century. All populations in this region were severely depleted by 1966, when the species received global protection from the International Whaling Commission (IWC). But despite the ban on hunting Humpbacks, illegal catches by Soviet whalers in particular continued until 1971. About 21,000 Humpbacks were recorded caught by modern whaling in the North Pacific during the 20th century, of which about 14,000 were taken in the eastern North Pacific.

In British Columbia, commercial whaling for Humpbacks began in the Strait of Georgia in the late 1860s. A succession of small companies, parties and individuals pursued them from sailing schooners and shore bases at Whaling Station Bay on Hornby Island, Whaletown Bay on Cortes Island, and Blubber Bay on Texada Island. The first phase of this industry was short-lived, ending in 1873 with a minimum of 81 Humpbacks having been caught. Whaling for Humpbacks in the Strait of Georgia resumed in 1907, when the Pacific Whaling Company opened a station at Page's Lagoon (now Piper's Lagoon) in Nanaimo. The whaling season started in November 1907, and by February 1908 at least 112 Humpbacks had been taken. The local Humpbacks appear to have been depleted quickly, and whaling from this station ended following the winter 1909–10 season. But between

1905 and 1911 the Pacific Whaling Company constructed and began operating whaling stations at Sechart in Barkley Sound and Cachalot Inlet in Kyuquot Sound on the west coast of Vancouver Island, and at Rose Harbour and Naden Harbour on Haida Gwaii. Between 1905 and 1943, when the last of these stations closed, 4811 Humpback catches were recorded, more than any other species. Over 80 per cent of this catch had been taken by 1917, with a maximum one-year catch of 1022 Humpbacks in 1911 (see Appendix 3). Although Humpback Whales were clearly depleted in BC waters by the 1940s, the whaling station at Coal Harbour on northwestern Vancouver Island, which operated from 1948 to 1967, continued to take Humpbacks when they could find them. A total of 804 Humpbacks were processed at Coal Harbour, but this represented only 8 per cent of the more than 10,000 whale catches recorded at this station.

Taxonomy and Population Structure

The genus *Megaptera* has only a single species and is one of only two genera in the family Balaenopteridae (the rorquals). Although Humpback Whales vary in size and colour pattern between the northern and southern hemispheres, these differences are not considered sufficient to warrant division into subspecies. Humpback populations differ in genetic structure among and within oceans. Genetic differentiation within oceans, including the North Pacific, results from the strong influence of maternal fidelity to particular breeding and feeding grounds. Humpback Whales in BC differ in mitochondrial DNA haplotype frequencies between northern waters and southwestern Vancouver Island. These differences reflect the fact that most whales off the northern coast migrate to Hawaiian breeding grounds, while many whales off southwestern Vancouver Island migrate to Mexico.

Conservation Status and Management

When commercial whaling of Humpback Whales was finally banned by the IWC in 1966, populations in all oceans were severely depleted. In the North Pacific an estimated 1200–1400 whales remained from an initial population that likely exceeded 15,000. The pre-whaling abundance of Humpback Whales in BC waters is unknown, but based on the rate of catches from shore stations in the early 1900s, at least 4000 must have been present off Vancouver Island in 1905. Humpback Whales in most oceans, including the North Pacific, have shown strong recovery since commercial whaling ended. The SPLASH project during 2004–2006 resulted in an estimate for the North Pacific of about 21,000 Humpbacks, but this is considered to be a low estimate. Based on statistical modelling using photo-identification records in BC from 1992 to 2006, we estimate that Humpbacks have been increasing at about four per cent annually and that they numbered about 2145 in 2006.

Although Humpback Whales appear to have done very well since the cessation of whaling, they still face a variety of threats from human activities. Chief among these are injury or mortality from vessel strikes and entanglement in fishing gear. Globally, Humpback Whales are the second most commonly struck whale species, after the Fin Whale, and they are the most frequently reported cetacean struck by vessels in BC waters. Between 2001 and 2008, DFO's Pacific Marine Mammal Response Program received 21 confirmed vessel strike reports involving Humpback Whales. Of these, 15 were witnessed strikes while the remaining six were of individuals documented with fresh injuries consistent with recent blunt force trauma or propeller lacerations. Vessel strikes in BC have mostly involved small boats (less than 10 metres long) travelling at speeds in excess of 15 knots (28 km/h). No deaths of Humpbacks that could be attributed to vessel strikes have yet been documented in BC waters.

Entanglement in fishing gear can also be a serious threat. In BC waters there have been 40 reports of entangled Humpbacks since 1987, including four confirmed deaths. Far more entanglements likely occur than this total might suggest, as most probably go unnoticed and unreported. A study of distinctive scars on the whales' peduncles (tail stocks) found that about half of all adult Humpbacks in southeastern Alaska showed evidence of having been entangled. In BC, gear types involved in the 40 entanglement incidents were mainly gillnets and float lines from crab or prawn traps, but seine nets and lines from aquaculture facilities and herring ponds, longline fishing gear, and boat anchors have also been documented.

Other anthropogenic factors with the potential to affect Humpback Whales include acute noise from sources such as military sonar and marine seismic exploration, as well as chronic noise from shipping, which may mask the sounds whales use to communicate or navigate.

So far, these problems do not appear to be affecting the recovery of Humpback Whales in British Columbia. As the population continues to grow, Humpbacks are reoccupying inshore coastal waters where they were once common. One example is the summer habitat in waters of western Johnstone Strait, Blackfish Sound and eastern

Queen Charlotte Strait, where local whale-watching operators, biologist Alexandra Morton and researchers with the Marine Education and Research Society have been photo-identifying Humpbacks since they first started reappearing in the early 1980s. Encounters were initially sporadic, and only a few individuals were identified each year through the 1980s and 1990s. Numbers of individuals identified each year began to increase steadily, reaching 10–15 whales each year by the mid 2000s, 20–35 by the late 2000s and 67 in 2011. Many of these animals are well known to the locals, coming back each season. Adult females are returning with their calves, many of which in turn come back in the following summers.

The recovery of Humpback Whales in most of the world's oceans led to the species being downlisted in 2008 from Vulnerable to Least Concern on IUCN's Red List. Humpbacks in Canadian Pacific waters were listed as Threatened under the Species at Risk Act in 2003, and a Recovery Strategy and Action Plan are being developed by DFO to promote their recovery. In 2011 COSEWIC assessed the population in Canadian waters as Special Concern because of its current abundance and upward trend; this downlisting was implemented under SARA in 2014. The Humpback Whale is currently on the BC Blue List.

Remarks

The genus name *Megaptera* is derived from the Greek *mega* for "large" and *pteron* for "wing", referring to the Humpback's huge flippers. The species name *novaeangliae* refers to New England, origin of the type specimen. The large flippers are the Humpback's most distinctive feature, and recent research indicates that they play a major role in the animal's great agility in the water. The flippers are highly flexible and mobile at the shoulder, but it is their unique tubercles (bumps or knobs) along the leading edges that make the Humpback the champion of whale "aquabatics" (figure 49). Experiments by hydrodynamics specialist Frank Fish and colleagues have shown that tubercles serve to alter the flow of water over the flipper, which significantly reduces drag and increases lift. This discovery has led to new "bio-inspired" designs using tubercles that may be used in the future to improve the performance of aircraft wings, ventilation fans and windmills. Humpback Whales also have tubercles on their head, but it is not yet clear if these also play a hydrodynamic role.

Figure 49. The flippers of Humpback Whales are the longest of all the cetaceans and make the species exceptionally manoeuvreable.
Photo: J. Hildering, Queen Charlotte Strait, July 2013.

Selected References: Barlow et al. 2011; Calambokidis et al. 2008; Cavanagh 1983; Clapham 2009; Clapham and Mead 1999; COSEWIC 2011; Fish et al. 2011; Ford and Reeves 2008; Ford et al. 2009; Ford, Abernethy et al. 2010; Gabriele et al. 1996; Gregr et al. 2000; McMillan et al. 2008; Merilees 1985; Monks et al. 2001; Nichol et al. 2010; Pike and MacAskie 1969; Rambeau 2008; Robbins et al. 2011.

Sperm Whale *Physeter macrocephalus*

Other Common Names: Cachalot.

Description

The Sperm Whale, made famous by the Herman Melville's *Moby Dick,* is perhaps the most distinctive and widely recognized of the whales. It is the largest of the toothed whales with a massive blunt head that is rectangular in profile and makes up 25–35 per cent of the total body length. The species has marked sexual dimorphism, with males almost one-third longer than females and three times the weight. Sperm Whales have a long, very narrow lower jaw, which is barely visible from the side. This contains 18–26 pairs of large, conical teeth that fit into sockets in the upper jaw, which is toothless. There is a single, S-shaped blowhole near the front of the top of the head and offset to the left, rather than being centred as in other cetaceans. The flippers are relatively short but quite wide and spatulate in shape. Instead of a well-defined dorsal fin there is a thick, rounded dorsal hump, which is followed by several smaller bumps along the dorsal ridge of the tail stock. The flukes are broad and triangular viewed from above, with nearly straight trailing edges and a deep notch.

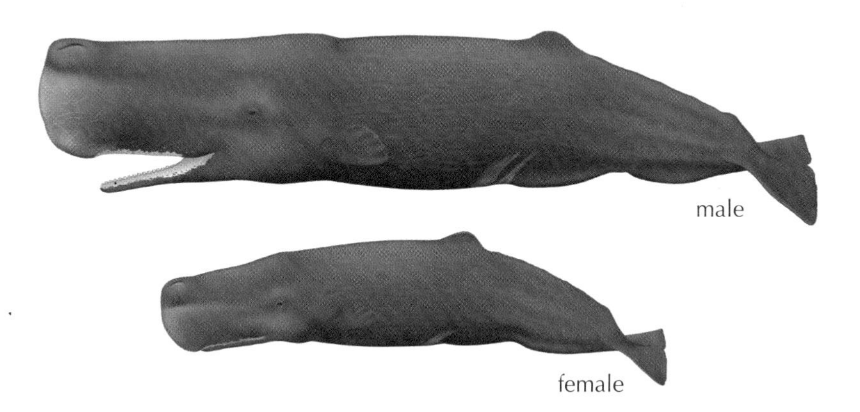

Sperm Whales are mostly black to brownish-grey, with white areas around the mouth and sometimes on the ventral surface, and darkly pigmented flukes on both dorsal and ventral surfaces. They often have whitish scars and scratches, especially on the head, which are the result of interactions with other Sperm Whales or injuries caused by the beaks of large squid, their main prey. The skin is creased with corrugations, particularly posterior to the flippers, giving the body a distinctly wrinkled or shrivelled appearance.

Measurements

length (metres):

male: 13.2 (8.2–19.5) n=3653
female: 10.7 (6.1–14.9) n=994
neonate: 3.5–4.5

weight (kg):

male: ~ 45,000
female: ~ 15,000
neonate: ~ 1000

Dentition

18–26 pairs in lower jaw only; any vestigial teeth in upper jaw do not erupt.

Identification

Sperm Whales are unlikely to be confused with any other large whales found in British Columbia waters. The low and bushy blow, angled forward and to the left, is distinctive at a distance in all but very windy conditions (figure 50). At a closer range, the uniquely squarish head and rounded dorsal hump are evident. Most sightings in BC – particularly north of Vancouver Island – are of adult males, usually alone but occasionally in groups of two to three. These males typically spend 7–10 minutes at the surface, moving slowly if at all, blowing every 10–30 seconds. Once their oxygen reserves are replenished, they submerge steeply, raising their tail flukes high into the air, and begin a deep, almost vertical dive that may last 30–40 minutes (figure 51). Sperm Whales stay

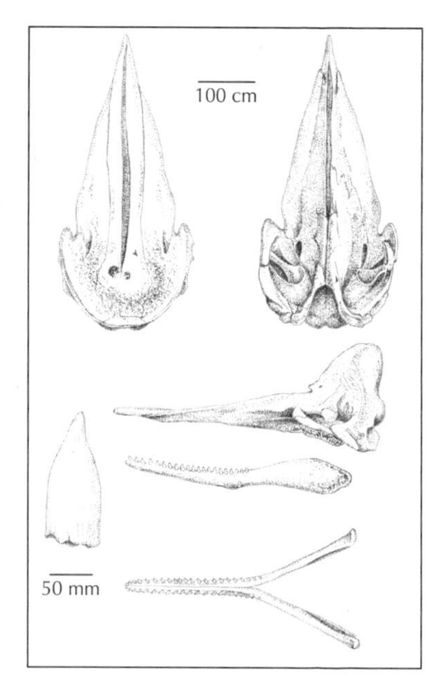

Figure 50. Male Sperm Whale off Kano Inlet, Haida Gwaii, blowing at the surface and showing the left-sided orientation of the blowhole. Photo: J. Towers, June 2009.

Figure 51. A Sperm Whale diving off the west coast of Gwaii Haanas National Park. Sperm Whales normally raise their flukes high above the surface when beginning their descent, and it is common to see Pacific Lampreys attached to the body or flukes. Photo: S. Olcen, July 2007.

underwater for so long that their presence in an area can better be determined by listening for their distinctive, loud clicks with a hydrophone. Larger groups of 10–15, usually females and their young, are most commonly found south of 51°N latitude.

Distribution and Habitat

Sperm Whales are second to Killer Whales as the most widespread cetaceans in the world. They inhabit most deepwater regions from the tropics to the edges of the polar pack ice, although only adult males routinely travel to colder latitudes beyond 50° latitude in each hemisphere. The greatest densities of Sperm Whales occur in productive waters along continental shelf breaks and on slopes, as well as in deep waters surrounding oceanic islands. Although female Sperm Whales are rarely found in waters shallower than 1000 metres, males are regularly seen in depths of less than 300 metres. Tagging and photo -identification studies have shown that female Sperm Whales have ranges spanning roughly 1000 km. Mature males have far more extensive ranges, migrating from high-latitude feeding grounds to tropical and subtropical waters where they temporarily join female groups for breeding. Young male Sperm Whales remain in breeding groups until they begin to mature, at which point they start moving to higher latitudes for feeding. This trend continues as males age, with the oldest individuals travelling the furthest into cold polar waters. But mature males do return periodically to lower latitudes to mate.

The distribution of Sperm Whales in BC waters was until recently known mostly from catches during the era of commercial whaling. Catcher boats based at whaling stations on the west coast of Vancouver Island and Haida Gwaii operated from May through September primarily in the open ocean to about 370 km (200 nautical miles) from shore. Locations of catches show considerable spatial segregation by sex off the BC coast. Breeding schools, which contain 10–15 females and their young, were found further from shore than males, and this separation increased over the whaling season as females moved further offshore and males inshore. What drives the females' offshore movement during summer is not clear. The attraction was once thought to be improved feeding opportunities, but UBC graduate student Ed Gregr and colleagues have suggested, based on a reanalysis of whaling data, that females move offshore for calving during July and August.

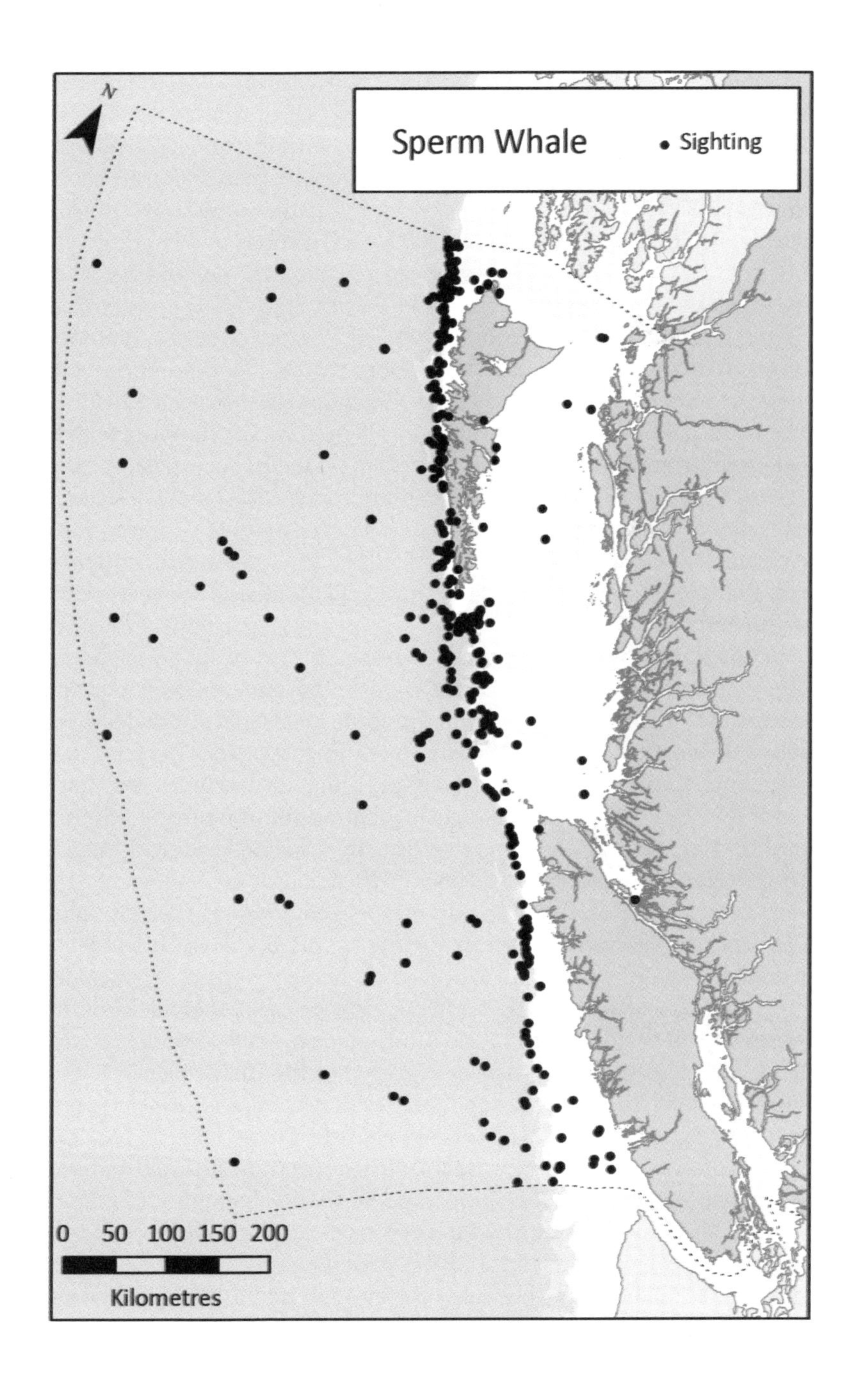
N
Sperm Whale
Sighting
0 50 100 150 200
Kilometres

Because whaling operations shut down during the months of October through April, the winter occurrence of Sperm Whales off the BC coast is poorly known. Since females and young are seldom found in waters with surface temperatures cooler than 15°C, breeding groups likely migrate to warmer waters well to the south of BC during winter. Males probably move toward lower latitudes in winter as well, but solitary Sperm Whales have been reported throughout the year by observers aboard weather ships at Ocean Station Papa, located at 50°N, 145°W, or about 1200 km west of Vancouver Island. Autonomous hydrophone moorings anchored to the sea floor about 500 km west of Vancouver Island and Haida Gwaii have recorded the distinctive loud clicks of Sperm Whales in all months of the year, with a peak during summer. In nearshore waters off the BC coast, adult male Sperm Whales are mostly seen in summer, but occasionally in winter as well. An autonomous hydrophone recorder that we deployed near Anthony Island at the southwest end of Haida Gwaii during 2009–10 also recorded Sperm Whale clicks during all months of the year, peaking in summer.

Whaling catches of Sperm Whales in BC were made primarily along the steep continental shelf edge or beyond, in deep oceanic waters (see figure 19). Occasionally, adult male Sperm Whales were taken by whalers in Hecate Strait, Dixon Entrance and Queen Charlotte Sound and in nearshore waterways along the northern mainland coast. Several whales were caught in the mouth of Milbanke Sound and one in the inshore waters of Squally Channel near Gil Island. More recently, between 2002 and 2010, the Pacific Biological Station's Cetacean Research Program has logged 75 Sperm Whale sightings during ship-based cetacean surveys off the BC coast. All of these were adult males – 68 solitary individuals and seven pairs. The whales were mostly scattered along the steep continental shelf break from the west coast of Haida Gwaii to the northwestern coast of Vancouver Island. Sightings of Sperm Whales reported to the BC Cetacean Sightings Network show a similar pattern, with several additional reports from north central Hecate Strait. In the late fall of 1984 I recorded the unmistakable clicks of a Sperm Whale from a fixed hydrophone anchored to the sea floor in Johnstone Strait at Telegraph Cove. Such visits to the inside passage by Sperm Whales appear to be very rare.

Sperm Whales show little genetic variation in different regions of the North Pacific, likely because of their high mobility. Tagging studies using implanted 'Discovery' tags during the whaling era showed large-scale movements by male Sperm Whales in BC. These 25-cm-long stainless steel tags inscribed with identification information were shot into the whale with a 12-gauge shotgun and were intended to be recovered when the whale was killed and processed (see figure 24). One whale tagged 130 km west of Estevan Point on Vancouver Island in the 1960s was later killed off Adak Island in the Aleutians. Another whale tagged by Japanese researchers in the Bering Sea was later recovered off Vancouver Island. In the summer of 2009, biologist Jan Straley and colleagues tagged three male Sperm Whales off Sitka, southeastern Alaska, with miniature satellite transmitting tags. Two whales moved south not long after being tagged, travelling through BC waters. One whale stayed close to shore and eventually rounded the tip of Baja California and went into the Gulf of California. The other whale travelled further offshore and was off Guatemala, still heading south, when the tag stopped transmitting. The third whale stayed near Sitka until mid October, then headed south, and was off Mexico when the tag went silent.

Natural History

Feeding Ecology

Sperm Whales feed mainly on mesopelagic and bathypelagic squid. A total of 55 species from 36 cephalopod genera have been documented as Sperm Whale prey. Many of these are small, weighing only a few hundred grams, but Sperm Whales also take larger squid, including the Giant Squid, which can reach lengths of 13 metres. Sperm Whales also eat a variety of deep-sea fishes.

The diet of Sperm Whales in British Columbia is known from analyses of the stomach contents of whales killed during the whaling period. Rowenna Flinn, a UBC student, analysed data that DFO scientist Gordon Pike collected from 697 Sperm Whales taken in the 1960s and found that the Robust Clubhook Squid was the dominant prey of both males and females. This is a large squid with mantle lengths reaching two metres (overall length four metres), although mantle lengths of one metre are more typical. Also important in the diet of both sexes were the smaller armhook squid and Ragfish, a deep-sea fish that reaches a length of two metres. Yelloweye Rockfish and other rockfish species were consumed frequently by males but less so by females, and both sexes fed occasionally on a variety of other fishes, including Pacific Spiny Dogfish, Pacific Lamprey, Longnose Skate and Pacific Hake.

Behaviour and Social Organization

The basic element of Sperm Whale society is the matrilineal family unit, which consists of about 12 females and their offspring and is commonly referred to as a breeding school. Most females spend their entire lives in the same unit with their close female relatives. Different units may temporarily travel together, forming larger associations that can persist for several days. Within the family unit there is communal care of the young, with females nursing calves that are not their own. Young whales are incapable of diving to the great depths achieved by adults, so individuals in the group take turns diving to forage so that calves are not left alone at the surface. Subadult males leave their family unit at an average age of six years and join with other young males of similar ages to form loose aggregations known as *bachelor schools*. As males grow older and larger, they travel in smaller and smaller groups and move to higher latitudes. When they reach physical maturity they are mostly solitary. Once they are in their late 20s, males start making return trips to the tropical and subtropical breeding grounds, where they roam among female units looking for mating opportunities.

The activities of Sperm Whales fall into two major categories: foraging and socializing/resting. Females and young spend about 75 per cent of their time foraging, and solitary males even more so. Sperm Whales make repeated deep dives to about 600 metres, on average, for durations of about 45 minutes, but some dives may be to more than 1000 metres for an hour or longer, while others are shorter and shallower. Between dives the whales spend an average of nine minutes at the surface. While foraging, Sperm Whales emit loud clicks at regular intervals of 0.5–1.0 second, presumably to orient themselves and detect prey. Periodically, clicks shift to a creak-type of sound emitted at a higher repetition rate as the whale acoustically examines a nearby object, likely a prey item.

Sperm Whales gather at or near the surface to engage in socializing/resting behaviour. They sometimes lie still and quiet and other times interact with behaviour that includes breaching, tail slapping, rolling and touching each other. During these social sessions, the whales often emit creaks and *codas*, which are short patterned sequences of clicks. These codas are the principal social signals of Sperm Whales, as they do not use the whistles or call types that are common to many of the other toothed whales. Groups of females are delineated into different acoustic clans with unique repertoires of codas. Codas are likely passed on from generation to generation by learning, and they may function as an acoustic "badge" of kin identity.

Life History and Population Dynamics

Sperm Whale calves are born following a gestation of 14–16 months. Although they begin to take solid food before reaching one year of age, they continue to be nursed for several years. Females become sexually mature at about nine years, when they are around 9 metres long, then their growth slows until they're physically mature at age thirty and an average of 10.6 metres long. Calving rates vary, but mature females have, on average, one calf every five years. Birth rates slow as females age and few give birth after about forty years old, when they become post-reproductive, or senescent.

Male Sperm Whales are only slightly larger than females during their first 10 years but then grow substantially larger over the next 20 years. They reach physical maturity at about age fifty and a length of about 16 metres. Males have a long pubescent period between ten and twenty years of age and only start being reproductively active when in their late twenties.

Sperm Whales are known to be attacked by Killer Whales, but this is probably a more significant cause of mortality for calves than adults. When attacked, female Sperm Whales group together into a defensive "rosette" formation. Some groups face inward with their tails out toward the attackers, while others face outward, using their jaws for defence. It is unlikely that Killer Whales would attack a mature male Sperm Whale unless it was weakened by age or injury.

Occasionally, for reasons that are often unclear, Sperm Whales strand themselves alive on shore, either alone or in groups. A group of 41 Sperm Whales became stranded near Florence, Oregon, on June 16, 1979, apparently after becoming disoriented in shallow water. All eventually died on the beach from overheating and suffocation caused by their own weight pressing on their heart and lungs. No mass strandings have been reported in BC, but 22 single strandings have been documented.

Exploitation

Beginning in the early 18th century, Sperm Whales were hunted over much of the world's oceans. For over 150 years whalers from several countries, particularly the US and UK, pursued Sperm Whales on the high seas using open whaleboats launched from square-rigged sailing ships. The open-boat hunt declined in the latter half of the 19th century as petroleum products were developed as substitutes for Sperm Whale oil. A resurgence in the hunt began after World War II, this time using modern catcher boats and factory ships that substantially reduced populations in many regions. Commercial hunting of Sperm Whales declined in the 1970s and 1980s and virtually ended with the International Whaling Commission moratorium in 1988. More than 300,000 Sperm Whales were killed in the North Pacific during the 20th century, with Soviet pelagic whaling operations responsible for killing about 160,000 from 1948 to 1979, mostly in the central and eastern North Pacific.

In British Columbia, Sperm Whales were never exploited by First Nations because they were considered too dangerous to hunt. Nuu-chah-nulth whalers on the west coast of Vancouver Island were quite familiar with the species and referred to it as *chichichwi'n*, or "teeth in middle (of body)". A total of 6019 Sperm Whales were reported killed by commercial whalers operating from shore stations in British Columbia from 1908 to 1967. Between 1933 and 1943 Sperm Whales comprised 79 per cent of the catch at the Rose Harbour and Naden Harbour whaling stations on Haida Gwaii. These were virtually all males, as female groups seldom ranged north as far as the nearshore hunting grounds of coastal whaling stations around Haida Gwaii. At the Coal Harbour whaling station on northwestern Vancouver Island, about 100–250 Sperm Whales were typically taken per year between 1948 and 1967 (see Appendix 3). This represented 20–32 per cent of the station's annual whale catch. More than 1000 Sperm Whales were brought into the Coal Harbour station in its last six years of operation, peaking at 304 in 1967, the year the station closed down. Although the catcher boats from Coal Harbour operated mostly off the west coast of Vancouver Island where female groups of Sperm Whales were often found, more than 90 per cent of the catch consisted of males due to size restrictions (minimum allowable length 35 ft, or 10.7 metres) and the prohibition of taking females with young. According to official reports, an additional 474 Sperm Whales were killed by the Japanese pelagic whaling fleet in the outer portion of Canada's EEZ from 1964 to 1975. The Soviet whaling fleet was also active in this area during the same period, but its catch data were falsified so it is unknown how many whales were taken.

Taxonomy and Population Structure

Although four species of Sperm Whales in the genus *Physeter* were described by Linnaeus in 1758, it has long been recognized that these all referred to the same species. The scientific name *Physeter catodon* was commonly used for the species in the past, but today most authorities agree that *Physeter macrocephalus* has precedence and it is used almost universally. No subspecies are recognized, but studies using molecular genetic techniques suggest that there may be discrete populations of Sperm Whales in the North Pacific.

Conservation Status and Management

Despite the long history of commercial whaling for Sperm Whales, most of the populations in many areas appear to have been less drastically depleted than many baleen whale populations. The global pre-whaling abundance of Sperm Whales likely exceeded a million, and this was reduced to perhaps a few hundred thousand by the late 1980s when commercial whaling was banned by the International Whaling Commission moratorium. Today, only relatively small numbers of Sperm Whales are killed annually by shore-based subsistence whalers in eastern Indonesia, and up to 10 are taken each year in the North Pacific by Japan under its "scientific" whaling program.

It has recently come to light that under-reporting and falsification of Sperm Whale catches by Soviet whalers during the 1960s and 1970s was more extensive than previously thought. A large proportion of their catch during these years was females, which was illegal at the time. This may have had a serious impact on Sperm Whale populations in the central and eastern North Pacific. There are now an estimated 80,000 Sperm Whales in the North Pacific, although this estimate is very rough. Their abundance off the west coast of North America appears to vary considerably from year to year. Biologists with the US National Marine Fisheries Service conduct periodic ship surveys as far out as 550 km (300 nautical miles) off California, Oregon and Washington to count cetaceans. In 2005 they estimated 3140 Sperm Whales in the survey area, but a repeated survey in 2008 yielded an estimate of only about 300. As there is no reason to suspect a decline in the population, analysts concluded that the difference was likely due to a difference in whale distribution between the two survey years. No abundance estimate is available for British Columbia.

Although the greatest threat to Sperm Whales – commercial whaling – has ceased, several other human-caused threats remain. Most Sperm Whale prey species are of little interest to fishermen, but Sperm Whales sometimes take fish off of fishing gear (mostly longline gear), an activity known as *depredation*. This depredation of longline catches is a relatively recent and increasing phenomenon in many regions. It's a growing problem in southeastern Alaska, where male Sperm Whales selectively remove Sablefish, also known as Alaskan Black Cod, from

Figure 52. A male Sperm Whale approaching a longline fishing boat south of Cape St James, Queen Charlotte Sound, to feed on discarded heads and entrails. Note the wrinkled appearance typical of Sperm Whales. Photo: J. Ford, May 2007.

longline gear. While it is usually more of a problem for the fishermen than the whales, some Sperm Whales become entangled in the fishing lines and die. In BC waters, Sperm Whales frequently loiter near long-line fishing boats, but seldom remove fish from the gear. Instead, they seem to be interested in feeding on non-target fish that are released as lines are retrieved, or on fish heads and other offal discarded from the boats (figure 52). In BC most Sablefish are caught using traps rather than longlines, which prevents the whales from removing the fish. Entanglement of Sperm Whales in drift gillnets is a cause of mortality in areas such as the Mediterranean, but it doesn't appear to be a threat in the eastern North Pacific where there is limited use of this fishing technique.

Sperm Whales also ingest marine debris that can cause choking or blockage of the gastrointestinal system. Various foreign objects including glass fishing floats, coconuts and shoes have been found in the stomachs of Sperm Whales. Gordon Pike, a whale biologist at the Pacific Biological Station from the late 1940s through 1960s, noted "incidental tidbits" such as a small stone, a piece of wood and an apple in the stomachs of Sperm Whales processed at the Coal Harbour whaling station. Sperm Whales are also at risk of ship strikes in areas of high traffic, although how often this happens is unknown.

The Sperm Whale is listed as Vulnerable on the IUCN Red List and considered Not at Risk in Canadian waters by COSEWIC. It is on the BC Blue List.

Remarks

The generic name for the Sperm Whale comes from the Greek *Physeter*, meaning "blower". The specific name is from the Greek *makros*, meaning "large" or "long", and *kephale*, meaning "head". The species got its common name from the oil-like wax in the head, called spermaceti, which was prized by whalers and used in candles, ointments and cosmetics. Early whalers evidently felt that this oil resembled sperm.

Selected References: Andrews et al. 2010; Carretta et al. 2011; Flinn et al. 2002; Ford, Koot et al. 2010; Gregr et al. 2000; Ivashchenko et al. 2013; Jefferson et al. 2008; Mellinger et al. 2004; Monks et al. 2001; Pike 1950; Pike and MacAskie 1969; Reeves and Whitehead 1997; Whitehead 2009.

Pygmy Sperm Whale *Kogia breviceps*

Other Common Names: None.

Description

The Pygmy Sperm Whale has a robust body with a shark-like head and a short, narrow lower jaw. It is very similar in appearance to the Dwarf Sperm Whale. It has a small, strongly falcate dorsal fin positioned well behind the middle of the back. Flippers are small and located far forward, close to the head. The head is blunt compared to most other toothed whales and dolphins, and it becomes increasingly squarish in older animals. The lower jaw ends well posterior to the snout and contains 12–16 (sometimes 10–11) pairs of fine, sharp teeth that fit into sockets in the upper jaw, which is toothless. The pair of longitudinal throat creases typically found in the Dwarf Sperm Whale is absent in the Pygmy Sperm Whale.

The body of the Pygmy Sperm Whale is dark bluish-grey or black on the dorsal surface, which blends to lighter shades on the sides and to white or pinkish-white ventrally. As in the Dwarf Sperm Whale, there is a lightly pigmented, crescent-shaped mark just behind the eye, sometimes referred to as a *false gill*.

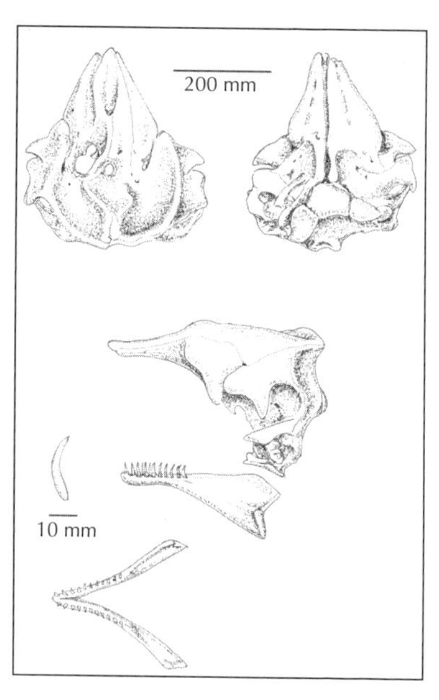

Measurements

length (metres):

adult: 3.5–3.7
neonate: ~ 1.2

weight (kg):

adult: ~ 450
neonate: ~ 50

Dentition

10–16 pairs in lower jaw only.

Identification

Both the Pygmy Sperm Whale and the similar Dwarf Sperm Whale are very difficult to spot in any but ideal sea conditions. They tend to lie motionless at the surface (called "logging"), show-

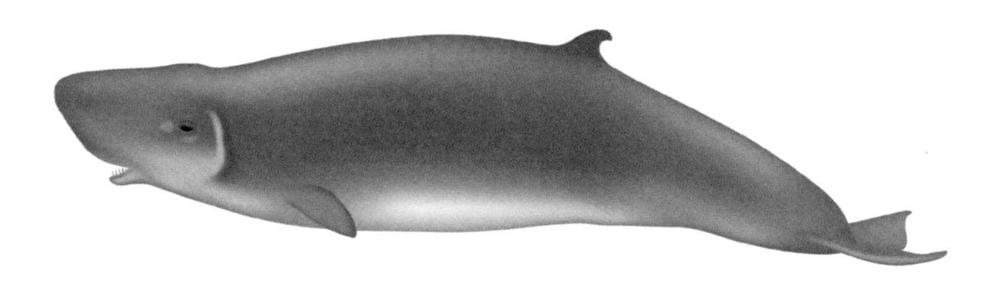

ing only their back and dorsal fin (figure 53), and they are difficult to approach for close examination. The two species are also extremely difficult to tell apart at sea. Adult Pygmy Sperm Whales are somewhat larger, growing to 3.8 metres compared to a maximum length of 2.7 metres in the Dwarf Sperm Whale. Also, the shape and position of the dorsal fin differs between the two species, with the Pygmy's fin relatively smaller and positioned further back on the body than the Dwarf's. The Pygmy Sperm Whale's dorsal fin also tends to be more strongly falcate, such that the tip is below the highest point of the fin.

Distribution and Habitat

Pygmy Sperm Whales are widespread in tropical to warm-temperate oceans of the world. They appear to prefer habitat farther offshore than the Dwarf Sperm Whale, generally occurring in deep waters seaward of the continental shelf and slope. In the eastern North Pacific, Pygmy Sperm Whales have occasionally been observed during ship-based cetacean surveys off California and Oregon. They are also known from strandings south of British Columbia, including three in

Figure 53. A Pygmy Sperm Whale logging at the surface in the Bahamas. This species has not been observed alive in BC waters.
Photo: C. Dunn, June 2010.

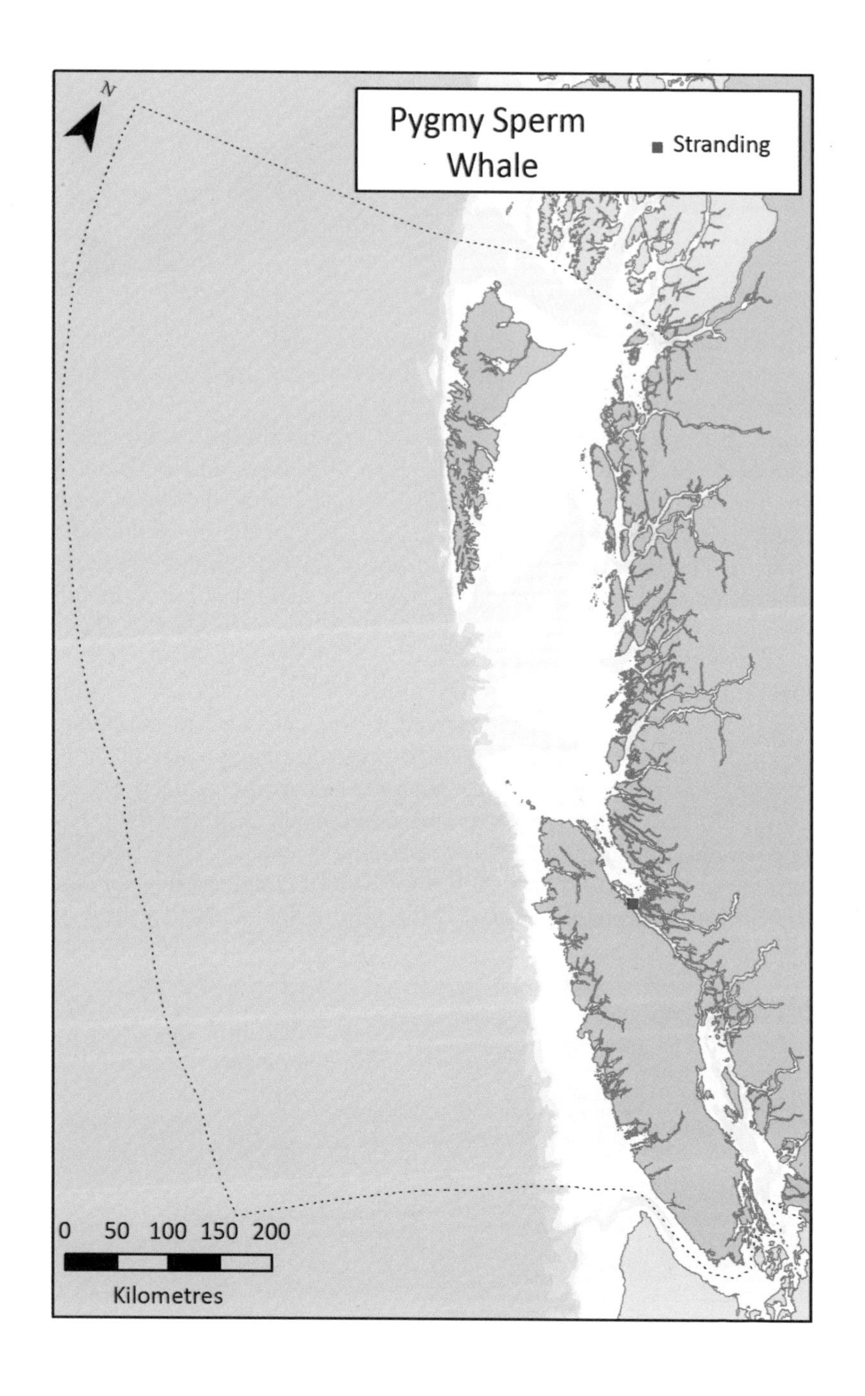
N
Pygmy Sperm Whale
Stranding
0 50 100 150 200
Kilometres

Washington. There is only one record, however, of the species in BC. This was the carcass of a mature male, 3.3 metres in length, found in a state of advanced decomposition on the beach at Hanson Island off northeastern Vancouver Island on July 31, 2003 (figure 54). The absence of more strandings, plus the lack of additional sightings in BC waters, suggest that these latitudes are at or beyond the northern limit of the species' normal range in the northeastern Pacific. There is, however, one record of a stranded Pygmy Sperm Whale farther north – at Yakutat, Alaska – also in July 2003. The species can be considered to be accidental in BC waters.

Natural History

Feeding Ecology

Pygmy Sperm Whales feed in deep water, primarily on squid and less frequently on deep-sea fish and shrimp. Their diet may be diverse – in South Africa, at least 67 prey species have been documented in the stomach contents of stranded Pygmy Sperm Whales. The stomach of the whale stranded on Hanson Island, BC, contained the beaks of 45 squid belonging to the family Gonatidae.

Figure 54. The only Pygmy Sperm Whale record for British Columbia is this mature male stranded on Hanson Island in July 2003. Photo: J. Borrowman.

Behaviour and Social Organization

Pygmy Sperm Whales are found singly or in small groups of up to about six individuals. Like Dwarf Sperm Whales, they are often seen lying motionless at the surface, exposing only the top of their head, back and dorsal fin (figure 53). They have very inconspicuous blows and dive without raising their flukes above the surface. Live-stranded animals produce echolocation-type clicks, but no other vocalizations have been recorded from the species.

Life History and Population Dynamics

Very little is known about the life history of Pygmy Sperm Whales. They reportedly calve between fall and spring, and gestation is reported to be 7–11 months. Some pregnant females have been seen with a calf, suggesting that births sometimes occur at intervals of only about a year. Sexual maturity is reached at lengths of 2.7–2.8 metres for females and 2.7–3.0 metres for males. Calves are about 1.2 metres long at birth. Little is known of the causes of natural mortality for Pygmy Sperm Whales, but they are occasionally preyed upon by Killer Whales and Great White Sharks. Two live-stranded Pygmy Sperm Whales in Japan had antibodies to the bacterium *Brucella*, which is the cause of brucellosis, a common disease in domestic animals. And a live-stranded animal in Taiwan was infected with the virus *Morbillivirus*, which is a known cause of mortality in marine mammals.

Exploitation

Like Dwarf Sperm Whales, Pygmy Sperm Whales have never been the targets of commercial whaling, likely due to their small body size, wariness and offshore range. But small numbers have been taken in shore-based cetacean hunts in the Lesser Antilles, Indonesia, Japan and the Philippines.

Taxonomy and Population Structure

Prior to the mid 1960s, all specimens of the genus *Kogia* were considered to belong to a single species, the Pygmy Sperm Whale. But a detailed examination of morphological evidence published in 1966 demonstrated that there are two species in the genus, the Pygmy and the Dwarf Sperm Whale. No subspecies have been described.

Conservation Status and Management

There is no reliable information on population sizes and trends for Pygmy Sperm Whales in any region. Because they are small and inconspicuous at sea, they are difficult animals to detect during ship-based surveys. Strandings are frequent in some parts of the world, including the southeastern US, South Africa and New Zealand, suggesting a fairly common distribution in these areas. There does not appear to be any major anthropogenic threat to Pygmy Sperm Whales at present. Small numbers are taken in shore-based cetacean hunts or as bycatch in gillnet fisheries, but this mortality is unlikely to have population-level effects.

The Pygmy Sperm Whale is considered Data Deficient by IUCN, and Not at Risk in Canadian waters by COSEWIC. It has no BC provincial listing as it is considered Accidental in BC waters.

Selected References: Baird et al. 1996, Carretta et al. 2011, McAlpine 2009, Reeves et al. 2002.

Dwarf Sperm Whale *Kogia sima*

Other Common Names: None.

Description

The Dwarf Sperm Whale is a small toothed whale with a robust, porpoise-like body that tapers rapidly to the tail flukes. It has a fairly high, falcate dorsal fin midway along the back. The small flippers appear to be positioned unusually far forward on the body. The head is conical with a short rostrum that becomes increasingly squarish with age. The small, narrow lower jaw ends well posterior to the rostrum and gives the mouth a distinctive, shark-like appearance. The Dwarf Sperm Whale has 7–12 pairs of sharp, thin teeth in the lower jaw and sometimes as many as three pairs of teeth in the upper jaw. Most, if not all, Dwarf Sperm Whales have several short longitudinal grooves on the throat.

Dwarf Sperm Whales are bluish-grey to black dorsally, which grades to lighter grey along the flanks and a dull white or pinkish on the ventral surface. A pale, crescent-shaped marking – often referred to as a *false gill*, as it resembles a fish's gill cover – is present between the eye and the insertion of the flipper.

Measurements

length (metres):
 adult: 2.5–2.7
 neonate: ~ 1.0
weight (kg):
 adult: ~ 275
 neonate: 20–40

Dentition

7–12 (rarely 13) pairs in the lower jaw; 0–3 pairs in the upper jaw.

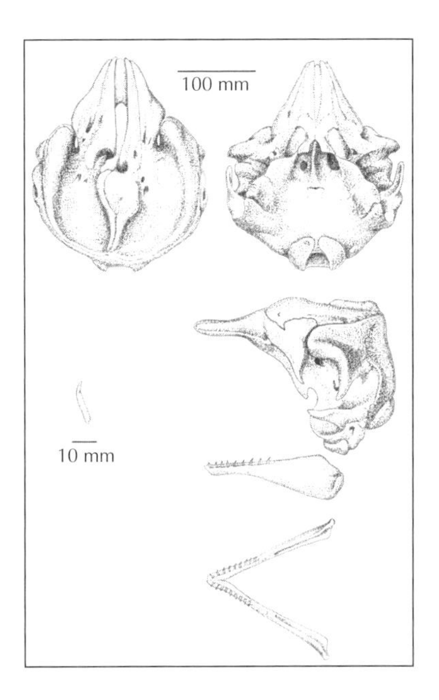

Identification

At a distance, Dwarf Sperm Whales might resemble porpoises or dolphins, but there are clear differences in appearance and behaviour. Dwarf Sperm Whales seldom swim at the surface, but instead tend to lie motionless, showing just their back and relatively large (for their size) and erect dorsal fin (figure 55). Because of this low profile, they are very difficult to detect in any but a flat sea. The Dwarf Sperm Whale is very similar in appearance to the Pygmy Sperm Whale, and the two cannot be reliably distinguished without close inspection. But neither species has been observed alive at sea in BC waters, and so future specimens are most likely to be stranded animals. Diagnostic features include body length (Dwarf Sperm Whales grow to less than 2.7 metres, Pygmy Sperm Whales more than 2.7 metres) and dorsal fin (triangular in shape,

Figure 55. A pair of Dwarf Sperm Whales moving slowly at the surface in the Gulf of California, Mexico. Photo: V. Shore, March 2009.

height more than five per cent of the total body length and located near midpoint of the back in Dwarf Sperm Whales; falcate in shape, less than five per cent of the total body length and positioned behind midpoint of the back in Pygmy Sperm Whales).

Distribution and Habitat

Dwarf Sperm Whales occur in tropical to warm-temperate waters worldwide, mostly in deepwater habitats, especially near continental slopes. Dwarf Sperm Whales appear to prefer warmer waters than Pygmy Sperm Whales, and tend to occur closer to the shelf break. They are not commonly observed at sea, likely due to their cryptic appearance and behaviour rather than rarity.

There is only a single record for the Dwarf Sperm Whale in British Columbia – a mature female that stranded alive at Pachena Bay on the west coast of Vancouver Island on September 24, 1981. Despite the efforts of volunteers from the Bamfield Marine Sciences Centre, who twice refloated the animal and pushed it into deeper water, it was found dead on the beach the following morning (figure 56). This is the northernmost stranding of the species in the world and should be considered an extralimital record. There are several stranding records for the species on California shores, but no confirmed sightings from cetacean surveys in coastal or offshore waters of the eastern North Pacific.

Figure 56. This Dwarf Sperm Whale stranded itself on the beach at Pachena Bay in September 1981 and died – it is the only specimen record of the species in BC. Photo: S. Suddes.

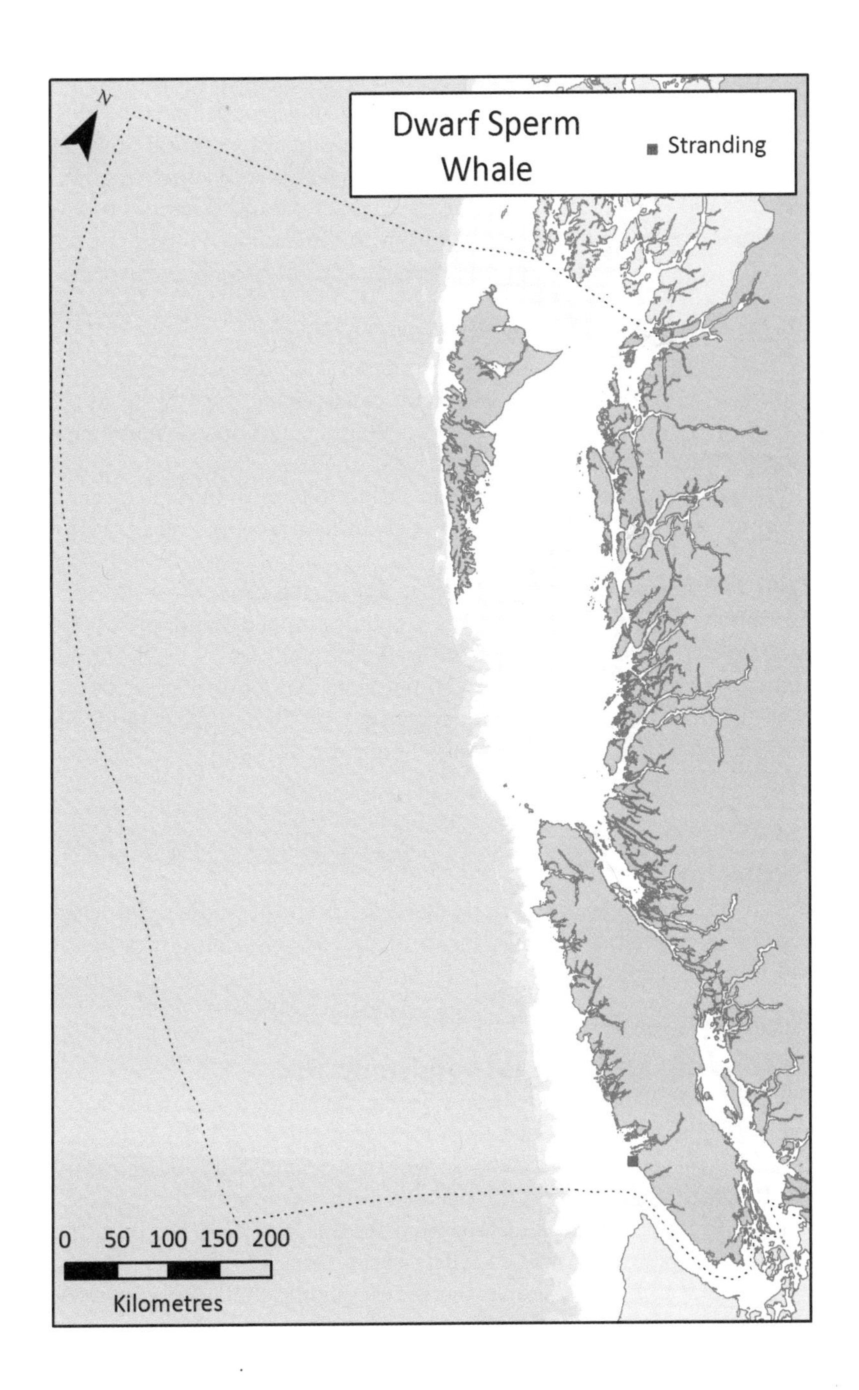
N
Dwarf Sperm Whale
Stranding
0 50 100 150 200
Kilometres

Feeding Ecology

Little is known of the feeding ecology of the Dwarf Sperm Whale. Based on examination of the stomach contents of stranded animals, their diet consists primarily of deepwater, pelagic squid and occasionally fish and crustaceans, including pelagic crabs and shrimp. The BC specimen had been feeding on squid from the families Gonatidae and Octopoteuthidae and on the mesopelagic shrimp *Notostomus japonicus*.

Behaviour and Social Organization

When Dwarf Sperm Whales are seen at sea, they are alone or in small groups of up to four individuals, with occasional groups of up to ten reported. They are usually observed floating motionless at the surface and rarely display energetic behaviour. When diving, they tend to sink vertically rather than roll at the surface like most other cetaceans. They are usually wary of boats and are difficult to approach.

Life History and Population Dynamics

Sexual maturity in both sexes occurs at an estimated length of 2.1–2.2 metres. Calving appears to be diffusely seasonal, with a peak during summer following a gestation estimated to be about a year. Dwarf Sperm Whales often bear scars suggestive of shark attacks, and this may be a cause of natural mortality. They are also known to be eaten by Killer Whales.

Exploitation

Dwarf Sperm Whales have never been targeted by commercial whalers, but they are occasionally taken in shore-based harpoon fisheries in Indonesia, Japan and the Lesser Antilles.

Taxonomy and Population Structure

Dwarf Sperm Whales and Pygmy Sperm Whales are so similar in appearance that they were once considered a single species. But a detailed review of the genus published in 1966 by a scientist at the Smithsonian Institution established the validity of the separate Dwarf species. The Latin name *Kogia simus* was used for this species in the past, but it has been changed to *sima* to reflect correct gender reference. Recent molecular studies have shown significant genetic differences between Dwarf Sperm Whales in the Atlantic and the Indo-Pacific region, suggesting that further splitting of the species may be warranted in the future.

Conservation Status and Management

Although they are seldom observed at sea, Dwarf Sperm Whales appear to be fairly common in some tropical regions. Vessel surveys have yielded estimates of about 11,000 Dwarf Sperm Whales in the eastern tropical Pacific and about 19,000 in the Hawaiian Islands Exclusive Economic Zone. There are no confirmed sightings off the North American west coast north of Mexico despite intensive vessel surveys for cetaceans, suggesting that the species is rare in the region.

Both Dwarf and Pygmy Sperm Whales are occasionally entangled in fishing gear and are known to ingest plastic flotsam such as bags and balloons, which may cause intestinal blockage and death. It is uncertain if these threats or the occasional directed takes in some areas are at a level that could affect populations.

Both IUCN and COSEWIC consider this species Data Deficient, signifying that there is insufficient information on its status to assign a conservation designation either globally or in Canadian waters. It is considered Accidental in BC.

Remarks

Both Dwarf Sperm Whales and Pygmy Sperm Whales have a sac in the lower intestine that contains about 12 litres of a thick, reddish-brown or chocolate-coloured liquid, which is expelled into the water when the whale is startled. This may serve as a distraction that aids in escaping from a predator, as is the case with the "ink" of octopus.

Selected References: Carretta et al. 2011, McAlpine 2009, Nagorsen 1985, Nagorsen and Stewart 1983, Reeves et al. 2002, Willis and Baird 1998b.

Baird's Beaked Whale *Berardius bairdii*

Other Common Names: North Pacific Bottle-nosed Whale, Giant Bottlenose Whale (both obsolete).

Description

Baird's Beaked Whale is the largest ziphiid (members of the family Ziphiidae), reaching lengths of 10–12 metres, with females growing slightly larger than males. It has a long, slender body with a small, triangular to falcate dorsal fin set far behind mid-back, the tip of which is usually rounded. The small flippers fit into shallow 'pockets' or indentations on the body. These whales have a long beak that rises abruptly to a prominent, bulbous melon. The genus Berardius is unusual in that the blowhole is oriented backwards compared to other odontocetes. When the blowhole is closed, its margin resembles a C in shape – in Baird's Beaked Whale, the horns of the C are pointed towards the tail whereas in all toothed whales they point towards the rostrum (an exception is the Sperm Whale, which has a unique blowhole configuration). The lower jaw has two pairs of teeth that erupt in both adult males and females.

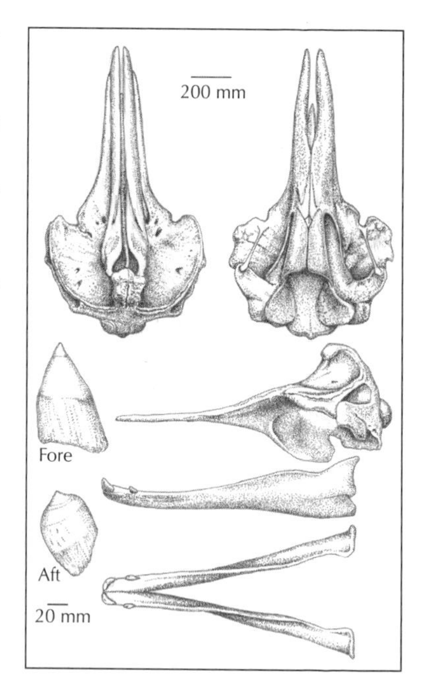

The anterior pair is larger and situated at the tip of the lower jaw. Because the lower jaw extends beyond the upper, this pair of teeth is exposed when the mouth is closed and whale barnacles of the genus *Conchoderma* may be attached to one or both of these front teeth. The posterior pair of teeth is smaller and peg-like, and concealed in the closed mouth. As in all beaked whales, there is a well-defined pair of grooves formed by folds in the skin and blubber along the throat that converge anteriorly into a V shape. These grooves extend from near the tip of the ventral side of the lower jaw to behind the gape. The caudal peduncle (tail stock)

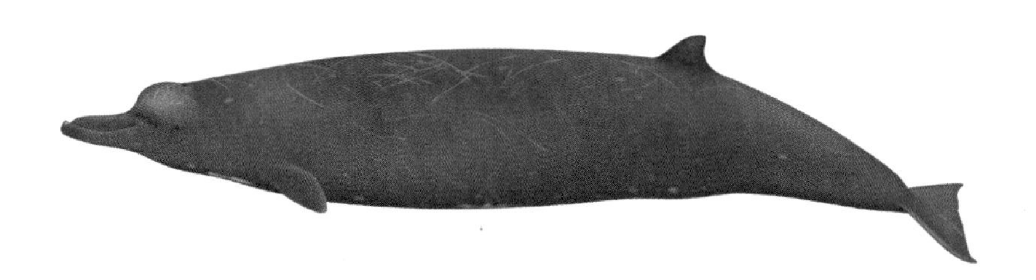

is deep and laterally compressed, and the flukes are relatively broad for the size of the animal (25–30 per cent of body length).

The coloration of both sexes is similar, ranging from slate grey to black, but patches of diatoms often give the skin a brownish shade. There are whitish, cloudy patches in the natural coloration ventrally, especially on the throat, genital area and around the umbilicus. The bodies of adults, both males and females, are usually criss-crossed with numerous white, linear scars that appear to be caused by tooth rakes during social interactions. Whitish oval scars thought to be caused by Cookiecutter Sharks and possibly Pacific Lampreys are often present as well.

Measurements

length (metres):

male: 10.2 (5.5–11.3) n=37
female: 10.1 (8.8–11.3) n=2
neonate: ~ 4.6

weight (kg):

adult: ~ 8000-11,000
neonate: unavailable

Dentition

Two pairs near the tip of the lower jaw adults (males and females).

Identification

Baird's Beaked Whales are the most conspicuous and readily identifiable of the beaked whales found in BC waters and, as a result, are the most often sighted. They are almost twice the size of Cuvier's, Hubbs' and Stejneger's Beaked Whales and have a more prominent beak, steep forehead and bulbous melon. When observed at sea, Baird's Beaked Whales are typically found in closely knit schools of 3–10 individuals, surfacing in unison and producing low, bushy blows (figure 57).

Figure 57. Three of four Baird's Beaked Whales observed eight kilometres west of Cape St James, Haida Gwaii, in May 2007. Photo: B. Gisborne.

Distribution and Habitat

Baird's Beaked Whales are endemic to cool-temperate waters of the North Pacific Ocean. They have been sighted in virtually all deep-water areas of the North Pacific north of 30°N, particularly near continental slopes, submarine escarpments and seamounts in waters 1000–3000 metres deep. The southernmost records are from the Gulf of California, Mexico (22°N), and the northernmost from around St Paul's Island in the Bering Sea (60°N). The species is migratory, arriving in continental slope waters during summer when sea surface temperatures are highest.

In British Columbia waters, whalers reported seeing Baird's Beaked Whales throughout the whaling season of April through early October, although most were sighted during July and August. Baird's Beaked Whales were never a primary whaling target in BC as they were comparatively uncommon, smaller and less valuable than the larger baleen whales and Sperm Whales. Still, over the course of commercial whaling in BC, 41 Baird's Beaked Whale were killed, almost half of these (19) in August. Since the end of the whaling period, 24 reliable sightings of Baird's Beaked Whales have been recorded in BC waters (see Appendix 2). Whaling kills and sightings are distributed widely along the continental slope and in oceanic waters between southern Haida Gwaii and the northern half of Vancouver Island. There are only three stranding records for the species in BC. Two of these were on the west coast of Vancouver Island, at Cox Bay and Long Beach, and the third near Rose Spit on Haida Gwaii.

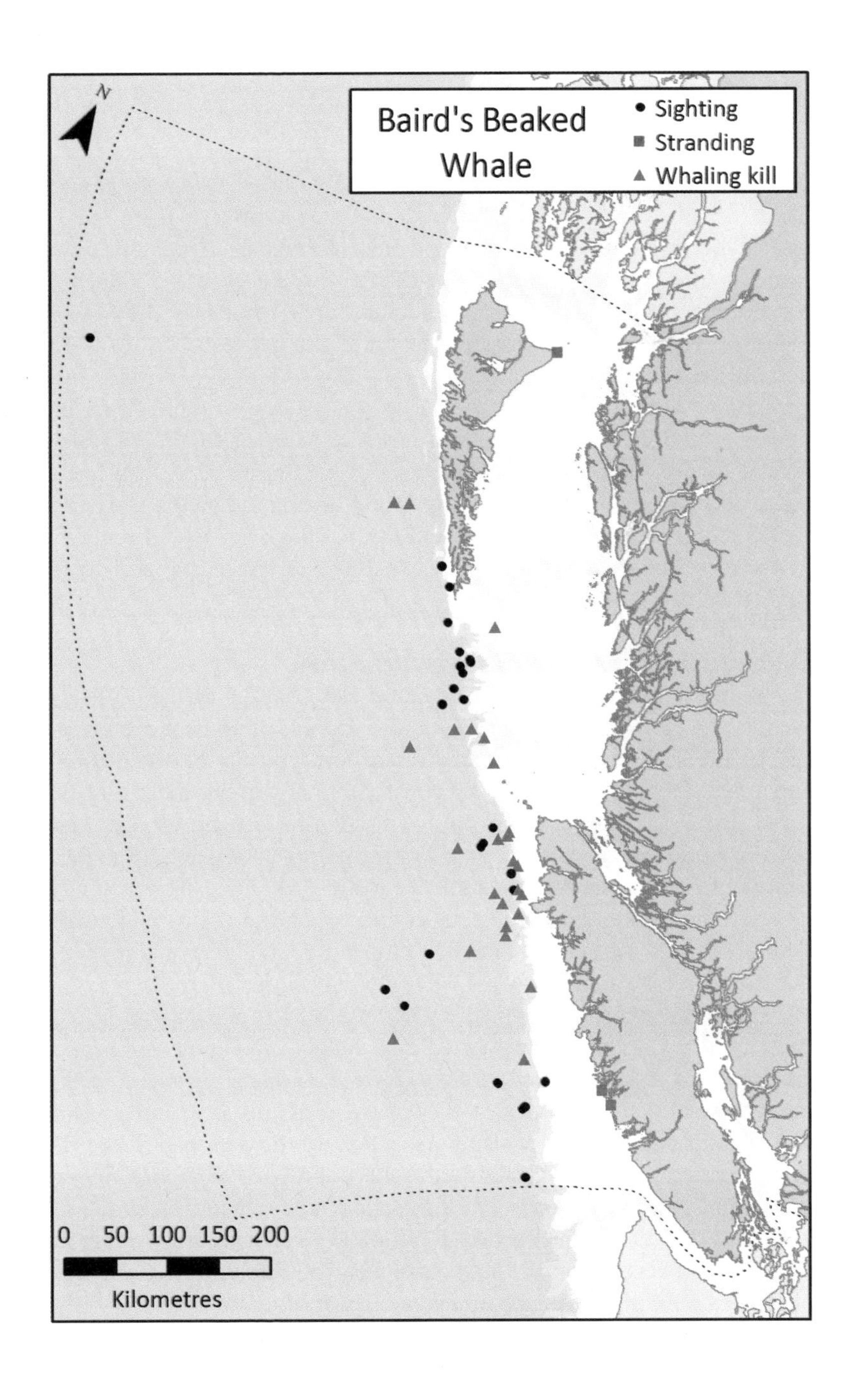
Baird's Beaked Whale
Sighting
Stranding
Whaling kill
N
0 50 100 150 200
Kilometres

Natural History

Feeding Ecology

Baird's Beaked Whales feed on a wide variety of deepwater squid and fishes. Examination of Baird's Beaked Whale stomachs from whaling catches in Japan indicated that fish were far more important than squid in the diet of whales taken off the Pacific coast of Honshu, but the reverse was true for whales taken further north in the Sea of Okhotsk. Off the Pacific coast of Japan, eight species of fishes were documented, representing two families – the codlings (Moridae) and grenadiers (Macrouridae) – as well as 30 species of squid from 14 families. In BC, Gordon Pike and Ian MacAskie of the Pacific Biological Station examined the stomachs of 13 Baird's Beaked Whales taken at the Coal Harbour whaling station during the 1950s and early 1960s. Three stomachs were empty but the others contained remains of squid (seven stomachs), small rockfish (seven stomachs) and skate egg-cases (two stomachs).

Behaviour and Social Organization

Unlike the smaller species of beaked whales, Baird's Beaked Whales are fairly gregarious and can be found in schools of 50 or more. Average group sizes in most regions are smaller, typically between three and seven. Details of social structure are not known, but there may be some sexual segregation. A group of 11 Baird's Beaked Whales that mass stranded near La Paz, Baja California, in 2006 were all males. Adults of both sexes carry scars from wounds inflicted during social interactions by the exposed teeth at the tip of the rostrum. Baird's Beaked Whales occasionally breach, flipper-slap, tail-slap and spyhop, but they rarely raise their flukes when diving (figure 58).

Like all beaked whales, Baird's Beaked Whales are accomplished deep divers. One male harpooned by whalers off Vancouver Island was reported to have "dived straight down at an amazing speed", taking with it 500 fathoms (almost 1000 metres) of line. A similarly rapid descent was observed by researchers in Japan, who attached data-loggers to Baird's Beaked Whales with a harpoon barb to record diving behaviour. Immediately after being tagged, one whale descended at more than 13 km/h to a record depth of 1777 metres, apparently in response to being tagged. More commonly, the whales dove to depths of about 1500 metres during an average 45-minute dive time.

Baird's Beaked Whales have a more diverse vocal repertoire than has been reported for other beaked whales, which likely reflects their greater sociality. They produce whistles at frequencies of 4–8 kHz and

Figure 58. A Baird's Beaked Whale breaching near a troller at Barkley Canyon, 100 km southwest of Bamfield, Vancouver Island.
Photo: W. Szaniszlo, July 2012.

short, irregular bursts of clicks, presumably for echolocation, with peak energy around 22–25 kHz.

Life History and Population Dynamics

Females reach sexual maturity at 10–15 years of age and a length of 9.8–10.7 metres, and males at 6–11 years and 9.1–9.8 metres. Once mature, females ovulate consistently every two years for the rest of their lives. After a gestation of around 17 months – the longest of all marine mammals – a female will give birth to a calf about 4.6 metres long. Females live an average of 54 years and males to more than 80 years. This higher mortality rate for females leads to a male-biased sex ratio of older whales in a population, and could explain the preponderance of males in whale catches – of 30 Baird's Beaked Whales taken in BC waters for which sex was recorded, 27 were male. Still, it seems unlikely that different longevities of the sexes could account for such an extreme male bias in the catches. Possibly other factors were involved, such as different distribution patterns of sexually segregated groups or sex-related behaviour that makes males more catchable than females. Because females are slightly larger than males, selection for larger animals by whalers would not help explain the bias of the catches toward males.

Killer Whales are likely the only predators of Baird's Beaked Whales. In Japan, as many as 40 per cent of individuals examined at whaling stations bore scars or injuries from Killer Whale teeth on their flukes and flippers.

Exploitation

Baird's Beaked Whale is one of the few beaked whales that has been exploited commercially. It has been hunted in Japanese waters from shore-based whaling stations since at least the early 17th century. Catches in Japan peaked following World War II at about 300 per year but have declined in recent years and are now regulated by a quota system. Baird's Beaked Whales were never heavily exploited in the northeastern Pacific. While Nuu-chah-nulth whalers of the west coast of Vancouver Island knew of a whale they called *tsitsilhn'i*, which was likely Baird's Beaked Whale, they did not hunt it because its oil and blubber were reported to cause diarrhea. Shore-based whaling operations in British Columbia reported killing 41 Baird's Beaked Whales between 1915 and 1966, and a similar number were taken by whalers in California, Washington and Alaska combined over the same period.

Taxonomy and Population Structure

Baird's Beaked Whale is very similar morphologically to Arnoux's Beaked Whale, which is in the same genus but inhabits circumpolar waters of the Southern Ocean. Although the decision to divide the genus into the two species has been questioned in the past, recent evidence from mitochondrial DNA supports the recognition of separate species.

Conservation Status and Management

The overall abundance of Baird's Beaked Whales is not certain, but some regional estimates are available. There are an estimated 7000 whales in Japanese waters, and about 50–60 are taken there annually. Large-scale ship-based surveys off California, Oregon and Washington between 1991 and 2008 resulted in estimates of 900–1300 Baird's Beaked Whales for this region, extending out to 555 km (300 nautical miles) from shore. Abundance appears to be stable.

This species, like other beaked whales, is likely to be vulnerable to disturbance or injury from loud human-caused noises such as those generated by navy sonar and seismic exploration. Ship strikes also threaten this species. A Baird's Beaked Whale stranded in Washington in 2003 appeared to have been struck by a ship.

Baird's Beaked Whale is listed as Data Deficient by IUCN and as Not at Risk in Canadian waters by COSEWIC. The species is on the BC List as Unknown due to insufficient information.

Remarks

Although whalers in British Columbia caught Baird's Beaked Whales as early as 1915, the species was not formally described in the province until 1953 when Gordon Pike, whale research scientist at the Pacific Biological Station from 1946 to 1968, published a description of two specimens brought into the whaling station at Coal Harbour. The genus *Berardius* is named after August Bérard (1796–1852), who was captain of the French corvette *Le Rhin* and brought the type specimen of Arnoux's Beaked Whale back from New Zealand in 1846. Baird's Beaked Whale was described in 1883 by Leonhard Stejneger and named after Spencer Fullerton Baird, past Secretary of the Smithsonian Institution.

Selected References: Balcomb 1989; Carretta et al. 2011; Dawson et al. 1998; Ford, Abernethy et al. 2010; Kasuya 2009; Minamikawa et al. 2007; Monks et al. 2001; Moore and Barlow 2013; Pike 1953b; Pike and MacAskie 1969; Reeves and Mitchell 1993; Walker et al. 2002.

Hubbs' Beaked Whale

Mesoplodon carlhubbsi

Other Common Names: Arch-beaked Whale.

Description

Hubbs' Beaked Whale is a medium-sized whale with a robust body that tapers at the ends to a small head and tail. Its dorsal fin is moderately falcate and positioned well behind the middle of the back. The height of the fin in adults is consistently 22–23 cm. The flippers are smaller than those of most other odontocetes and fit into depressions in the body wall behind the points of insertion. These "flipper pockets" are found in most species of beaked whales. The size of the tail flukes is typical of a medium-sized whale; as in other beaked whales, the flukes lack a median notch.

The head shape of Hubbs' Beaked whales is similar to that of other species in the genus, with a long, narrow rostrum that slopes up to a moderately bulbous melon. There is a prominent pair of ventral throat grooves. The mouthline is arched, giving it a noticeably S-shaped curve when viewed from the side. Adult males have a prominent pair of teeth at the peak of this arch, about midway along the mouthline. These teeth are exposed outside the closed mouth and may project to the top of the rostrum or higher. The teeth do not erupt in females.

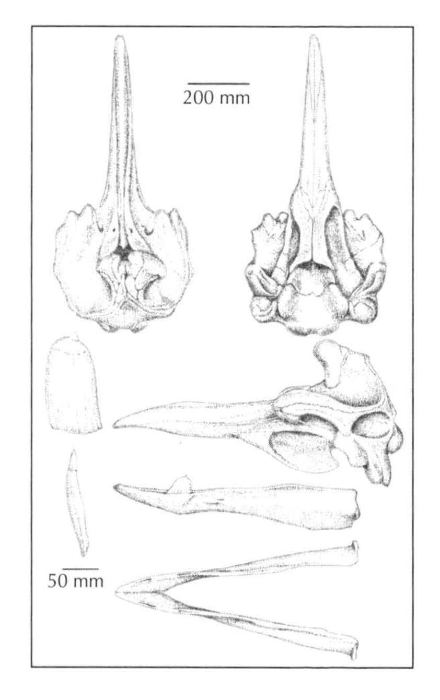

Adult males are mostly dark grey to black, with no obvious differentiation between coloration of dorsal and ventral surfaces. Many males show heavy linear scarring on their body, particularly on the flanks. Females and subadults are a medium grey that fades to a light grey or white on the ventral surface, and they

lack the scars seen on adult males. In both sexes, the flukes are lighter on the ventral surface than on the dorsal surface. The most distinctive feature of Hubbs' Beaked Whales is the pigmentation of the head. In adult males the rostrum and beak (anterior portion of the mandibles) are a bright white. In adult females and subadults the beak is lighter than the rest of the head but the contrast is not as great as in males. Adult males also have a distinct patch of white on the top of the melon, resembling a white cap or "toque".

Measurements

length (metres):

adult: 5.0 (4.7–5.3) n=5

neonate: ~ 2.5

weight (kg):

adult: ~ 1500

neonate: unavailable

Dentition

One pair of laterally compressed teeth midway along the mandible; erupted through the gum only in adult males.

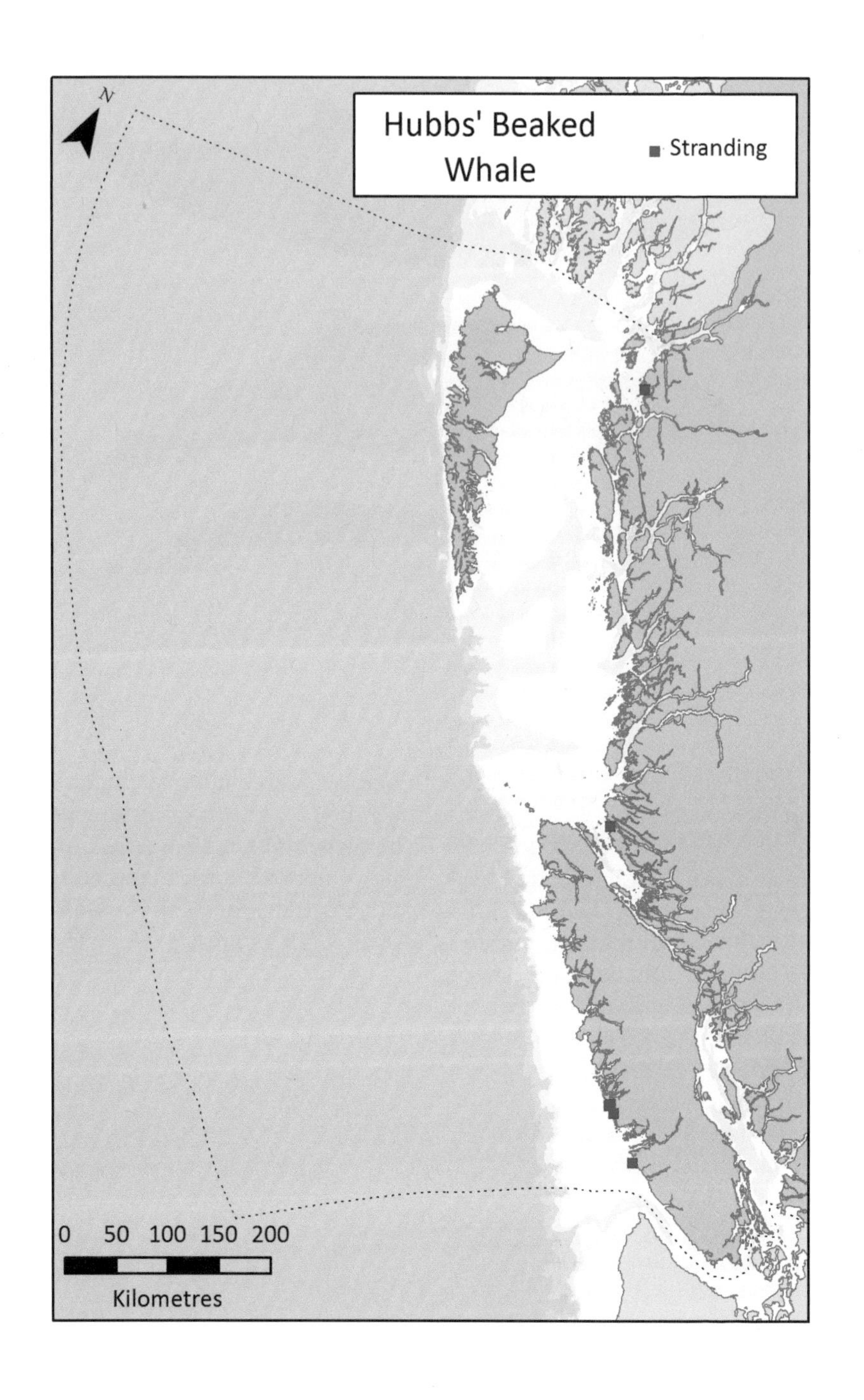
N
Hubbs' Beaked Whale
Stranding
0 50 100 150 200
Kilometres

Identification

Beaked whales, particularly the smaller *Mesoplodon* species, are difficult to detect and identify at sea, so confirmed sightings of Hubbs' Beaked Whales are rare. The bright white beak and "toque" on mature males easily distinguish them from the other mesoplodonts found in BC waters, Stejneger's Beaked Whales. But this feature is only evident in live or very fresh specimens. Cuvier's Beaked Whales, particularly males, can have a similar lightly pigmented head, but the beak is much shorter than that of Hubbs' Beaked Whales. Female and subadult Hubbs' Beaked Whales are difficult to differentiate from Stejneger's Beaked Whales using external features.

Distribution and Habitat

Hubbs' Beaked Whale appears to be restricted to temperate waters of the North Pacific, but its distribution is known primarily from strandings, as the species is rarely seen alive. It is a deep diving, oceanic species thought to have a continuous distribution across the North Pacific, but this has not been confirmed. Along the west coast of North America, strandings have been recorded from San Diego, California, (33°N) to Prince Rupert, BC (54°N). No strandings have been reported north of British Columbia, although numerous strandings of Stejneger's Beaked Whales have been recorded. This suggests that British Columbia may be at or close to the northern limit of the Hubbs' Beaked Whale's range. In the western Pacific, Hubbs' Beaked Whale strandings have taken place only on the northeastern coast of Honshu, Japan, at about 38°N latitude.

In British Columbia Hubbs' Beaked Whale has only been recorded in seven strandings (Appendix 2). The earliest record of the species in BC was a moderately decomposed male that drifted into Prince Rupert harbour in December 1962. Five additional strandings were recorded in the 1960s and 1970s: three along the west coast of Vancouver Island (figure 59), one at Skull Cove on Bramham Island south of Cape Caution and one without data on location or date of stranding. The most recent stranding was near Pachena Point on the west coast of Vancouver Island in February 1992. There have been no confirmed sightings of live Hubbs' Beaked Whales in BC.

Natural History

Feeding Ecology

Hubbs' Beaked Whale is a deep-diving species that feeds on mesopelagic squid and fishes. Squid of the genera *Gonatus*, *Onychoteuthis*, *Octopoteuthis*, *Histioteuthis* and *Mastigoteuthis*, as well as the Pacific Viperfish, have been identified from the stomach contents of stranded animals.

Behaviour and Social Organization

Nothing is known about the behaviour and social organization of Hubbs' Beaked Whales as they are so rarely seen alive. But the extensive scarring on adult males appears to be mostly the result of fighting. These scars are up to two metres long and often occur as parallel pairs that are likely caused by the exposed teeth of other males. Aggression among males, perhaps to determine dominance for breeding, is likely the cause of these long scars.

Life History and Population Dynamics

Little information is available on breeding, calving, growth and maturity. Examination of fetuses from strandings suggests that calving takes place in summer.

Figure 59. A female Hubbs' Beaked Whale stranded at Long Beach, Vancouver Island, in January 1969. Note the lack of a median notch in the trailing edge of the tail flukes, typical of beaked whales. Photo: P. Whittington.

Exploitation

Hubbs' Beaked Whales have been taken occasionally in cetacean hunts in Japan and rarely as bycatch in a drift gillnet fishery for billfish off California. There is no record of the species being exploited in British Columbia.

Taxonomy and Population Structure

The first specimen of this species was collected and described by prominent marine biologist Carl L. Hubbs at Scripps Institution of Oceanography, La Jolla, California, in 1945. He originally identified the specimen as Andrews' Beaked Whale, *Mesoplodon bowdoini*, a species known only from New Zealand, South Australia and various subantarctic islands. Subsequently, ziphiid specialist Joseph Moore determined from skull characteristics that it was a distinct species and in 1963 named it after Hubbs. Recent molecular evidence from DNA confirms the species' close relationship with *M. bowdoini* and also shows a more distant relationship to the sympatric Stejneger's Beaked Whale than was previously thought.

Conservation Status and Management

Nothing is known about the abundance of Hubbs' Beaked Whales or their conservation status. The paucity of sightings suggests that the species is rare, but like all mesoplodonts, Hubbs' Beaked Whales are very inconspicuous at sea.

The species is considered Data Deficient by IUCN Red List and Not at Risk by COSEWIC. Its BC conservation status is Unknown.

Remarks

All beaked whales are deep divers that feed mostly on squid and occasionally fish and crustaceans at depths of 1000 metres or more. But until recently it has been a mystery how these animals catch their prey. Teeth are virtually absent in most beaked whales, except for one or two pairs that protrude through the gums only in mature males and appear to function as tusks in intraspecific fighting. In some mesoplodonts these erupted teeth can restrict the opening of a male's mouth to no more than a few centimetres. From anatomical studies of stranded beaked whales and behavioural observations of live stranded animals, the late John Heyning, a whale scientist at the Los Angeles County Museum of Natural History, and his colleague James

Mead at the Smithsonian Institution determined that beaked whales are almost certainly suction feeders – they catch prey by sucking it into their mouths and swallowing it whole. All beaked whales have at least one pair of throat grooves that converge anteriorly toward the rostrum. These deep grooves allow the floor of the oral cavity to stretch ventrally and expand its capacity considerably. This expansion, together with rapid retraction of the tongue in a piston-like motion, creates a sudden drop in oral cavity pressure that sucks prey into the mouth.

Selected References: Heyning and Mead 1996, Houston 1990a, Mead et al. 1982, Willis and Baird 1998a.

Stejneger's Beaked Whale

Mesoplodon stejnegeri

Other Common Names: Bering Sea Beaked Whale, Sabre-toothed Beaked Whale.

Description

Stejneger's Beaked Whale is a medium-sized whale with a robust body that is typical of the genus, tapering at the ends to a relatively small head and tail. It has a falcate dorsal fin situated well behind the mid-back, and a distinct keel on the tail stock between the anus and the tail flukes. As in other mesoplodonts, the tail flukes lack a median notch. The flippers are small and fit into flipper pockets on each side. Stejneger's Beaked Whale has a long, well-defined beak that slopes smoothly back to a relatively flat melon. The mouthline curves upwards, forming a prominent arch. In males, two teeth on the lower jaw protrude from the leading edge of this arch on the mandibles and tilt forward and inward so as to pinch or constrict the opening of the upper jaw. Teeth do not erupt from the gums in females or subadults.

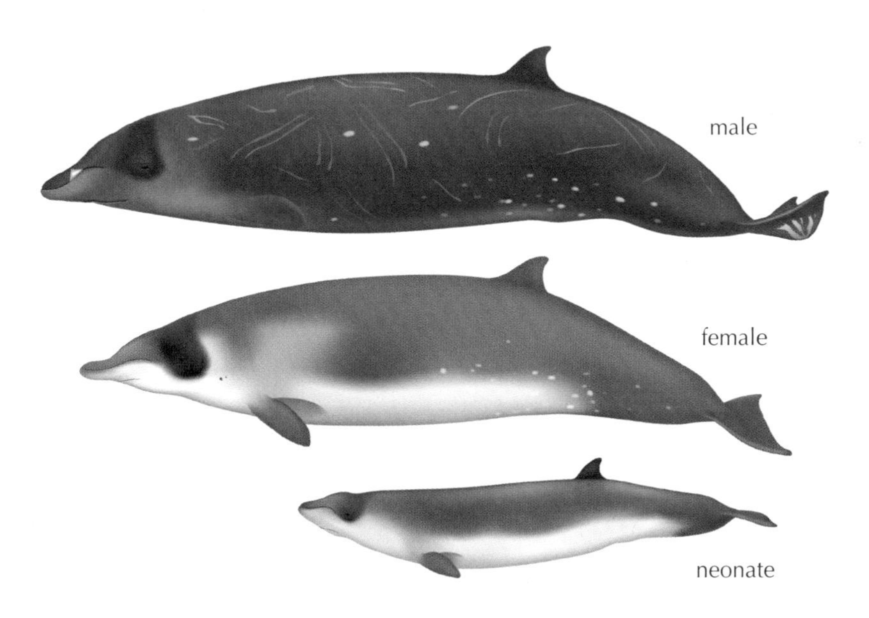

Male Stejneger's Beaked Whales are typically dark grey to black over most of the body, although they can appear brownish when observed at sea. This brownishness is likely due to a film of diatoms that is often found on the bodies of cetaceans. Females and subadults are lighter, and their coloration grades to a much lighter shade on the undersides. Both sexes usually have a prominent black cap on the top of the head, which extends downward like goggles over the eyes. The undersides of the tail flukes may exhibit light-grey or white striations that radiate out from the posterior margin, resembling a starburst pattern. Adults usually have many circular or oval-shaped white scars, four to eight centimetres in diameter, especially on the rear half of their bodies. These are likely caused by bites of Cookiecutter Sharks and possibly Pacific Lampreys. Adult males have many linear scars from wounds presumably inflicted by the exposed teeth of other males during aggressive interactions.

Measurements

length (metres):

adult: 5.2–5.7

neonate: 2.2

weight (kg):

adult: ~ 2000

neonate: ~ 80

Dentition

One pair of laterally compressed teeth, midway along the mandible; erupted only in adult males.

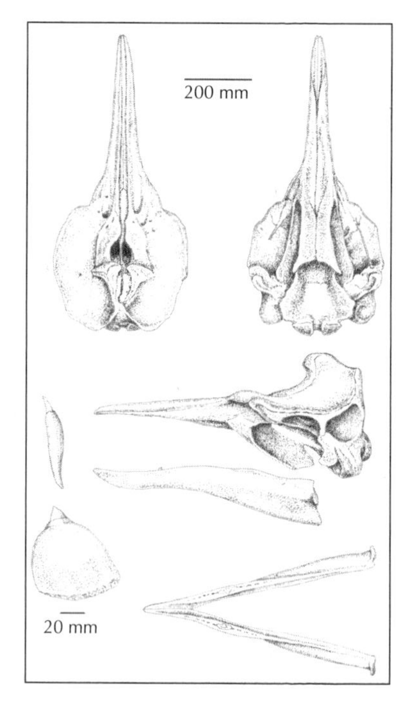

Identification

Identification of beaked whales at sea can be very difficult. Stejneger's Beaked Whales in BC waters are likely to be confused with Hubbs' Beaked Whales, the only other mesoplodont known to occur in the region. But a Stejneger's Beaked Whale has a less prominent bulge ahead of the blowhole than a Hubbs', and the latter has a light-coloured

Figure 60. A male Stejneger's Beaked Whale stranded on the east coast of Graham Island in April 1998. Note the extensive scarring from wounds caused by the teeth of other males. The whale's exposed teeth can be seen in the close up of its head (right). Photos: C. Tarver.

or white melon. Adult male Hubbs' Beaked Whales have a distinctly white rostrum and the two teeth on the mandibles are considerably less exposed than in Stejneger's Beaked Whales. Whales of either species could be confused with small Common Minke Whales at a distance, as all three have a dorsal fin set fairly aft on the back, but a Minke Whale's dorsal fin tends to be more falcate than that of a Stejneger's Beaked Whale (figure 60). Helpful tips on identifying beaked whales can be found on a website called The Beaked Whale Resource (www.beakedwhaleresource.com).

Distribution and Habitat

Stejneger's Beaked Whales are rarely seen at sea, so their general distribution has been inferred from strandings. They are endemic to the cold-temperate waters of the North Pacific Ocean, the Sea of Japan and deep waters of the southwestern Bering Sea. On the west coast of North America strandings have been documented as far south as Monterey Bay, California, and to Honshu, Japan, on the Asian coast. Most sightings of live animals have been in the Aleutian Islands, Alaska, and were closely associated with the steep slope of the continental shelf in waters 700–1500 metres deep. Scars from Cookiecutter Shark bites indicate that these whales may move seasonally to warmer waters, as these sharks are not known to range further north than about 38°N.

Stejneger's Beaked Whale was first documented in British Columbia from a beach-worn partial skull found in August 1953 near Port McNeill on the northeastern coast of Vancouver Island. Since then, four additional strandings have been recorded in BC, two on the west coast of Vancouver Island and two on Haida Gwaii (see Appendix 2). Only one sighting of what was probably a living Stejneger's Beaked Whale has been reported in western Canadian waters. This was a female beaked whale, most likely a Stejneger's Beaked Whale, found alive but disoriented in shallow water in Newcastle Channel, south of Departure Bay, Nanaimo, in August 1979 (figure 61). It was observed and photographed by Michael Bigg, marine mammal scientist from the nearby Pacific Biological Station, and it eventually swam

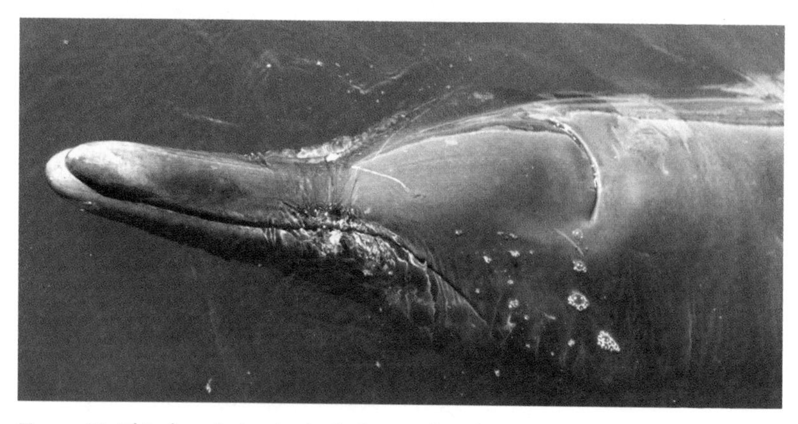

Figure 61. This female beaked whale wandered into Departure Bay, Nanaimo, on August 5, 1979. She was found disoriented in shallow water, but swam off after being pushed into deeper water. Photo: M. Bigg.

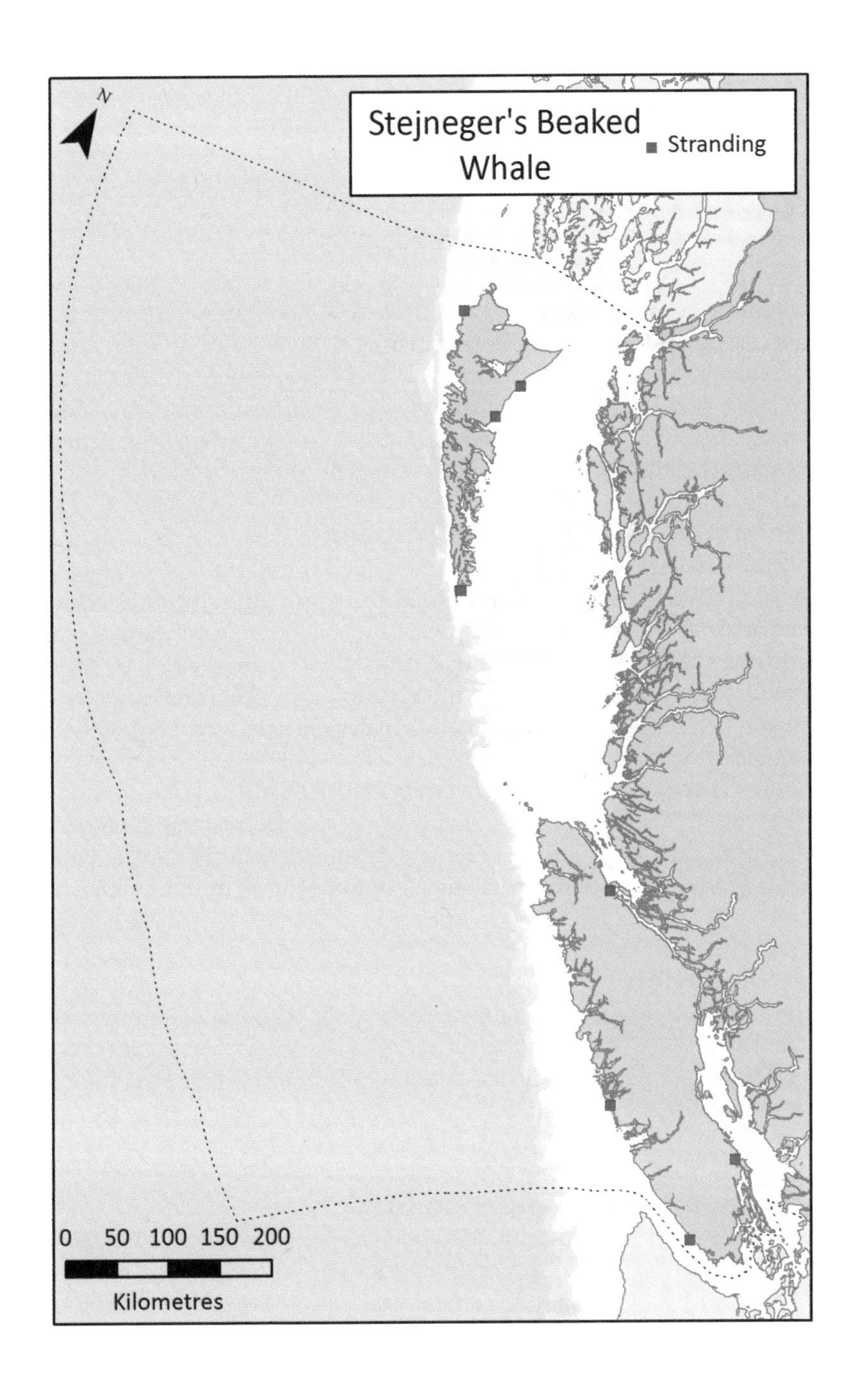

Stejneger's Beaked Whale
Stranding
N
0 50 100 150 200
Kilometres

off. This animal appears to have strayed far from its normal deepwater oceanic habitat and was most likely sick or injured.

Natural History

Feeding Ecology

Stomach contents of Stejneger's Beaked Whales stranded in Alaska and the Sea of Japan were dominated by remains of deepwater squid belonging to two families, Gonatidae and Cranchiidae. At least some of these squid species do not undergo daily vertical migrations, indicating that the whales were feeding at depths of more than 200 metres. No fish remains were found in the stomachs of these specimens.

Behaviour and Social Organization

Stejneger's Beaked Whales have been seen travelling in groups of 3 or 4 to as many as 15. Animals tend to swim tightly together at the surface and to swim and dive in unison, suggesting a cohesive social structure. The presence of long, linear scars on the bodies of adult males suggests that aggressive interactions with other males are frequent, perhaps related to access to females for breeding.

Life History and Population Dynamics

Very little is known about the life history of this species. Based on sizes and presumed growth rates of fetuses, it appears that the calving period is protracted, with births taking place from spring through summer.

Exploitation

Stejneger's Beaked Whales have occasionally been hunted along with Cuvier's Beaked Whales in Japan, and they have also been caught there incidentally in salmon gillnets. There is no record of their exploitation in the northeastern Pacific.

Taxonomy and Population Structure

Beaked whales in the genus *Mesoplodon* have long been challenging to taxonomists. Historically, species have been distinguished by skull characteristics, especially the shape, size and position of the teeth of adult males. However, because of anatomical similarities, especially in females and young individuals, misidentifications have been common and molecular genetic analyses have been playing an increasingly important role in species identification. At present 14 species of mesoplodonts are recognized in the world's oceans, and new species continue to be found. The most recent, *M. perrini*, was identified in 2002 and is known only from five strandings in California.

Conservation Status and Management

As with all mesoplodonts, there is little information on the overall abundance or trends of Stejneger's Beaked Whale. Given the infrequency of sightings, the species appears to be rare. Based on ship surveys from 1991 to 2008 the abundance of mesoplodonts off the US west coast (all species combined) appears to be declining. Although there is no directed hunting of Stejneger's Beaked Whales, they, like other beaked whales, may be particularly vulnerable to the effects of human-caused noises, particularly from military mid-frequency sonar and from air guns used in seismic surveys. Several strandings involving multiple beaked whales have been linked to such sound sources, and necropsies have suggested that the deaths were caused by gas-bubble disease, induced by behavioural responses to noise.

Stejneger's Beaked Whale is listed as Data Deficient by IUCN Red List and as Not at Risk in Canada by COSEWIC. Its BC Provincial List designation is Unknown due to a lack of information.

Remarks

Stejneger's Beaked Whale is one of the better-known members of the genus *Mesoplodon*. It was described in 1885 by Frederick W. True from a beach-worn skull collected in 1883 at Bering Island in the western Bering Sea by Leonhard Stejneger, renowned zoologist at the Smithsonian Institution.

Selected References: Houston 1990b, Loughlin et al. 1982, Loughlin and Perez 1985, Moore and Barlow 2013, Pike and MacAskie 1969, Walker and Hanson 1999, Willis and Baird 1998a.

Cuvier's Beaked Whale *Ziphius cavirostris*

Other Common Names: Goose-beaked Whale.

Description

Cuvier's Beaked Whale is a robust, small to medium-sized toothed whale with a spindle-shaped body typical of beaked whales. The beak is shorter than those of other ziphiids (members of the family Ziphiidae) and slopes steeply but smoothly up to the melon. This shape, together with the distinctively curved mouth-line, gives the whale's head profile the appearance of a goose beak. The dorsal fin is relatively small and falcate, and it is positioned about two-thirds of the way along the body. The flippers are small and narrow, and can be tucked into pockets along the body wall, as seen in beaked whales of the genus *Mesoplodon*. The tail flukes are proportionately large and lack a distinctive notch along the trailing margin. Adult males have a pair of conical teeth that protrudes from the tip of the lower jaw; these teeth slant forward and are exposed when the mouth is closed (figure 62). A pair of throat grooves converges, but does not meet, anteriorly.

Coloration of male Cuvier's Beaked Whales is a dark slate grey over most of the body, with a distinctively white head. This white coloration continues over the dorsal surface as far as the dorsal fin. Females vary in colour from a dark grey to a reddish-brown, with a slight lightening around the head. Both sexes commonly have light oval scars that are likely from Cookiecutter Shark bites. Long, linear scars are found only on males and appear to be the result of raking by the exposed teeth of other males during fighting.

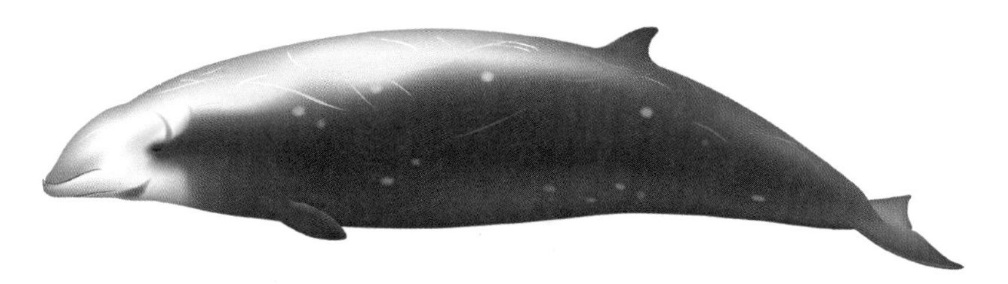

Measurements

length (metres):

adult: 5.7 (4.7–6.0) n=7

neonate: ~ 2.7

Figure 62. A surfacing adult male Cuvier's Beaked Whale shows the exposed pair of teeth at the tip of its lower jaw. Photo: C. Dunn, Bahamas, June 2010.

weight (kg):
adult: 2200–2900
neonate: ~ 250–300

Dentition

Two exposed teeth at the tip of the lower jaw in adult males only.

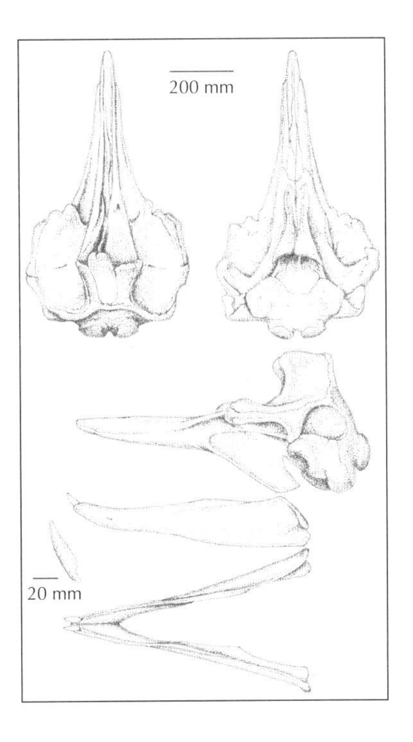

Identification

Like other similarly sized beaked whales, Cuvier's Beaked Whales swim alone or in small groups, avoid vessels and have a low and diffuse blow, which makes them difficult to detect at sea. With their whitish head, short beak and steeply sloped melon, Cuvier's Beaked Whales are fairly distinctive in appearance but can be confused with other members of the family Ziphiidae. In BC waters adult male Hubbs' Beaked Whales also have white on their heads, but this is confined to the anterior portions of the rostrum and the top of the melon. Both Hubbs' Beaked Whales and Stejneger's Beaked Whales have a longer beak than Cuvier's Beaked Whales. Baird's Beaked Whales also have a far more prominent rostrum, have no white on their head and are considerably larger. At a distance, mature Risso's Dolphins, which become increasingly whitish with age and have numerous linear scars on their body, could be confused with male Cuvier's Beaked Whales. But Risso's Dolphins are smaller, and have a differently shaped head and a much taller, erect dorsal fin.

Distribution and Habitat

Stranding records indicate that Cuvier's Beaked Whale is the most cosmopolitan of the ziphiids. The species is found in deep, cool-temperate to tropical waters of all the world's oceans and most seas. There are far more stranding reports of Cuvier's Beaked Whales than of any other species of beaked whale, suggesting that they are also the most abundant. Their distribution is strongly associated with continental slope waters with depths of 1000–2000 metres. But Cuvier's Beaked Whales also occur in other oceanic regions, including waters over abyssal plains with depths of more than 5000 metres.

Records for British Columbia include twenty strandings, five sightings, and one incidental catch (see Appendix 2). Strandings are distributed almost evenly between Haida Gwaii (eleven; figure 63) and Vancouver Island (eight), with one on the mainland coast near Bella Bella. Strandings are documented in all seasons but are most common in summer. Three sightings were made during Cetacean Research Program surveys off the west coast of Vancouver Island: two

Figure 63. An adult female Cuvier's Beaked Whale found stranded at Balcom Inlet, Gwaii Haanas National Park Reserve, on August 28, 1988. Photo: B. Falconer.

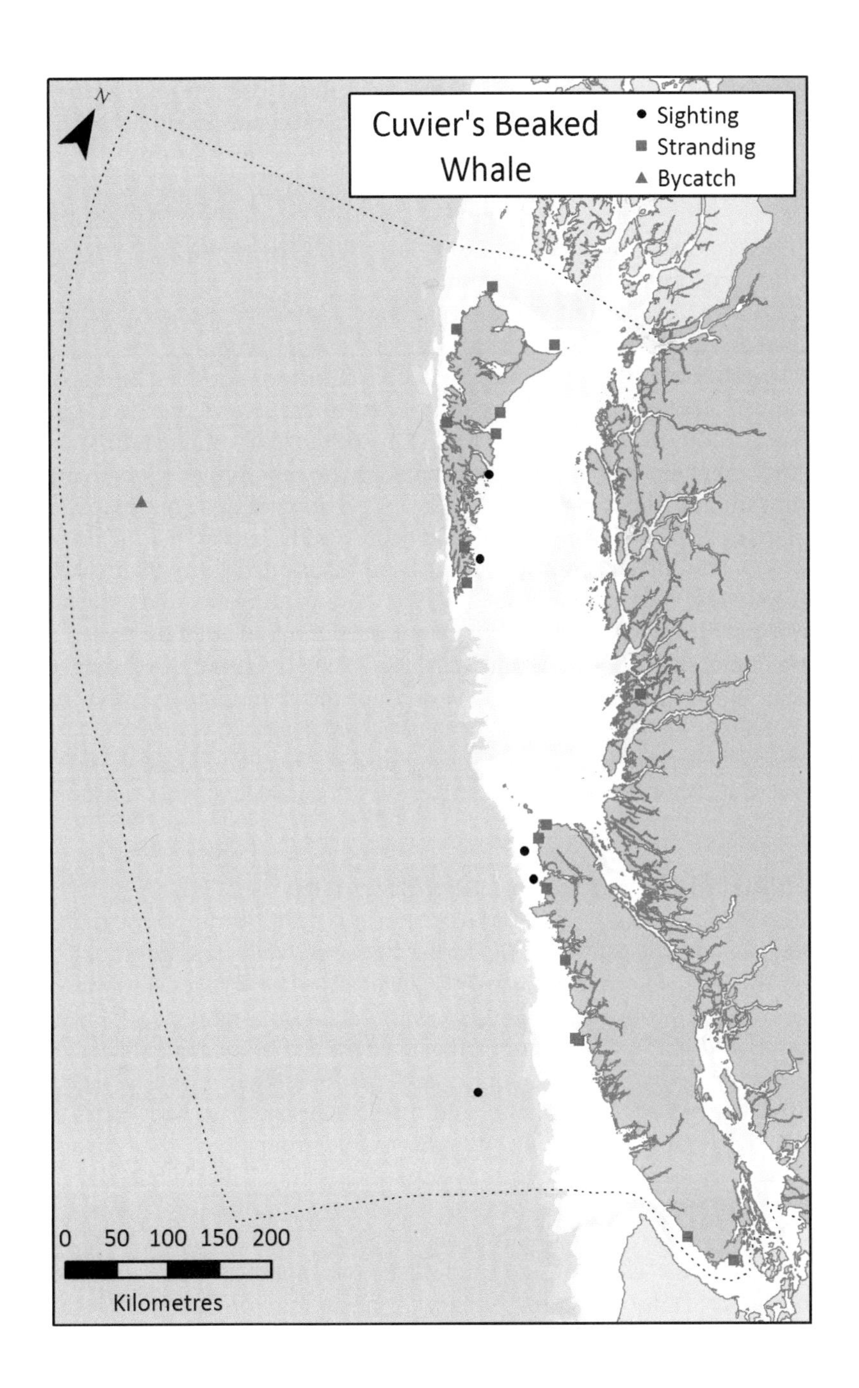
Cuvier's Beaked Whale
Sighting
Stranding
Bycatch
N
0 50 100 150 200
Kilometres

offshore of Quatsino Sound and the third about 100 km southwest of Tofino. The other two were sightings of single animals in Haida Gwaii, one near Skedans Island in June 1989 and one off the east side of Moresby Island in August 2005.

Natural History

Feeding Ecology

Cuvier's Beaked Whales forage in deep waters for a variety of mesopelagic and bathypelagic prey species. Their diet is known only from the remains of prey found in the stomach contents of stranded animals. These remains are typically dominated by oceanic squid belonging to the families Gonatidae, Cranchiidae and Histioteuthidae but may also contain the remains of deepwater shrimp and fishes.

One stomach that I retrieved from a stranded Cuvier's Beaked Whale at Langara Island in July 1991 contained beaks from at least 693 squid, as well as a few fish vertebrae. A group of students in a marine mammal course at Bamfield Marine Sciences Centre determined that 92 per cent of these squid were from the genus *Gonatus*, and the remainder were made up of the genera *Chiroteuthis*, *Cycloteuthis*, *Taonius*, *Teuthowenia/Galiteuthis*, *Gonatopsis* and *Neoteuthis*. The fish vertebrae could not be identified but likely were from a deepwater, offshore species.

Behaviour and Social Organization

Recent field studies in the Mediterranean, near Hawaii and the Bahamas, and off southern California have provided new information on the behaviour and sociality of Cuvier's Beaked Whales. They tend to travel in groups three or four, with occasional sightings of ten or more. Resightings of photographically identified individuals show that group composition can be quite stable over a period of several years, especially among adult males, and that groups can have high levels of fidelity to fairly small areas. Whether stable groups consist of maternally related individuals, as in species such as the Killer Whale, is not yet clear.

The recent development of data-logging tags that can be attached temporarily to whales using suction cups or small barbs has provided new insights into the extraordinary diving abilities of Cuvier's Beaked Whales. Group members dive in unison for an average of about one hour (maximum recorded duration: 137 minutes) typically to depths averaging 1400 metres (maximum recorded depth: 2993 metres).

These are the longest and deepest dives ever recorded for a mammal. Following these deep dives, the whales spend an hour or so close to the surface, diving to no more than 200 metres, during which time they presumably recover from the oxygen debt they've accumulated. Suction cup tags are also capable of recording the sounds produced by tagged animals. The recordings show that when Cuvier's Beaked Whales reach depths of 400–500 metres during their descent on a deep dive, they start emitting nearly continuous echolocation clicks at a rate of about two per second. These clicks are likely used to search for food, and when a prey item is detected they shift to a faster repetition rate and the clicks turn into a buzz. The whales stop clicking once they begin their ascent to the surface and remain silent until their next deep dive.

Life History and Population Dynamics

Very little is known about reproduction in Cuvier's Beaked Whales. They have been reported to reach sexual maturity at average lengths of 5.8 metres for females and 5.5 metres for males. There is no information on seasonality of calving.

Exploitation

Cuvier's Beaked Whales have been taken occasionally in hunts for Baird's Beaked Whales in Japan, but no catches have been reported in recent years. There is no history of exploitation of this species in the northeastern Pacific.

Taxonomy and Population Structure

Despite the extensive range of Cuvier's Beaked Whales, only a single species is recognized in the genus. Recent studies have shown strong variation in mitochondrial DNA haplotype frequencies across different ocean regions, indicating low rates of movement, at least for females, between areas.

Conservation Status and Management

The overall population of Cuvier's Beaked Whales is not known, but as with most beaked whales, they do not appear to be abundant. Photo-identification studies off Hawaii and southern California and in the Mediterranean have resulted in a few dozen individuals being identified in each study area. The predicted density, based on ship surveys in the eastern tropical Pacific, is about 4.5 whales per 1000 square kilometres. Surveys undertaken in 2005 and 2008 by the US NOAA out to 550 km (300 nautical miles) off the west coast of the mainland US resulted in an estimate of 2143 Cuvier's Beaked Whales.

The most serious human threat to Cuvier's Beaked Whales appears to be exposure to intense mid-frequency military sonar, which has been linked to strandings of multiple individuals in various regions of the world. A stranding involving two Cuvier's Beaked Whales in Baja California was associated with intense underwater seismic air gun pulses used in a nearby geophysical survey. Two stranded individuals in Croatia and France had ingested plastic bags, which contributed to their deaths.

Cuvier's Beaked Whale is considered of Least Concern by IUCN and as Not at Risk by COSEWIC. It is on the BC Yellow List.

Remarks

Ziphius cavirostris was first described by the renowned French naturalist G. Cuvier, from a partial cranium collected near the village of Fos, France, in 1804. Although once known as the Goose-beaked Whale because of the head's appearance in profile, the common name Cuvier's Beaked Whale is now more widely used and accepted. In Japan, the species is known as *akabokujira*, meaning "baby-faced whale."

Selected References: Baird et al. 2006; Cowan and Hatter 1940; Falcone et al. 2009; Ferguson et al. 2006; Ford, Abernethy et al. 2010; Heyning 1989; Heyning and Mead 2009; Houston 1991; McSweeney et al. 2007; Mitchell 1968; Schorr et al. 2014; Tyack et al. 2006; Willis and Baird 1998a.

Long-beaked Common Dolphin

Delphinus capensis

Other Common Names: Long-beaked Saddleback Dolphin.

Description

The Long-beaked Common Dolphin is a small delphinid that reaches a maximum size of about 2.5 metres. It is very similar in appearance to the Short-beaked Common Dolphin; the two were only recognized as distinct species in the mid 1990s. The two species share the same basic body morphology, although the Long-beaked Common Dolphin tends to be more slender. It has a tall, slightly falcate dorsal fin, and its flippers curve backwards and are pointed at the tips. Both species have a relatively long beak, but as its name implies, the Long-beaked Common Dolphin has a longer beak, which is almost 10 per cent of its total body length. Long-beaked Common Dolphins have a somewhat flatter melon and the angle where it joins the beak is less acute than in the Short-beaked Common Dolphin.

The colour pattern of the Long-beaked Common Dolphin is similar to that of the Short-beaked Common Dolphin, but generally less distinct. This criss-crossing pattern of shades forms an hourglass motif on the dolphin's sides, which is unique to these two species. The back portion of the hourglass is dark brownish-grey while the anterior portion between the eye and dorsal fin, known as the thoracic patch, is a tan or yellowish-tan colour. Both species also have a light grey stripe extending from the area immediately above the flipper to the anal region, which separates the whitish belly from the flank patterns, but on the Long-beaked Common Dolphin the stripe is wider toward the flipper.

Measurements

length (metres):

male: 2.0–2.5

female: 1.9–2.2

neonate: 0.8–1.0

weight (kg):

adult: <235

neonate: unavailable

Dentition

47-67 pairs in both the upper and lower jaws.

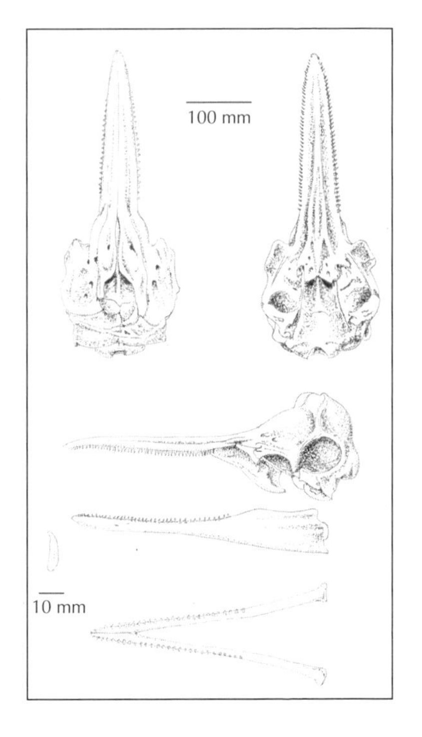

Identification

Long-beaked and Short-beaked Common Dolphins are rare in BC waters. The unusual criss-cross/hourglass pattern on the sides and the tan to yellowish-tan thoracic patch of both species should distinguish them from the commonly sighted Pacific White-sided Dolphin and the rarely seen Striped Dolphin. The two Common Dolphin species are easily confused at sea but can be distinguished by the distinct but more muted colour pattern of the Long-beaked species. The Long-beaked Common Dolphin has wider chin-to-flipper stripes that may merge with the dark lip patch, making the lower jaw appear darker than that of the Short-beaked species. Off California, both species travel in schools of a few to several thousand. Sightings of Long-beaked Common Dolphins in BC have been of single animals or groups of up to four.

Distribution and Habitat

Long-beaked Common Dolphins are found in warm-temperate to tropical waters in many ocean regions. They tend to be found closer to shore than the Short-beaked Common Dolphins, generally within 180 km of the coast. In the eastern Pacific they typically occur from central California (Point Conception) to northern Chile, including the Gulf of California. They are also found along the western Pacific coasts of Korea and southern Japan, along the east coast of South America and along the west and south coasts of Africa. A very long-

Figure 64. One of a pair of Long-beaked Common Dolphins in False Creek, Vancouver, in September 2002. Photo: A. Carson.

beaked form of the species (*D. capensis tropicalis*) occurs in coastal waters around the rim of the Indo-Pacific, from the Red Sea to Taiwan and south central China.

The first record of Long-beaked Common Dolphins in BC is from February 2, 1993, when an adult male stranded alive on Mayne Island and died shortly afterwards (see Appendix 2). The next record is a series of sightings of one to four dolphins in the late summer and fall of 2002 at Goose Bay in Rivers Inlet, Port McNeill and in Vancouver harbour (figure 64). In September 2003 a pair of Long-beaked Common Dolphins was observed on several occasions in Victoria harbour. This pair later became entangled in a salmon gillnet in Alberni Inlet on October 7, 2003; one animal was released alive but the second died and was recovered for post-mortem examination. No further sightings have been reported in BC waters since 2003, although two Long-beaked Common Dolphins were observed in southern Puget Sound from June through fall 2011, and an adult female – likely one of the two individuals seen in 2011 – was found stranded in March 2012.

The records of Long-beaked Common Dolphins in BC represent the northernmost occurrence of the species in the eastern North Pacific. As these dolphins are rarely seen north of Point Conception, California, the BC records should be considered extralimital. The sightings in 2002 and 2003 coincided with warm water conditions off the BC coast caused by a weak El Niño event, so it may be that warm ocean temperatures in some years cause a northward shift in their distribution off California, and some individuals may stray even further north.

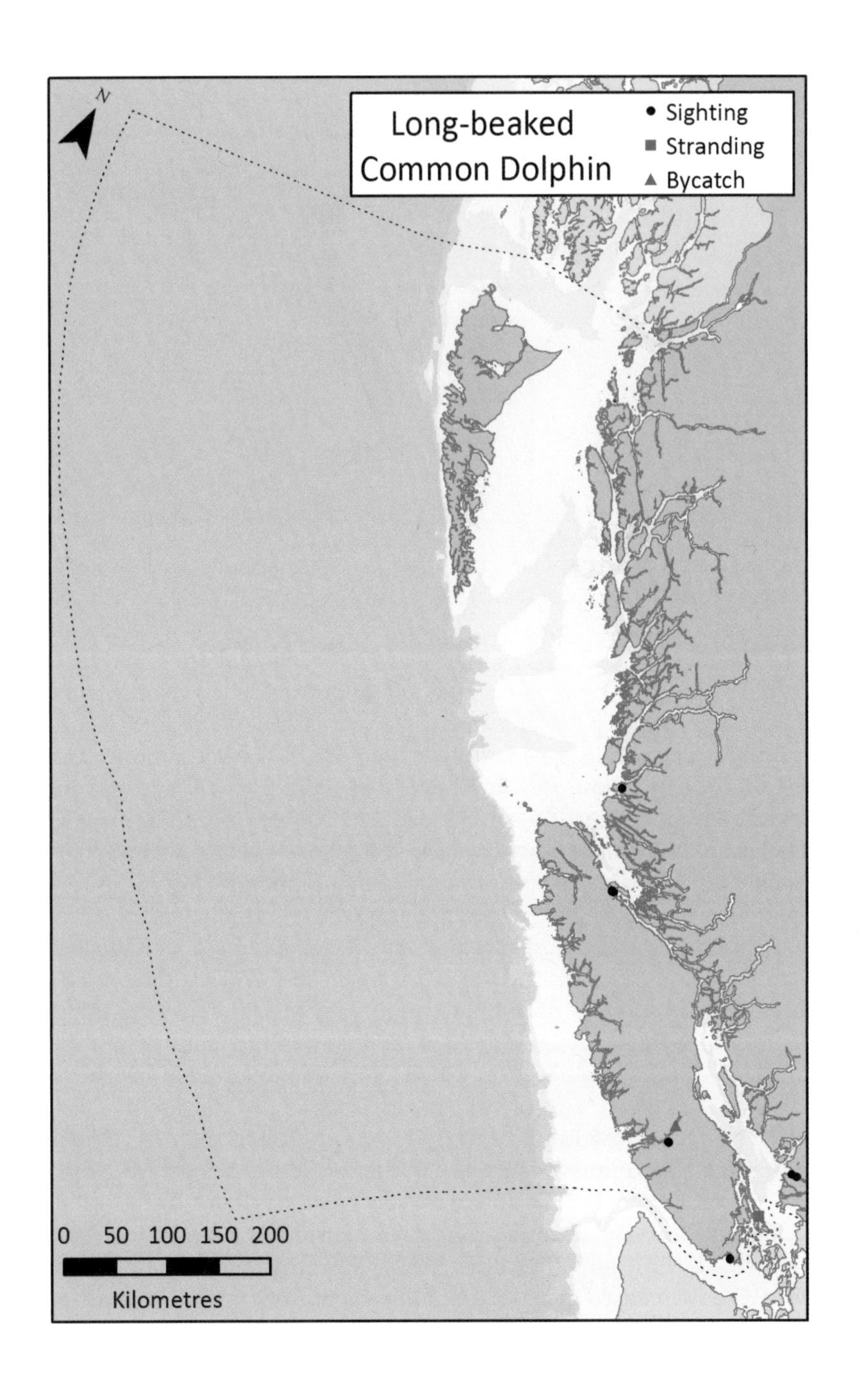
Long-beaked
Common Dolphin
Sighting
Stranding
Bycatch
N
0 50 100 150 200
Kilometres

Natural History

Feeding Ecology

Long-beaked Common Dolphins feed on squid, as well as a wide variety of small schooling fishes, such as sardines, anchovies and hake. The dolphin entangled in Alberni Inlet in October 2003 had the remains of four Pacific Hake and six Pacific Herring in its stomach. One comparative study of the diets of the two Common Dolphin species off southern California found considerable overlap in prey species. But since the Long-beaked species prefers warmer, shallower and more coastal waters than the Short-beaked species, there were some differences related to their habitat preference.

Behaviour and Social Organization

Long-beaked Common Dolphins are very gregarious and can be found in schools of up to several thousand. Average school size observed during surveys off southern California was 318. These large schools are thought to be composed of smaller subunits of about 20–30 individuals, which may be closely related. There may also be sex and age segregation within schools. Long-beaked Common Dolphins are very energetic and swift swimmers, often porpoising at high speed and making a variety of aerial jumps, breaches and leaps (figure 65). They regularly and eagerly approach moving vessels to ride the bow and stern waves.

Figure 65. Long-beaked Common Dolphin in the Gulf of California, Mexico. Photo: V. Shore, February 2011.

Life History and Population Dynamics

Most studies of the life history of Common Dolphins were conducted before the Long-beaked and Short-beaked species were distinguished and so there is no information relating to these species separately. Common Dolphins are born at a length of 80–93 cm following a gestation of 10–11.7 months. Sexual maturity in the eastern Pacific is estimated to occur at 7–12 years of age for males and 2-4 years for females. Maximum lifespan is estimated to be about 30 years. Long-beaked Common Dolphins are preyed upon by Killer Whales and possibly sharks.

Exploitation

Long-beaked Common Dolphins are sometimes taken for human consumption in Japan, Taiwan and Venezuela. There is growing concern about large numbers of Long-beaked Common Dolphins killed in Peru for human food and shark bait, and there are recent anecdotal reports that many dolphins, including Long-beaked Common Dolphins, are being killed for bait in coastal fisheries off Baja California. There are no records of the species being exploited in BC.

Taxonomy and Population Structure

Until fairly recently there was much uncertainty regarding the taxonomic status of the genus *Delphinus*, especially in the eastern North Pacific. In 1873 American zoologist W.H. Dall proposed a new species, *Delphinus bairdii*, from coastal waters of southern California, which he considered distinct in beak length and coloration from *D. delphis*. The validity of this new species was challenged by some and supported by others, and until this uncertainty was resolved, the names *D. bairdii*, *D. delphis bairdii* and *D. delphis* were applied inconsistently to Common Dolphins in the eastern North Pacific. It was not until 1994 that J. Heyning and W. Perrin provided a clear description of two distinct species in the region, and this has more recently been supported by genetic analysis. The Long-beaked form of Common Dolphin was originally proposed as a new species, *D. capensis*, in 1828 by British zoologist J.W. Gray from a specimen collected near the Cape of Good Hope, South Africa. This name has taken precedence to that later proposed by Dall. In the original *Mammals of British Columbia* Cowan and Guiguet classified the only specimen of Common Dolphin from BC known at the time as *Delphinus bairdii*, but in fact this stranded individual was a *Delphinus delphis*.

Conservation Status and Management

Although not as abundant as the Short-beaked Common Dolphin, the Long-beaked Common Dolphin is relatively numerous. There are an estimated 62,000 off southern California and 69,000 in the Gulf of California. Although the species is not considered at risk, some populations may be threatened by human activities, particularly in places where dolphins are caught for human food and fishing bait, and where they are killed accidentally in net fisheries.

Because of uncertainty about the status of local or regional populations, the IUCN Red List considers the species Data Deficient. Its status in British Columbia has yet to be ranked by COSEWIC or the BC Conservation Data Centre because of its apparently extralimital occurrence.

Selected References: Barlow 2010, Barlow and Forney 2007, Ford 2005, Guiguet 1954, Hammond et al. 2008a, Heyning and Perrin 1994, Jefferson et al. 2008, Perrin 2009, Pike and MacAskie 1969, Reeves et al. 2002.

Short-beaked Common Dolphin
Delphinus delphis

Other Common Names: Short-beaked Saddleback Dolphin.

Description

The Short-beaked Common Dolphin is a small delphinid very similar in appearance to the closely related Long-beaked Common Dolphin. Both species have the same basic body structure with tall, slightly falcate dorsal fins and slender, curved flippers with pointed tips. The body of the Short-beaked Common Dolphin, however, is somewhat more robust than the slender Long-beaked species and, as its name indicates, it has a somewhat shorter beak. It also tends to have a more rounded, bulging melon, and the angle at which the melon merges into the beak is more acute than in the Long-beaked species.

Both species of Common Dolphins have a criss-cross colour pattern that forms an hourglass motif on their sides. The pattern, unique to these species, is more distinct on the Short-beaked Common Dolphin. The back portion of the hourglass, the *cape*, is dark brownish-grey while the anterior portion between the eye and dorsal fin, the thoracic patch, is light tan or yellowish-tan. The grey stripe that extends from the flipper to the anal region in the Long-beaked species is reduced or absent in the Short-beaked Common Dolphin, and the white on the belly extends up over the flipper. A dark stripe that runs from chin to flipper is narrower on the Short-beaked than on the Long-beaked Common Dolphin.

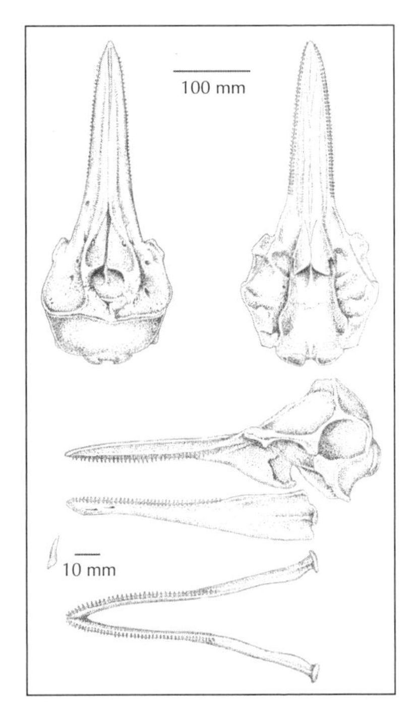

Measurements

length (metres):

male: 1.7–2.3
female: 1.6–2.2
neonate: 0.8–0.9

weight (kg):

adult: ~ 200
neonate: unavailable

Dentition

41–57 pairs in both the lower and upper jaws.

Identification

While Short-beaked and Long-beaked Common Dolphins occur in British Columbia waters, both species are very rare here and considered accidental. In fact, Short-beaked Common Dolphins are known in BC waters only from three stranded specimens. The unusual criss-cross or hourglass pattern on the sides and the tan to yellowish-tan thoracic patch of both species should distinguish them from the commonly sighted Pacific White-sided Dolphin and the rare Striped Dolphin. The two Common Dolphin species are easily confused at sea, but aside from the difference in beak lengths, the overall coloration pattern of the Short-beaked species is bolder than that of the Long-beaked species. And in the Short-beaked species, the chin to flipper stripe is reduced, and the white of the belly extends above the flipper. While there have been no confirmed sightings in BC waters, off California, dolphins of both species may travel in schools of a few to several thousand.

Distribution and Habitat

The Short-beaked Common Dolphin is primarily an oceanic species found in tropical to cool-temperate waters of the Pacific and Atlantic Oceans. It ranges from nearshore waters to thousands of kilometres offshore and is most abundant in areas with steep bathymetry that creates productive upwelling conditions. In the northeastern Pacific, Short-beaked Common Dolphins are normally found to as far north as northern California.

The species is known in BC from only three stranded animals (see Appendix 2). The first was a single adult specimen found dead on a Victoria beach on April 8, 1953. This animal was first identified by BC Provincial Museum biologist Charles Guiguet as *Delphinus bairdii*, now known as the Long-beaked Common Dolphin. But in a 1994 monograph by American cetologists John Heyning and William Perrin describing unequivocally the existence of two distinct *Delphinus* species in the northeastern Pacific, the authors suggested that the specimen was a Short-beaked Common Dolphin, basing their opinion on a photograph in Guiguet's report. I subsequently measured the skull of the specimen, which is in the Royal BC Museum's collection and confirmed it to be of the Short-beaked species. The second BC record is an adult female found floating in Nuchatlitz Inlet, west coast of Vancouver Island, on November 25, 1994 (figure 66), and the third is a beach-worn skull found by a tour group in the Skedans Islands, Haida Gwaii, in July 2011. This is the northernmost record of the genus in the eastern North Pacific.

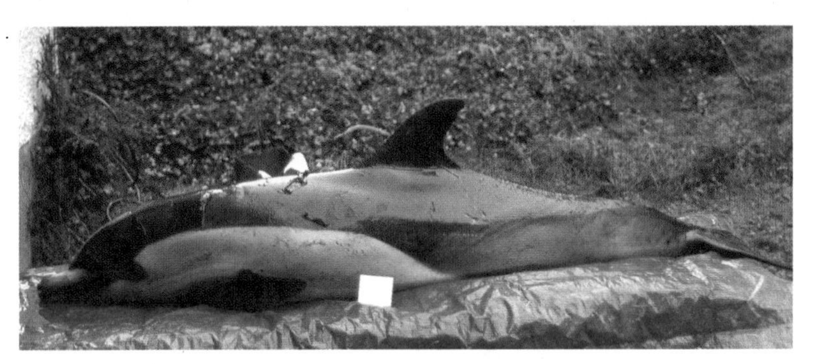

Figure 66. This Short-beaked Common Dolphin was found floating dead at Nuchatlitz Inlet, Vancouver Island, on November 25, 1994. It is one of only three records of this species in British Columbia. Photo: S. Wischniowski.

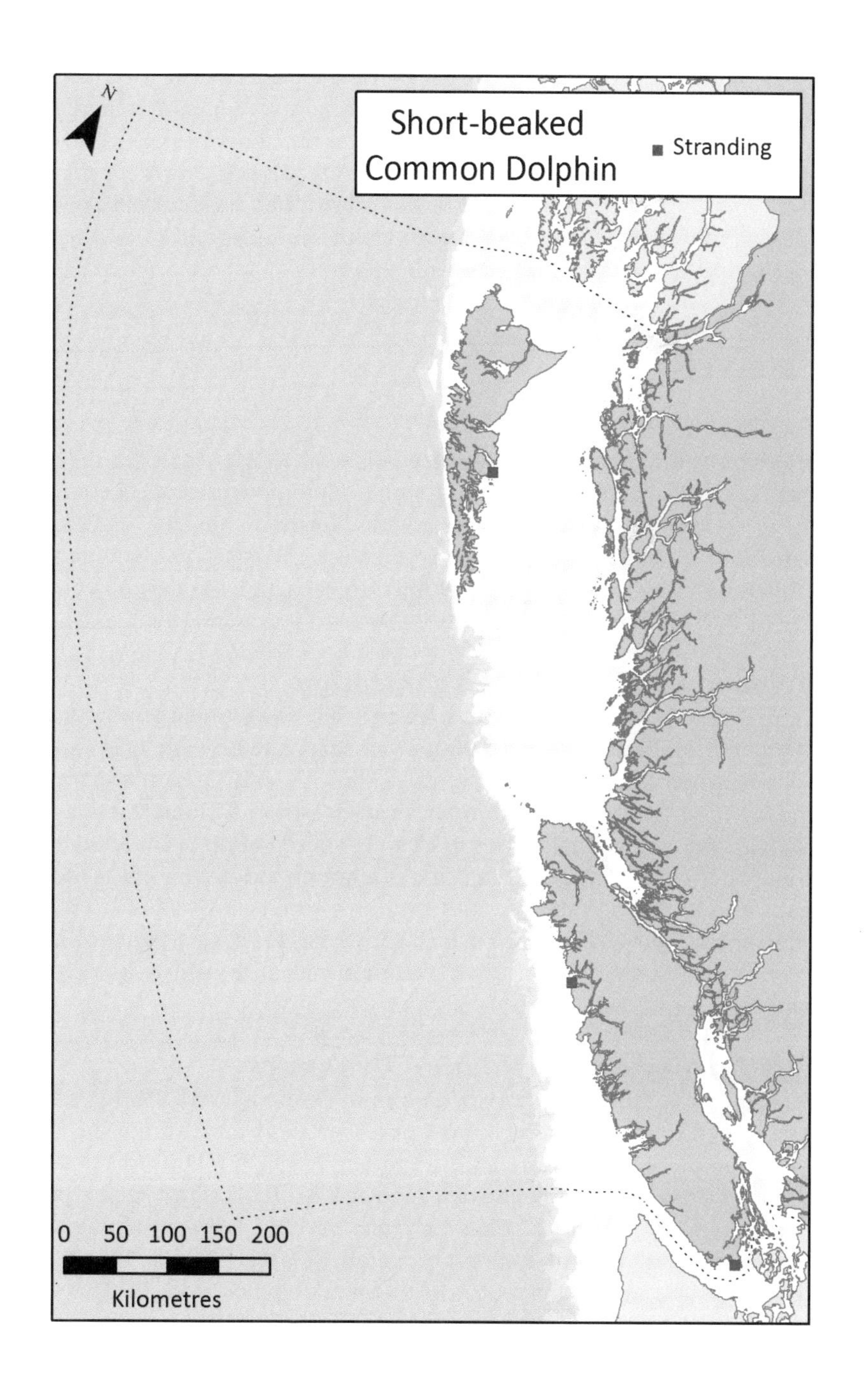
N
Short-beaked
Common Dolphin
Stranding
0 50 100 150 200
Kilometres

It's interesting that there are no sighting records of Short-beaked Common Dolphins in BC waters yet there are several such records for Long-beaked Common Dolphins. Overall, Short-beaked Common Dolphins are far more abundant in the eastern North Pacific and typically range considerably farther north than the Long-beaked species, to waters off southern Oregon. But the occurrence of either species in BC waters must be considered rare.

Natural History

Feeding Ecology

Short-beaked Common Dolphins feed on a wide variety of small (less than 20 cm) schooling fishes and squid. Common prey species off the west coast of North America include Northern Anchovy, Pacific Sardine and Pacific Herring. The stomach of the specimen stranded in Victoria in 1953 was reportedly filled to capacity with partially digested Pacific Herring.

Behaviour and Social Organization

Short-beaked Common Dolphins are typically found in large schools of several hundred to over 10,000, often segregated by age and sex. Schools observed during surveys off California and Oregon usually contained about 100–200. Short-beaked Common Dolphins are active, fast-moving animals that can swim at speeds of up to 40 km per hour. They frequently porpoise out of the water and display energetic leaps and breaches. They are also avid bow-riders and wake-surfers, often approaching vessels and even large whales for a ride (figure 67). Like most oceanic dolphins, they communicate with high-frequency whistles.

Life History and Population Dynamics

Most studies of the life history of Common Dolphins were conducted before the Long-beaked and Short-beaked species were distinguished. Common dolphins are born at a length of 80–93 cm following a gestation of 10–11.7 months. Sexual maturity in the eastern Pacific is estimated to occur at 7–12 years for males and 2–4 years for females. The maximum lifespan is probably about 30 years. Killer Whales and sharks are significant predators of small dolphins, and this likely includes the Short-beaked Common Dolphins. Predation on Short-beaked Common Dolphins by False Killer Whales has been observed in the eastern tropical Pacific.

Figure 67. A Short-beaked Common Dolphin surfing in a ship's wake in California waters. Photo: B. Pitman, September 2009.

Exploitation

Short-beaked Common Dolphins are not currently a target of directed catches for human consumption, although a major hunt for the species took place in the Black Sea until the early 1980s. They are frequently the victim of accidental capture and mortality in net fisheries using various types of gear.

Taxonomy and Population Structure

Until fairly recently, all Common Dolphins were considered to belong to a single species, *Delphinus delphis*, although several different species and subspecies have been proposed over the years. Then, in 1994, cetologists John Heyning and William Perrin provided clear evidence of two species, the widely distributed, oceanic Short-beaked Common Dolphin and the more regional, nearshore Long-beaked Common Dolphin. These are now generally accepted, at least for the North Pacific.

Conservation Status and Management

Despite its rarity in BC waters, the Short-beaked Common Dolphin is one of the most abundant cetaceans in the world. There are almost three million in the eastern tropical Pacific and about 350,000 off the

US west coast, estimated from ship surveys between 1991 and 2005. A major human-related threat to the species is bycatch in net fisheries. Large numbers died in tuna purse seines in the eastern tropical Pacific until this was mitigated in the 1990s, and there are also incidental catches in various gillnet and pelagic trawl fisheries throughout their range. Reduced prey availability and habitat degradation are considered to be the main current threats to Short-beaked Common Dolphins in the Mediterranean, where the species is listed by IUCN as Endangered.

Otherwise, the species is assessed by IUCN as Least Concern. Short-beaked Common Dolphins, which are common off Canada's east coast, are considered Not at Risk by COSEWIC. The species has not been given any conservation status in BC as it is considered accidental in the province.

Selected References: Barlow and Forney 2007, Ford 2005, Guiguet 1954, Heyning and Perrin 1994, Perrin 2009, Pike and MacAskie 1969.

Short-finned Pilot Whale

Globicephala macrorhynchus

Other Common Names: Pacific Pilot Whale, Pothead, Blackfish.

Description

The Short-finned Pilot Whale is a large member of the family Delphinidae, the ocean dolphins. Its body is long but robust with a thick tail stock. The melon is exaggerated and blunt, and is more bulbous in adult males than females. A sloping mouthline angles steeply upward toward the eye. The dorsal fin is well forward on the body, approximately one-third of the way back from the head. It is relatively low and rises at a shallow angle from a distinctively broad base into a strongly falcate or hooked form. Adult males grow significantly larger than females and their dorsal fin becomes disproportionately large and hooked. The flippers are long and falcate, reaching about 16 per cent of the total body length.

Short-finned Pilot Whales are mostly all black or dark brownish-grey, which is why they are sometimes referred to as Blackfish (a common name also used in the past for Killer Whales, especially

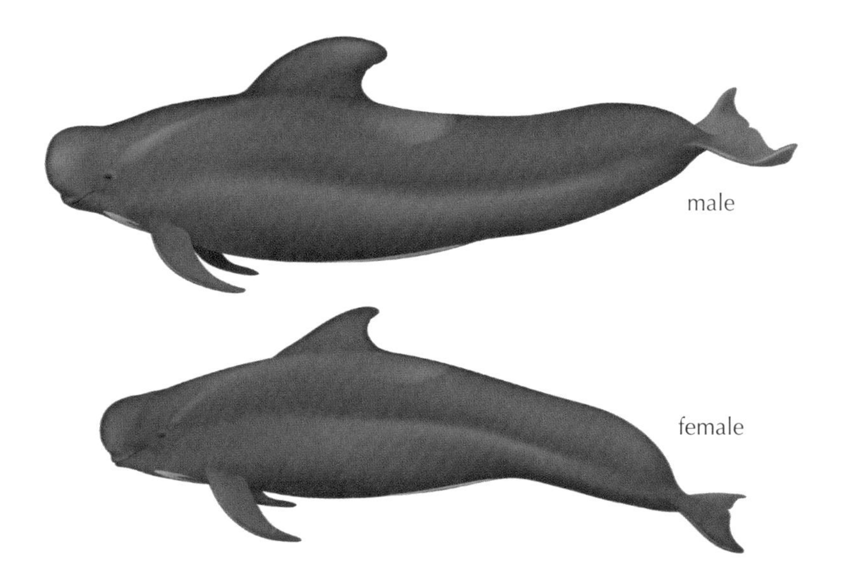

on the BC coast, and for several other odontocete species). There are light-grey patches under the throat and in the uro-genital region, which are connected by a thin mid ventral stripe. There may also be a grey saddle patch behind the dorsal fin, and often a faint grey streak that begins at the eye and extends upwards and back toward the dorsal fin.

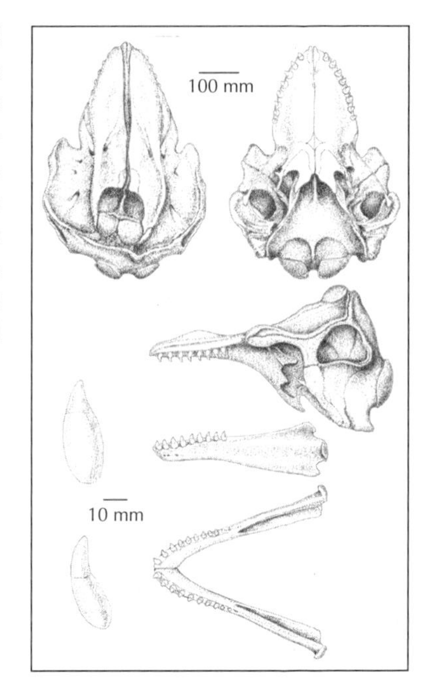

Measurements

total length (metres):

male: ~ 7.2
female: ~ 5.5
neonate: 1.4–1.9

weight (kg):

adult: 1000–3500
neonate: 60–100

Dentition

7–9 pairs of sharp, conical teeth in both the lower and upper jaws.

Identification

In BC waters Short-finned Pilot Whales could be confused with False Killer Whales, which are similar in size and colour. But the Short-finned Pilot Whale's head is blunter and its dorsal fin is lower, wider and set further forward on the body (figure 68). Both species are rarely seen along the BC coast. Killer Whales are similar in size but have distinctive white patches and differ in body shape. Risso's Dolphins are also similar in size and shape to Short-finned Pilot Whales, but they are usually light grey to whitish in colour, have numerous visible scars on their body and have a much taller dorsal fin. Short-finned Pilot Whales in the eastern Pacific have been seen in mixed groups with other species, such as Sperm Whales, Pacific White-sided Dolphins and Risso's Dolphins.

Distribution and Habitat

Short-finned Pilot Whales are found in warm-temperate to tropical waters of the Pacific, Atlantic and Indian oceans. In the eastern

Figure 68. Short-finned Pilot Whales are mostly black and have a low-profile dorsal fin that is broader at the base, relative to body size than in Killer Whales. Photo: J. Towers, Gulf of California, Mexico, December 2009.

Pacific, they have been documented from the northern Gulf of Alaska down the coast of the Americas as far as Peru, as well as from the Hawaiian Islands. They occur mostly in oceanic habitat, particularly in continental shelf break and slope waters, and other areas of high bottom relief.

Short-finned Pilot Whales appear to be rare visitors to the BC coast, at least in continental-shelf waters. A 1929 report by BC naturalists G.H. Wailes and W.A. Newcombe stated that the species was common in BC waters, but evidently they were confusing Pilot Whales with Killer Whales, which were also known as Blackfish at the time and are certainly numerous along the BC coast. There are only ten confirmed records of Short-finned Pilot Whales in BC waters. Three are of incidental catches of a total of six individuals during an experimental Canadian drift-net fishery for Flying Squid that took place in 1986 and 1987. These catches were well offshore – between 275 and 350 km west of Vancouver Island and southern Haida Gwaii. There is only a single stranding record for the species in BC, a 4.52-metre female at Metchosin, near Victoria on southern Vancouver Island, examined on November 23, 1977 (figure 69).

There are six reliable sighting records of Short-finned Pilot Whales in BC waters, all in deep waters beyond the shelf break (see Appendix 2). Group sizes ranged from 2 to 15, except for one observation of approximately 150 animals about 300 km west of Cape St James on the southern tip of Haida Gwaii. Several unconfirmed sightings of Pilot Whales have been reported in nearshore waters off the west coast of

Figure 69. An adult female Short-finned Pilot Whale stranded near Metchosin in November 1977. Photo: M. Bigg.

Vancouver Island, but these are not included here as they lack credibility. No Short-finned Pilot Whales were observed in more than 2800 cetacean sightings during 35,000 km of shipboard surveys undertaken by the Pacific Biological Station's Cetacean Research Program from 2002 to 2010, but most of this survey was restricted to continental shelf and slope waters. The distribution of Short-finned Pilot Whales off California is affected by El Niño events, so it is possible that occasional sightings of the species off the BC coast are associated with a general northward shift in distribution during warm-water years. But the Short-finned Pilot Whale should be considered a rare visitor to coastal BC waters.

Natural History

Feeding Ecology

The diet of Short-finned Pilot Whales consists primarily of squid and, to a lesser extent, fish. Their appearance in nearshore areas of southern California is associated with the seasonal spawning migrations of squid. Short-finned Pilot Whales have also been observed chasing Short-beaked Common Dolphins in the eastern tropical Pacific, but it is unclear whether they prey on this species to a significant extent.

Behaviour and Social Organization

Short-finned Pilot Whales are gregarious and normally travel in schools of 15 to 50 that have a matrilineal structure. These kin groups are

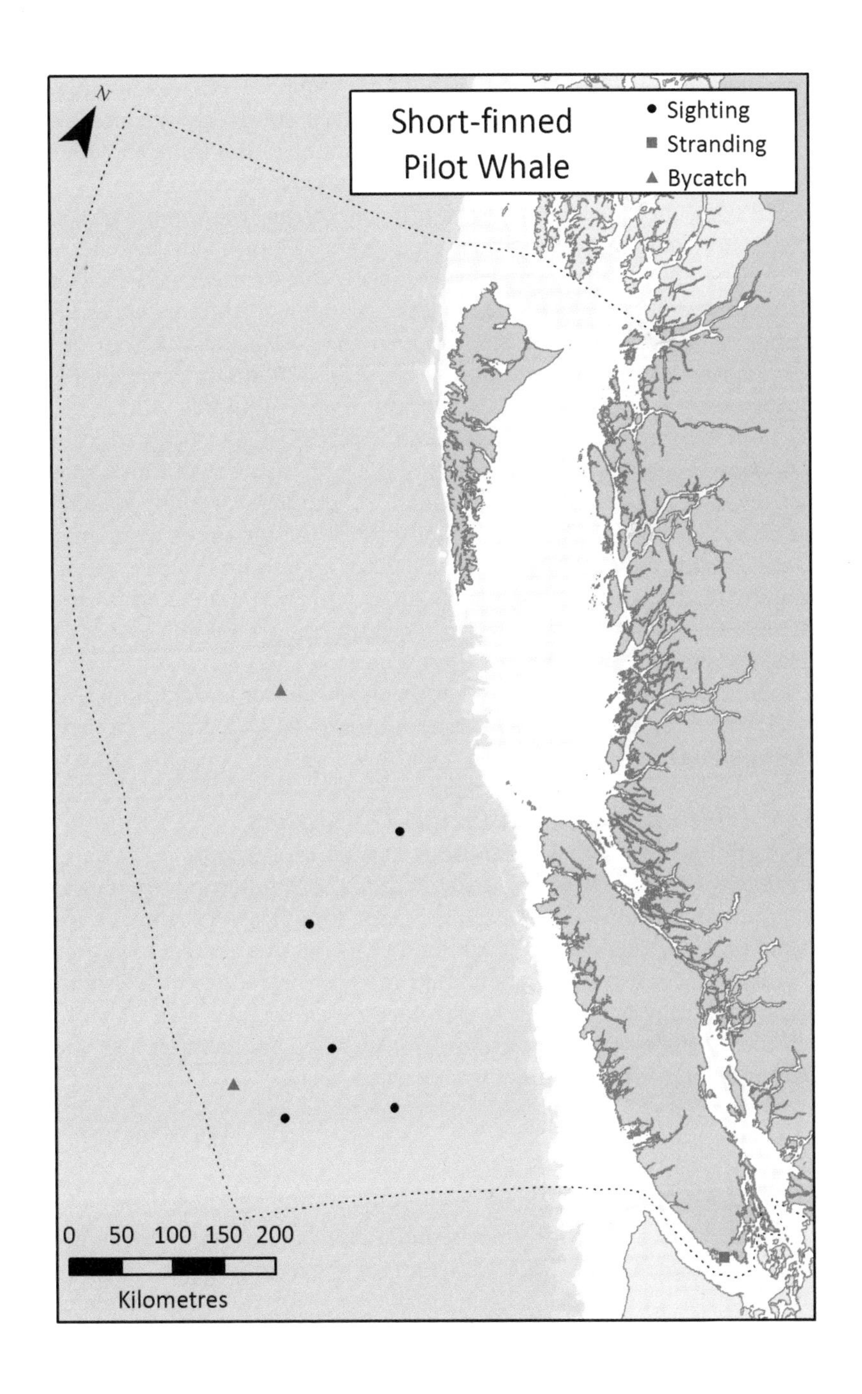
N
Short-finned
Pilot Whale
Sighting
Stranding
Bycatch
0 50 100 150 200
Kilometres

stable over time, with whales born into the group staying well into maturity, if not for their entire lives. In this regard, their social structure is similar to that of Resident Killer Whales, although not quite as stable. Genetic evidence indicates that adult males leave their natal group to breed with unrelated females, but it is not known what proportion of them stay away or return to their natal group. Mature females tend to be more numerous than adult males, likely due to the male's shorter lifespan as well as the permanent dispersal of many adult males. Gatherings of several hundred Short-finned Pilot Whales have been observed, which likely are temporary associations of multiple matrilineal groups taking advantage of localized prey concentrations.

While foraging in the open ocean, a school of Short-finned Pilot Whales will spread out and form a broad front, often more than a kilometre wide, with individuals and small subgroups swimming roughly parallel to each other but separated by tens or hundreds of metres. Occasionally, large aggregations of 150 or more animals may occur in good feeding areas. They forage most actively at night, and during the day schools typically gather into tightly knit groups and rest at or near the surface. Short-finned Pilot Whales are very vocal, producing a wide array of complex whistles and pulsed calls for social contact, as well as echolocation clicks for orientation and finding objects in their surroundings.

Life History and Population Dynamics

Short-finned Pilot Whales mature at around 8 or 9 years for females and 13 to 17 years for males. A single calf is born following a gestation of about 15 months, and the calf is nursed for at least 2 years. The interval between calves averages about 5 years but may extend to almost 8 years in older females. Females become post-reproductive at around age 40 but may continue to suckle young – either their own or those of other mothers in the kin group – for up to 15 additional years. The potential longevity of females is at least 63 years.

Exploitation

Short-finned Pilot Whales are hunted in various parts of their wide range, including Japan, St Vincent and St Lucia in the Caribbean, Indonesia and Sri Lanka. Most of the hunting, except in Japan, involves small-scale ("artisanal") fisheries that take relatively small numbers of animals each year for food. Up to several hundred a year are taken in harpoon and drive fisheries in Japan. There is no record of Short-finned Pilot Whales being hunted in BC waters.

Taxonomy and Population Structure

Although Short-finned Pilot Whales are currently treated as a single species with no subspecies, there is evidence suggesting there may actually be two or more species or at least subspecies. Two distinct forms exist off Japan – northern and southern – which differ in cranial and external morphology, genetic structure and life history. The taxonomic status of these forms has yet to be resolved. In the past, the Short-finned Pilot Whale in the eastern North Pacific was considered by some biologists to be a species distinct from that in the tropical Pacific, and it was named *Globicephala scammonii* after Charles Scammon, the famous whaling captain of the 19th century. More recent studies have shown that this distinction is not warranted, and the name *G. scammonii* is no longer used.

Conservation Status and Management

Short-finned Pilot Whales are reasonably abundant, although some regional populations are depleted as a result of long-term exploitation. There are estimated to be around 600,000 in the eastern tropical Pacific and 9000 in Hawaiian waters. The species was once common off southern California, but after a strong El Niño event in 1982–83 it virtually disappeared from the region for about a decade. Few sightings have been made off the US west coast in recent surveys, and only a few hundred animals are estimated to occur in those waters today. Short-finned Pilot Whales are sometimes taken as bycatch in drift or seine net fisheries, although this mortality does not seem to be sufficient to threaten populations in the eastern North Pacific. Pilot whales are also thought to be sensitive to loud anthropogenic sounds such as those associated with navy sonar and seismic exploration. Mass strandings have been linked to such sounds, but conclusive cause-and-effect evidence is lacking.

The IUCN has designated the Short-finned Pilot Whale as Data Deficient, because of taxonomic uncertainty about the pilot whale in the North Pacific and the related difficulty of determining the conservation status of populations. COSEWIC designated it as Not at Risk in 1993. The species is on the BC Yellow List.

Selected References: Baird and Stacey 1993; Carretta et al. 2011; Ford, Abernethy et al. 2010; Jefferson et al. 2008; Olsen 2009; Pike and MacAskie 1969; Reeves et al. 2002; Stacey and Baird 1993; Taylor et al. 2008; Wailes and Newcombe 1929.

Risso's Dolphin *Grampus griseus*

Other Common Names: Grampus.

Description

Risso's Dolphin is a distinctive-looking, small to medium-sized toothed whale. The anterior half of the body is robust, but the posterior half tapers to a relatively slender tail stock. The head is broad and bulbous with a squarish profile, and the blunt front part of the melon has a distinct vertical cleft or furrow. The flippers are long, falcate and pointed. The dorsal fin is erect, pointed and quite tall for the size of the animal. Dentition is unusual in that there are few teeth in the forward portion of the lower jaw and typically none in the upper jaw.

Coloration of Risso's Dolphins is quite variable and striking, and it changes dramatically with age. Calves have a light ventral surface and grey to dark brownish-grey dorsal surface. As the dolphins age, their body colour lightens – except for the dorsal fin and flippers, which remain mostly dark – and they accumulate many white scars on the back and flanks, presumably from the teeth of other Risso's Dolphins and possibly from the beaks and suckers of their squid prey. Due to light pigmentation and extensive scarring, older animals can be almost white in appearance. There is little sexual dimorphism in body size or coloration in Risso's Dolphins.

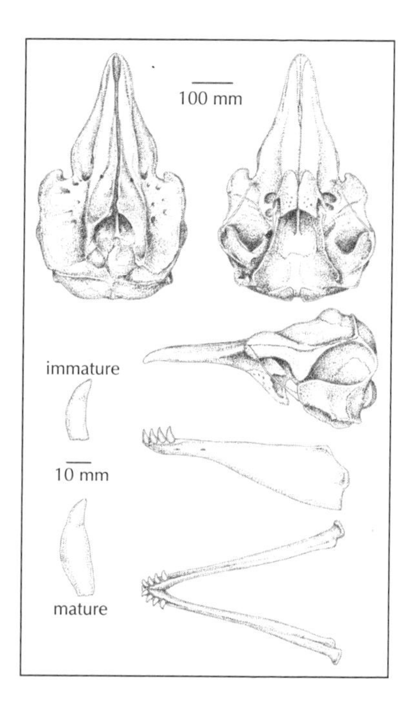

Measurements

length (metres):
- adult: 3.6–4.0
- neonate: 1.1–1.5

weight (kg):
- adult: ~ 500
- neonate: ~ 60

Dentition

2–7 pairs in the front of the lower jaw; none or 1–2 pairs of vestigial teeth in the upper jaw.

Identification

Risso's Dolphins are easy to identify at sea if viewed at close range. They travel in groups that vary widely in size, from 5 to 100 or so. Adults are mostly light in colour with extensive scarring, and they have a distinctively blunt head with a vertical cleft in the front of the melon, which may only be visible at very close range (figure 70). In BC waters they could be confused with Killer Whales at a distance

Figure 70. Risso's Dolphins off the west coast of Vancouver Island. The characteristic cleft on the front of the melon is visible on the animal on the right. Photo: B. Gisborne, August 2007.

Figure 71. A Risso's Dolphin with a very young calf off the west coast of Vancouver Island. The adult's dorsal fin is tall and resembles that of a Killer Whale, but it is mostly light to dark grey rather than black.
Photo: B. Gisborne, September 2013.

because of their relatively tall dorsal fin, but Risso's dorsal fin is lighter in colour (figure 71). Short-finned Pilot Whales are somewhat similar in body shape but are larger and much darker than Risso's Dolphins.

Distribution and Habitat

Risso's Dolphins are widely distributed in deep waters of the continental shelf break and slope in tropical to temperate areas of the southern and northern hemispheres. In the North Pacific they range as far north as the Gulf of Alaska and the Kamchatka Peninsula. But they favour warmer temperatures, with most sightings taking place in waters 15–20°C, and rarely in waters of less than 10°C.

Until recently Risso's Dolphins were considered rare in BC. The first record of the species in the province was a 3.6-metre individual shot by a local resident in May 1964 at Big Bay on Stuart Island near Campbell River. This specimen was the first record for Risso's Dolphins in coastal waters of the northeastern Pacific north of California, where the species is commonly observed. The second record for BC was a 2.6-metre male that washed ashore on Vargas Island off the west

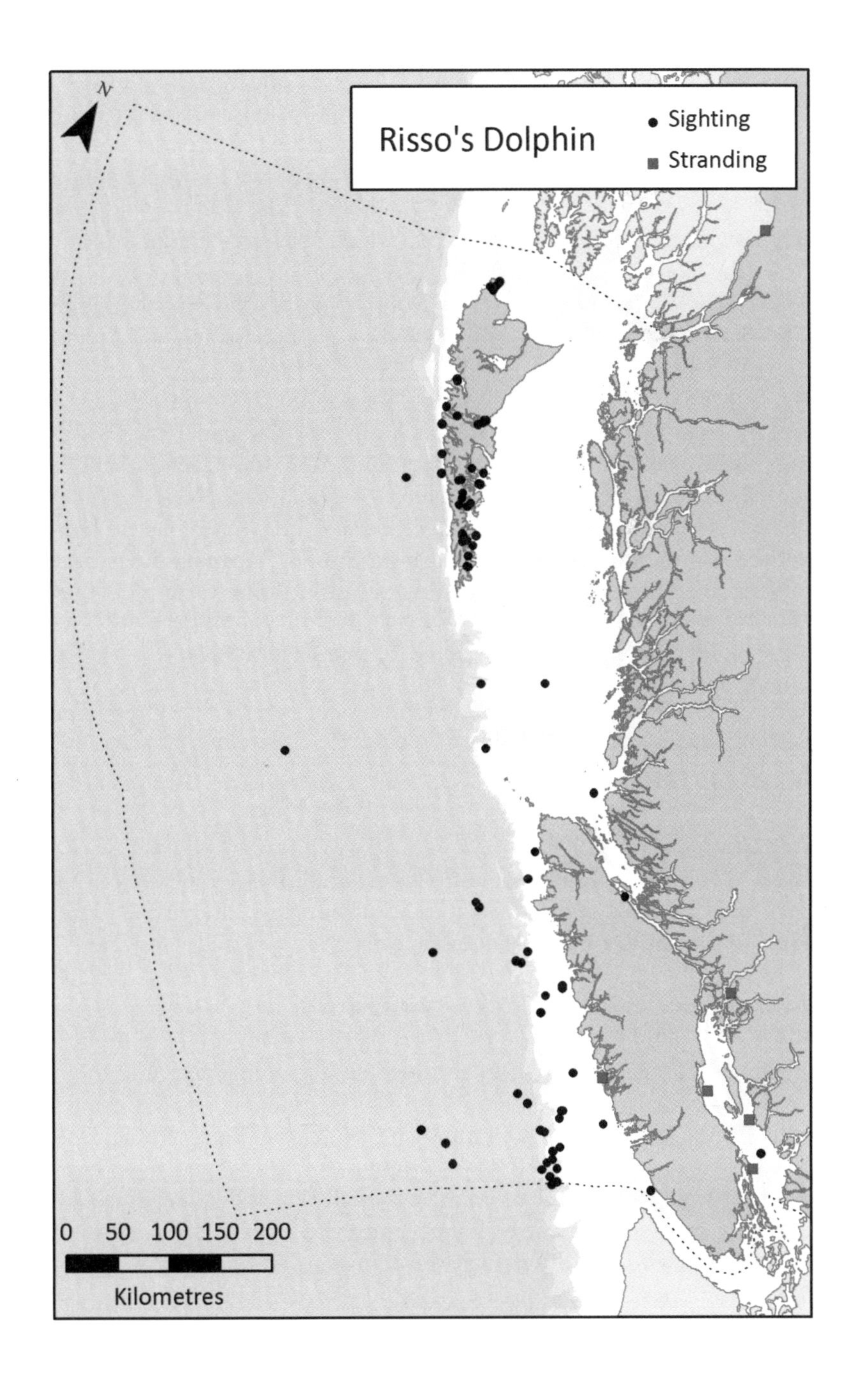
Risso's Dolphin
Sighting
Stranding
N
0 50 100 150 200
Kilometres

coast of Vancouver Island on April 17, 1970, and was examined by biologist David Hatler. The third and most northerly record for BC is of an individual that stranded alive but later died at Stewart on November 15, 1977. The first confirmed sighting of live Risso's Dolphins in BC was made by biologist Tom Reimchen, who observed a group of 14 swimming through Parry Pass, which separates Langara Island from the northwestern corner of Graham Island, Haida Gwaii, on March 27, 1978. Later that year a group of about 10 Risso's Dolphins was seen on four occasions in the Strait of Georgia during late August to early September, 1978.

Since the 1970s there have been five additional stranding records and 82 reported sightings of Risso's Dolphins. The majority of sightings have been made by tour operators and Parks Canada staff in the nearshore waters of Gwaii Haanas National Park Reserve. Other records are from around Langara Island, off the west coast of Vancouver Island, and in Queen Charlotte Sound. Most sightings have taken place during summer, possibly because of the warming of surface waters and also increased observer effort during this time of year. Overall, Risso's Dolphin sightings in BC waters have been recorded in nine months of the year.

Natural History

Feeding Ecology

Risso's Dolphins feed almost entirely on squid. There is evidence that they feed mostly at night, likely to take advantage of nocturnal migrations to the surface by some squid species.

Behaviour and Social Organization

Risso's Dolphin is a gregarious species that is usually found in moderate-sized groups of 10 to 100, although aggregations of several thousand have been reported. They can also be found in mixed groups with Pacific White-sided Dolphins and Northern Right Whale Dolphins. Most sightings of Risso's Dolphins in BC waters have been of groups of 25 or fewer animals, although one school reported off the west coast of Vancouver Island comprised about 150 dolphins. Risso's Dolphins are often seen travelling and surfacing slowly, but they sometimes become highly energetic – breaching, porpoising and spyhopping. Unlike some other dolphins, they seldom ride the bow or wake of vessels. Little is known of this species' social structure, but there is some evidence from photo-identification studies in California and the Azores that they form stable clusters of pairs or groups of 3–12 individuals, and that adult males often travel together in sexually segregated groups. Risso's Dolphins emit a variety of burst-pulse signals for social signalling as well as clicks for echolocation.

Life History and Population Dynamics

The life history of this species has been little studied. Sexual maturity is thought to be attained at ages of 8–10 years for females and 10–12 years for males, which corresponds to body lengths of 2.6 to 2.7 metres in both sexes. Gestation is estimated to be 13–14 months and intervals between calves average 2.4 years. Calving appears to peak in fall and winter off California. Lifespan is at least 35 years.

Exploitation

Risso's Dolphins are taken in drive hunts in Japan and until recently in the Solomon Islands, and by harpooning in Indonesia, Japan and the Lesser Antilles. There is no record of the species being hunted by First Nations or by commercial whalers in British Columbia.

Taxonomy and Population Structure

Genetic studies have shown that Risso's Dolphins are most closely related to False Killer Whales, Pygmy Killer Whales, Melon-headed Whales and Pilot Whales. No subspecies are recognized. Risso's Dolphins are known to have hybridized with Common Bottlenose Dolphins, in captivity and in the wild.

Conservation Status and Management

Risso's Dolphins are widespread and fairly abundant, and do not appear to be at risk globally. Several hundreds are taken annually in drive fisheries in Japan and in directed hunts in Sri Lanka. Some are killed incidentally as bycatch in drift-net and seine fisheries in Taiwan, Philippines, Peru, the Mediterranean Sea, and the North Atlantic. The species has been assessed as Least Concern by the IUCN Red List and as Not at Risk in Canada by COSEWIC. It is on the BC Yellow List.

Remarks

The common name for the species comes from Antoine Risso, a French naturalist whose description of the type specimen formed the basis of the species' formal description by natural historian Georges Cuvier in 1812. But researchers often refer to the species simply by its generic name, *Grampus*. Note that Killer Whales were formerly assigned to the same genus, and occasionally Grampus has been used to refer to them as well, especially historically. The name *grampus* is derived from Old French and Latin terms meaning "grand fish".

Selected References: Baird and Stacey 1991a, Guiguet and Pike 1965, Hatler 1971, Leatherwood et al. 1980, Reimchen 1980, Stroud 1968, Taylor et al. 2008.

Pacific White-sided Dolphin

Lagenorhynchus obliquidens

Other Common Names: Lag, Striped Dolphin (obsolete).

Description

Pacific White-sided Dolphins – or *Lags* as they are often called by field researchers – are stocky animals with a short, thick rostrum. The beak is differentiated from the melon only by a shallow crease. The dorsal fin is relatively large for the size of the animal and varies in shape. Typically, it is moderately falcate, with a sharp to slightly rounded tip, but in adult males it may become lobate (broadly rounded) and noticeably hooked compared to females. The flippers are long and curved, with slightly rounded tips.

The colour pattern of Pacific White-sided Dolphins is complex and bold. The back is dark grey to black and the belly is pearl white with a distinct black border. The flanks are light grey except for a broad swath of black that sweeps diagonally backward past the dorsal fin to the genital region. This black band is overlain by a pair of thin grey lines, or "suspender stripes", that extend forward from the grey flank. These stripes are easily seen from above when the dolphin is riding bow waves. The beak and lips are black with a narrow stripe extending back toward the flipper. The dorsal fin and flippers are two-tone – the leading edges are black but lighten to grey on the trailing edges.

Measurements

length (metres):

male: ~ 2.5
female: ~ 2.4
neonate: 0.92–1.00

weight (kg):

male: ~ 200
female: ~ 145
neonate: ~ 15

Dentition

23–36 pairs in both the upper and lower jaws.

Identification

Pacific White-sided Dolphins are among the easiest cetaceans to identify in BC waters. They tend to be seen in groups of 10 to 200 and occasionally in large schools of more than 1000, either in a compact formation when travelling or widely dispersed when foraging. When swimming quickly they "porpoise" or "dolphin leap", with their bodies mostly or completely clearing the water when they surface to breathe (figure 72). They are well-known for a variety of spectacular aerial manoeuvres, including repetitive breaches, belly flops, cartwheels, and flipper and tail slaps.

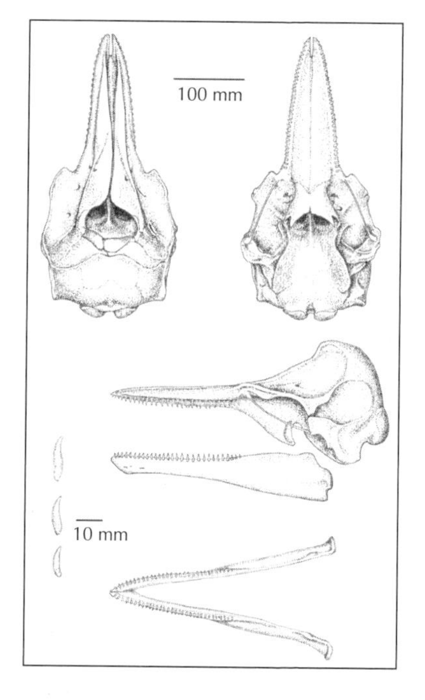

The colour pattern of the Pacific White-sided Dolphin is more striking than that of any other small cetacean in the region except the Striped Dolphin, the Long-beaked Common Dolphin and the Short-beaked Common Dolphin, all of which have similar body markings but are rare in BC waters. The most likely species to be confused with the Pacific White-sided Dolphin is the Dall's Porpoise, which is similar in size and has a bold white patch on its side (in fact, the Pacific White-sided Dolphin's sides are grey, not white). But Dall's Porpoises are generally found in groups of

Figure 72. Fast-swimming Pacific White-sided Dolphins frequently clear the surface when travelling. Photo: J. Ford, Queen Charlotte Strait, August 2005.

fewer than 10, and they rarely leap clear of the water's surface, even when swimming rapidly. Pacific White-sided Dolphins sometimes associate with Northern Right Whale Dolphins and Risso's Dolphins in mixed schools, but this is seen more frequently in California than in British Columbia. The colour of Pacific White-sided Dolphins can sometimes be muted by a thick growth of diatoms on the skin, which gives them a yellow or brownish hue and obscures some of their markings. This is most frequently observed on dolphins in offshore waters.

Distribution and Habitat

The Pacific White-sided Dolphin is endemic to temperate to cold waters of the North Pacific, where it is one of the most abundant cetacean species, and occupies a range of habitat types. These dolphins are widely distributed in pelagic waters across the North Pacific, most commonly from about 35°N to 47°N but extending north to the Aleutian Islands. In continental slope waters of the western North Pacific, they range from southern China to the Kamchatka Peninsula, including the Sea of Japan and southern Okhotsk Sea. In the eastern North Pacific, they occur from the southern Gulf of California northward to the Gulf of Alaska, including offshore, continental shelf and inshore waters.

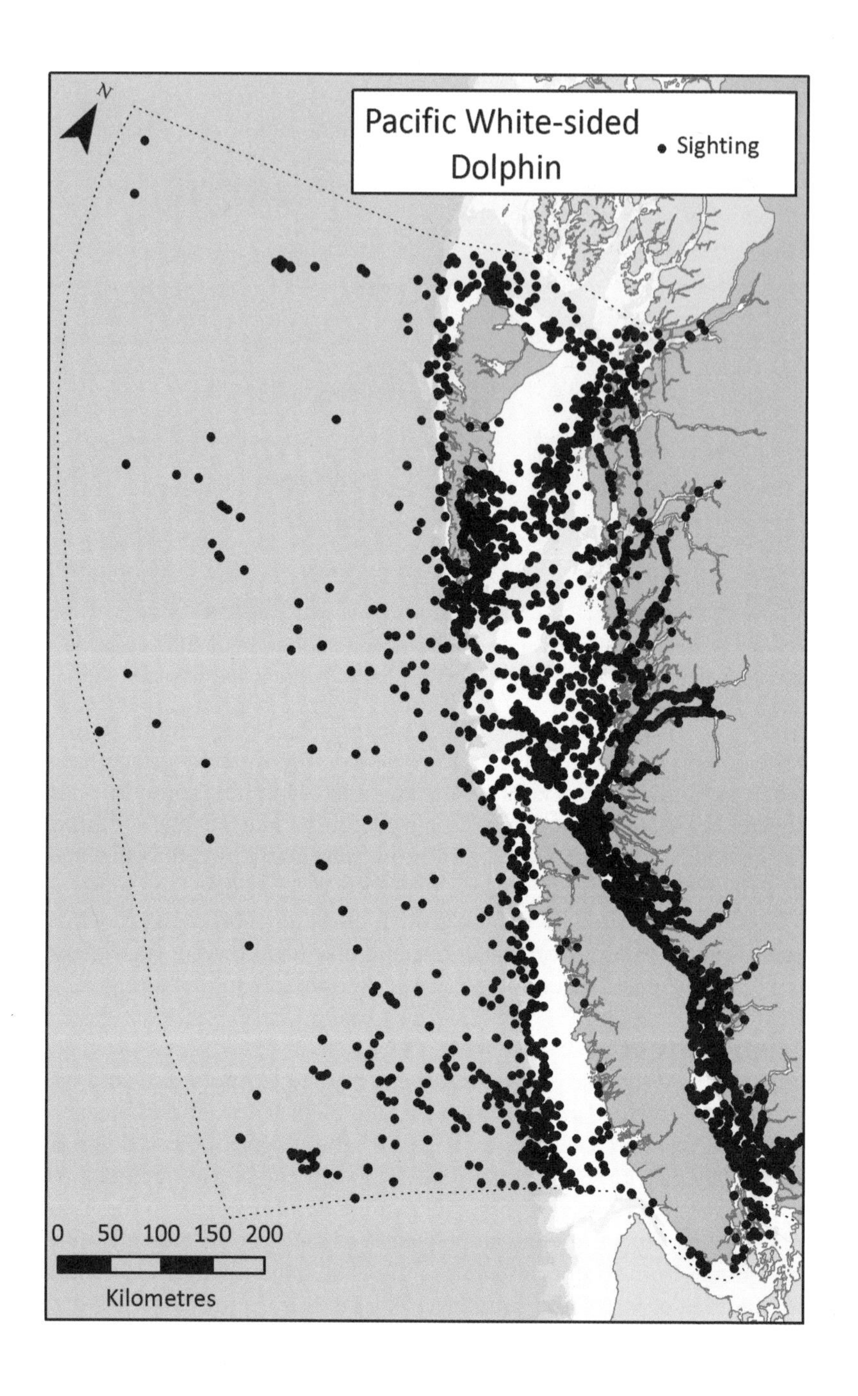
Pacific White-sided Dolphin
Sighting
N
0 50 100 150 200
Kilometres

Pacific White-sided Dolphins are found throughout coastal and offshore waters of British Columbia. They are frequently seen around Haida Gwaii, particularly in the waters of Gwaii Haanas National Marine Conservation Area and near Langara Island. They are also widespread in Dixon Entrance, Hecate Strait and Queen Charlotte Sound. Off the west coast of Vancouver Island, sightings of Pacific White-sided Dolphins are concentrated along the productive continental shelf and slope. The species is also common throughout the many narrow channels and fjords of the central and northern mainland coast. Interestingly, the occurrence of Pacific White-sided Dolphins in these inshore waters has changed dramatically over the past three decades. Over much of the past century these dolphins were rarely seen in inside coastal passages. But starting in the early 1980s they began to appear with increasing regularity, first in the waters of Fitz Hugh Sound and adjoining Fisher, Burke and Dean channels, then in Rivers Inlet, Queen Charlotte Strait, Johnstone Strait, and in the passes and channels of the Broughton Archipelago. Around 1998 they began to be seen in the Strait of Georgia, and the frequency of sightings has increased steadily since then. Pacific White-sided Dolphins are now frequent visitors to the Sunshine coast, the western Strait of Georgia near Gabriola Island, and Howe Sound. Several sightings of large groups (about 1000 individuals) of Pacific White-sided Dolphins in the Strait of Georgia in 2013 drew considerable public and media interest. The reason for the increased use of inshore waters by these dolphins in recent decades is not clear, but it is likely due to changing distribution and abundance of their prey.

Pacific White-sided Dolphins are present in BC coastal waters throughout the year, though their distribution appears to shift offshore somewhat during summer. Biologist Alexandra Morton studied dolphins in the Broughton Archipelago during the 1980s and 1990s and found that they were most abundant from October through January, with a peak in November. Despite increased effort and better sighting conditions, the dolphins were seldom seen during summer months. In recent years, however, they have been seen throughout the summer in nearby Johnstone Strait.

Natural History

Feeding Ecology

More than 60 species of fishes and 20 species of cephalopods have been documented in the diet of Pacific White-sided Dolphins. They are opportunistic feeders, and their prey varies by location and season.

In offshore waters, their diet is dominated by squid and lanternfish (myctophids) whereas forage fish species such as anchovies, sardines and herring are more important in coastal areas. In the 1990s University of BC graduate student Kathy Heise studied the diet of Pacific White-sided Dolphins by recovering and identifying prey fragments, especially fish scales, from the vicinity of foraging dolphins off northeastern Vancouver Island and in Fitz Hugh Sound on the central BC coast. She found that the dolphins were feeding primarily on Pacific Herring 15–23 cm in length and on juvenile and adult salmon up to 60 cm long, including Sockeye, Pink and Coho (figure 73). Biologist Alexandra Morton also collected prey remains in the Broughton Archipelago and found the dolphins were feeding mostly on Pacific Herring and Capelin. Food remains in the stomachs of dolphins found stranded or drowned as bycatch in fishing nets have provided further insight into diet. Heise examined 11 such stomachs and found Pacific Herring, Chum Salmon, Sablefish, Walleye Pollock, squid and shrimp fragments.

Behaviour and Social Organization

Pacific White-sided Dolphins are very social and gregarious, and they are normally found in large schools. In 377 sightings made during Cetacean Research Program cetacean surveys along the BC coast from 2002 to 2008, the average estimated group size was 60 dolphins, within a range of just 2 to about 3000. About 90 per cent of sightings were of 100 or fewer dolphins, and only 5 per cent were of schools of more than 200. The largest schools we've seen were in open waters around Haida Gwaii. Researcher Kathy Heise noted an average group size of about 100 dolphins in Queen Charlotte Strait and Fitz Hugh Sound, while Alexandra Morton found groups of 11 to 50 to be most common in the Broughton Archipelago.

The Pacific White-sided Dolphin is a fast-swimming, energetic species known for its acrobatic aerial displays. Nineteenth-century whaling captain and naturalist Charles M. Scammon described these dolphins as "tumbling over the surface of the sea, or making arching leaps, plunging again on the same curve or darting high and falling diagonally sidewise upon the water, with a spiteful splash, accompanied by a report that may be heard at some distance." When moving rapidly, individuals in a school leap or "porpoise" clear of the surface, often with multiple animals in the air at a time. Pacific White-sided Dolphins are eager surfers and often ride on the bow waves and stern wake of all sizes of vessels, from small speedboats to cruise ships. As they race through the water, the dolphins may show all manner of

Figure 73. A group of Pacific White-sided Dolphins in Blackfish Sound foraging for salmon, one of which can be seen here.
Photo: B. Hill, August 2011.

leaps and jumps, including backflips, somersaults and belly slaps. Once, one accidentally leapt three metres onto the deck of a moving research vessel.

When foraging, Pacific White-sided Dolphins spread out widely, with individuals and subgroups moving in different directions and surfacing and diving independently. When travelling, schools become more tightly spaced and highly integrated, with all members moving in the same direction at the same speed. Biologist Nancy Black studied the activity budget of Pacific White-sided Dolphins in Monterey Bay, California, and found that schools spent 24 per cent of their time feeding, 33 per cent milling, 22 per cent travelling, 18 per cent socializing and only 3 per cent resting.

Occasionally, dolphin schools move extremely rapidly as a tight unit in a single direction, frothing the water in their headlong dash. This is sometimes referred to as *squalling*, since at a distance it resembles an intense wind squall that whips the sea surface into whitecaps. In many cases, squalling dolphins are fleeing from Transient (Bigg's) Killer Whales, their main predators. This appears to be a successful anti-predator strategy unless the pursuing Killer Whales manage to chase the school into a confined bay or inlet. In their panic to escape, some dolphins can become stranded in the shallows or on shore if it is a sloping beach. Dead dolphins left in the aftermath of a Transient Killer Whale attack have been found in such locations as Higgins Pass, Codville Lagoon and Kent Inlet on the central BC coast. Although Pacific White-sided Dolphins are quick to flee from approaching Transient Killer Whales, they are regularly seen near or even mixed in with groups of Resident Killer Whales. It's clear that they recognize the differences between these two ecotypes of killer whales and don't consider fish-eating Residents a threat. In fact, these dolphins may harass Resident Killer Whales on occasion, perhaps taking pieces of fish that have been broken up for sharing among the whales.

Although Pacific White-sided Dolphins are highly social, little is known of their social structure. Large schools no doubt include males and females and a whole range of ages, but small groups of mature males with large, hooked dorsal fins are sometimes observed. This species is quite vocal under water, making a wide variety of high-pitched chirps, trills, rasps and whistles. When lowering a hydrophone into the water near a dolphin school, one generally hears loud buzzes and sweeping tones generated by rapidly emitted echolocation clicks as the dolphins approach and acoustically investigate the unusual object hanging in their midst.

Life History and Population Dynamics

Pacific White-sided Dolphins reach sexual maturity at 7.5–11 years of age. Most calves are born from June through August after a 11–12 month gestation and are nursed by their mother for 8–10 months. The average calving interval may be greater than four years. Males can live up to 42 years and females 46 years.

Exploitation

Pacific White-sided Dolphins have never been the primary targets of commercial hunting, but some are taken periodically by harpooning and in drive fisheries in Japan. In British Columbia, Pacific White-sided Dolphins were hunted in the past by First Nations, although they were not as important as some other marine mammal species. Archaeologist Alan McMillan and co-workers found that Pacific White-sided Dolphins represented about four per cent of marine mammal remains identified in middens at the Nuu-chah-nulth village site Ts'ishaa in the Broken Group Islands of Barkley Sound. Teeth of Pacific White-sided Dolphins were found distributed throughout more than 2000 years of accumulated midden layers examined at the Kwakwaka'wakw village known today as Hopetown in the Broughton Archipelago.

Taxonomy and Population Structure

Only a single species of Pacific White-sided Dolphin is recognized, although cranial morphometric analysis and mitochondrial DNA evidence have revealed the presence of two forms along the west coast of North America: one north of 36°N latitude and one south of 32°N in the waters around Baja California. The taxonomic significance of these two forms has yet to be established.

Conservation Status and Management

The Pacific White-sided Dolphin is one of the most abundant cetaceans in the North Pacific. Estimates of 900,000 to 1,000,000 animals have been made from vessel surveys in the oceanic waters of the central north Pacific, but these may be inflated by the tendency for dolphins to approach vessels. Abundance off the US west coast varies considerably with ocean conditions, and estimates have ranged from 13,000 to 122,000, with an average of 24,000 from 1996 to 2001. In 2004 and 2005, researchers Rob Williams and Len Thomas conducted line-transect vessel surveys in coastal BC waters and estimated 25,900 Pacific White-sided Dolphins. This estimate may also be on the high side due to dolphin attraction to the survey vessel. Biologist Alexandra Morton used photo-identification of individual dolphins to develop a catalogue of over 1000 animals in the Broughton Archipelago from 1987 to 2005. Graduate student Erin Ashe then used this catalogue to estimate annual abundance of dolphins in that area using mark-recapture statistical modelling and found numbers varied from an estimated 355 in 1992 to 2047 in 2000.

In the past, the greatest threats to Pacific White-sided Dolphins were the high-seas drift gillnet fisheries for salmon and squid, which operated through much of the central and western Pacific from the 1970s through the 1980s. These fisheries may have killed about 100,000 Pacific White-sided Dolphins as bycatch during that period. Although a United Nations moratorium ended these high seas fisheries in 1993, some bycatch of dolphins continues in net fisheries off Japan and along the west coast of North America, but this mortality does not appear to be sufficient to have population-level effects.

Pacific White-sided Dolphins are considered Least Concern by IUCN (Red List) and Not at Risk in Canadian waters by COSEWIC. They are on the BC Yellow List.

Remarks

Although Pacific White-sided Dolphins are abundant in British Columbia waters, until recently there were relatively few published accounts of the species in this region. The first recorded sighting was by Wilfred H. Osgood, who in his 1901 book on the natural history of the Queen Charlotte Islands described "a porpoise supposed to be this species". A skull collected at Estevan Point on the west coast of Vancouver Island in 1943 is the first specimen record for Canadian waters. The first recorded sighting in Canadian waters was of a school of at least 1000 Pacific White-sided Dolphins near Langara Island, Haida Gwaii, in June 1959, which warranted a published note by cetacean scientist Gordon Pike. Later, in their 1969 report "Marine Mammals of British Columbia", Gordon Pike and Ian MacAskie listed only 19 reliable sightings of the species in BC waters. By 2012 Pacific White-sided Dolphins had been reported to the BC Cetacean Sightings Network on 4800 occasions, which represents about 10 per cent of all cetacean sightings in the network's database.

Selected References: Ashe 2007, Black 1994, Heise 1997a and b, Lux et al. 1997, McMillan et al. 2008, Morton 2000, Pike 1960, Pike and MacAskie 1969, Stacey and Baird 1991b, Williams and Thomas 2007.

Northern Right Whale Dolphin

Lissodelphis borealis

Other Common Names: None.

Description

The Northern Right Whale Dolphin is an average-sized dolphin with an unusually long, streamlined body. It is somewhat compressed dorso-ventrally rather than laterally as in most dolphins, and the posterior half of the body tapers to an extremely slim tail stock. Most notably, this dolphin has no dorsal fin. The beak is short and demarcated from the melon with a shallow crease, and the mouthline is long and straight. The flukes and flippers are small, narrow and finely structured.

Northern Right Whale Dolphins are predominantly black with a distinct narrow band of white on the ventral side that extends from the throat to the flukes. This band is widest between the flippers, where it is visible in a lateral view of the animal. There is also a small white patch behind the tip of the lower jaw. The rear half of the flukes is grey dorsally and white ventrally. The calves have a muted colour pattern and are primarily brownish-grey rather than black. In some individuals the ventral white field extends up the sides, over the rostrum and onto the dorsal surface of the flippers (figure 74).

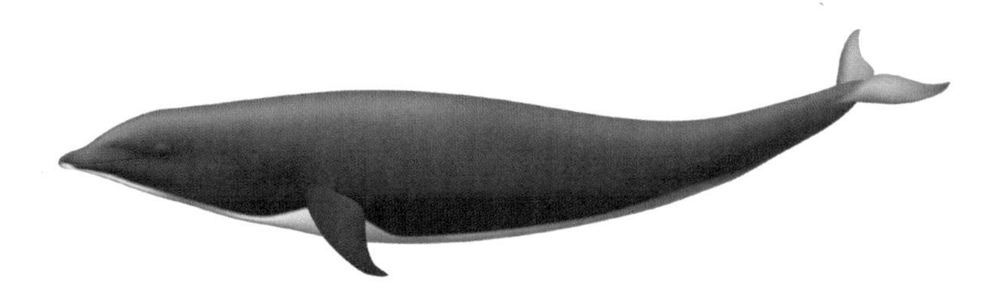

Measurements

length (metres):

male: ~ 3.1
female: ~ 2.3
neonate: 0.9–1.0

weight (kg):

adult: 60–115
neonate: unavailable

Dentition

37–56 pairs of fine, sharp teeth in both the upper and lower jaws.

Identification

Northern Right Whale Dolphins are easy to identify at sea because they are the only finless dolphins in BC waters. They are usually seen in groups of ten to fifty, though aggregations of several hundred have been reported. They are high-speed swimmers and, when travelling, make graceful, low-angle leaps that often expose their entire body (figure 75). At a distance it is possible to confuse these dolphins with porpoising sea lions or Northern Fur Seals. Northern Right Whale Dolphins are often seen in association with other cetaceans, most commonly with Risso's Dolphins and Pacific White-sided Dolphins.

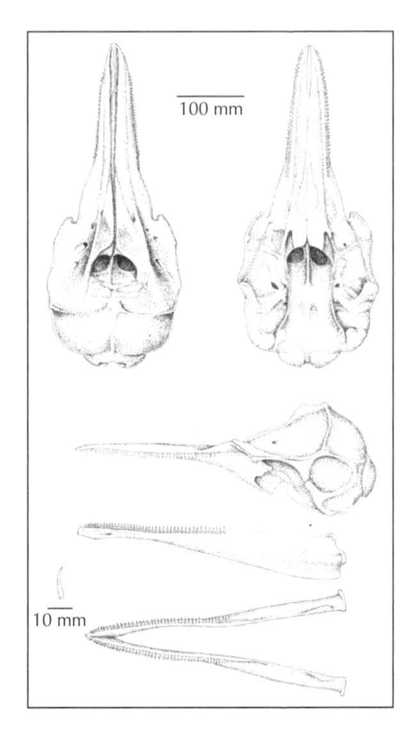

Distribution and Habitat

Northern Right Whale Dolphins inhabit warm to cool-temperate waters of the North Pacific between about 30°N and 51°N. Southern British Columbia appears to be the northern limit of the species' normal range in the eastern North Pacific. A few sightings have been recorded as far north as 55°N in the Gulf of Alaska and in the central Aleutian Islands, but these are considered extralimital. The habitat of the species is deep oceanic waters near or beyond the continental break and slope, and it is seldom seen close to shore unless the edge of the continental shelf is very near the coast.

The first record of the Northern Right Whale Dolphin in British Columbia waters was a school of some 200 animals encountered on February 13, 1970, about 100 km west of Ucluelet on the west coast of Vancouver Island. A specimen was taken from this school

Figure 74. The Northern Right Whale Dolphin in the foreground has normal colouration while the other displays an uncommon colour morph with more extensive white colouration. Photo: R. Palm, west of Tofino, Vancouver Island, September 1994.

Figure 75. Part of a large group of Northern Right Whale Dolphins off the northwestern tip of Vancouver Island. Photo: J. Ford, August 2007.

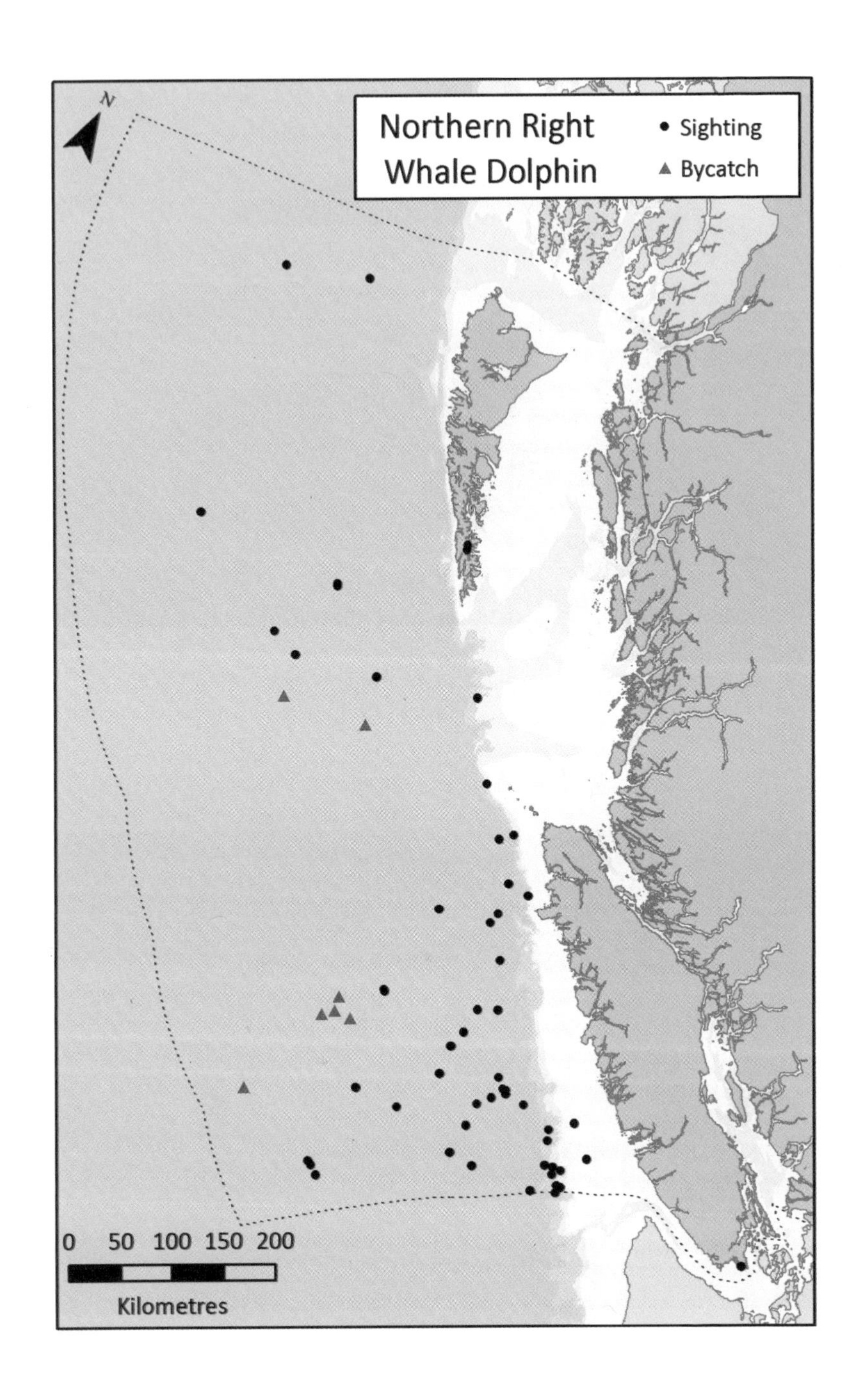

N
Northern Right
Whale Dolphin
Sighting
Bycatch
0 50 100 150 200
Kilometres

for the Royal BC Museum collection (RBCM 6982). Since then there have been 46 recorded occurrences of Northern Right Whale Dolphins in BC waters. Of these, 39 were sightings, most of which were in deep water beyond the continental shelf off the west coast of Vancouver Island. Several sightings have also been reported off the west coast of Haida Gwaii and in nearshore waters of Gwaii Haanas National Marine Conservation Area. One lone individual was seen repeatedly close to shore in McNeill Bay, Victoria, over several weeks in the summer of 1998. Seven reported occurrences involved incidental catches of Northern Right Whale Dolphins in a short-lived experimental drift-net fishery for Neon Flying Squid in 1986 and 1987. These catches were made 250–300 km off the west coast of Vancouver Island.

Off California, where the species is more common, there is a seasonal shift in the distribution of Northern Right Whale Dolphins northwards and offshore during summer, and southwards and inshore during winter. In BC waters most sightings have taken place during summer, but this is likely biased by the increased sighting effort at this time of year. There is probably an increased presence of Northern Right Whale Dolphins off our coast in warm-water years, such as during El Niño events.

Natural History

Feeding Ecology

The diet of Northern Right Whale Dolphins consists of fish and squid. A study of the stomach contents of Northern Right Whale Dolphins taken as bycatch in the high-seas Japanese drift-net fishery documented over 30 species of mesopelagic fishes, primarily lanternfish (myctophids) and squid. Stomach contents of dolphins stranded or collected in California were similar, consisting mostly of squid and lanternfish but also including representatives of the deepwater fish families Bathylagidae (deep-sea smelts), Melamphaidae (bigscale or ridgehead fishes) and Paralepididae (barracudinas). One stomach sample contained the remains of Pacific Saury, a pelagic schooling fish often found near the surface.

Behaviour and Social Organization

Northern Right Whale Dolphins are gregarious and typically found in groups of up to several hundred individuals – occasionally in gatherings of several thousand. In BC waters reported group sizes averaged 60 dolphins, with a range of 2 to about 400. Virtually nothing is known of their social structure. The swimming behaviour of Northern Right Whale Dolphins has two general modes: slow and fast. When moving slowly, individuals usually expose only the head and blowhole to breathe, creating very little surface disturbance. Fast-moving groups swim rapidly just below the surface, coming up to breathe during a series of low-angle leaps above the surface. These leaps are usually graceful, with the individual re-entering the water smoothly head-first. Occasionally the dolphins will belly-flop, side-slap or tail-slap upon re-entry. These percussive leaps are sometimes repeated dozens of times in a row.

Northern Right Whale Dolphins are quite timid compared to many other dolphins. They are usually wary of boats and avoid or flee from them, unless in the company of other species, such as Pacific White-sided Dolphins, that tend to approach vessels to ride their bow and stern wakes. This seems to embolden the Northern Right Whale Dolphins, and they will ride the waves alongside their more extroverted associates. When frightened, Northern Right Whale Dolphins flee explosively at high speeds of over 30 km per hour. Describing the collection of the first British Columbia specimen from a school of 200 dolphins in February 1970, RBCM biologists Charles Guiguet and W.J. Schick noted that "when the animal was shot, the entire school reacted violently by jumping, a spectacular display according to crew members".

Studies of the underwater acoustics of Northern Right Whale Dolphins indicate that their vocalizations consist exclusively of echolocation clicks and burst-pulse signals; they do not appear to produce whistles like most other ocean dolphins. Interestingly, complex patterns of burst-pulse signals were heard repeatedly from particular subgroups of individuals, suggesting that there may be individual or group-specific "signatures" in this species.

Life History and Population Dynamics

The life history details of Northern Right Whale Dolphins are known mostly from examination of animals taken as bycatch in high-seas drift-net fisheries in the central North Pacific. Both sexes reach sexual maturity at about 10 years of age, when males are about 215 cm and females 200 cm in length. Calves are born at a length of about 92 cm after a gestation of 12 months, and the calving interval is two years. The lifespan is at least 42 years. Killer Whales are very likely a source of natural mortality for Northern Right Whale Dolphins, but this species is one of the few oceanic dolphins that has not yet been documented as prey of these predators.

Exploitation

Northern Right Whale Dolphins have never been a target of major fisheries in any part of their range, although some are taken regularly in small-cetacean fisheries in Japan. There is no record of their exploitation in BC waters.

Taxonomy and Population Structure

The genus *Lissodelphis* consists of two species, the Northern Right Whale Dolphin and the Southern Right Whale Dolphin, *Lissodelphis peronii*, which is found in temperate waters of the southern hemisphere. The two species are remarkably similar except for their coloration – the southern species tends to have considerably more white on its head and sides than its northern cousin. It has been suggested that the two should be considered subspecies, but this has not been widely accepted.

Conservation Status and Management

Large numbers of Northern Right Whale Dolphins were killed incidentally in high-seas drift-net fisheries during the 1970s and 1980s, with mortalities in the central North Pacific estimated at 15,000–24,000 a year during the 1980s. High-seas drift-net fisheries ended under an international United Nations moratorium in 1993, but by then the abundance of Northern Right Whale Dolphins may have been reduced by as much as 30 per cent. Low levels of bycatch in drift-net fisheries continue in the Exclusive Economic Zones of several North Pacific countries, but no such fisheries are undertaken in Canadian waters. Current population abundance is estimated to be about 68,000 for the North Pacific. Off the US west coast, the estimate from ship surveys in 2005 and 2008 is around 8300 individuals, but no abundance estimate is available for BC waters.

The Northern Right Whale Dolphin is considered Least Concern on the IUCN Red List, Not at Risk in Canada by COSEWIC, and is on the BC Yellow List.

Remarks

The unimaginative common name for this graceful species is based on its lack of a dorsal fin, a feature shared with the Right Whale. The genus name for Northern Right Whale Dolphin is from the Greek word *lissos*, meaning "smooth", referring to the animal's finless back. The species name *borealis* is from the Latin word for "northern".

Selected References: Baird and Stacey 1991b, Carretta et al. 2011, Guiguet and Schick 1970, Jefferson et al. 1994, Leatherwood and Walker 1979, Rankin et al. 2007, Reeves et al. 2002.

Killer Whale *Orcinus orca*

Other Common Names: Orca, Blackfish.

Description

The Killer Whale is the largest member of the dolphin family (Delphinidae). Its robust body, tall and erect dorsal fin and unique coloration give it a very distinctive appearance. It has a relatively blunt head and a poorly defined beak compared to most other dolphins. Killer Whales are highly sexually dimorphic, with mature males up to 17 per cent longer and 40 per cent heavier than mature females. At maturity, the female's dorsal fin remains falcate and reaches a height of about 90 cm, but the male's fin tends to have a triangular shape (occasionally with a forward cant in old males) and can be twice the height of a female's. In both sexes the flippers are large and oval in shape, and they become disproportionately larger in adult males. The same is true of the tail flukes, the tips of which curl downwards in mature males.

The most distinctive feature of the Killer Whale is its striking coloration. The body is generally black dorsally and white ventrally. Above and behind the eye on each side of the head is a conspicuous elliptical white patch, referred to as the *postocular* or *eye patch.* On the rear flanks, the ventral white region forms patches that extend almost halfway to the dorsal ridge. At the rear base of the dorsal fin is a grey, variably shaped saddle patch. In neonates the normally white areas on the body have an orange hue, and the saddle patch is indistinct or absent for the first year of life. Considerable variation exists in the size and shape of white and grey patches among Killer Whale populations. In some populations, particularly in the southern hemisphere, Killer Whales have a faint grey cast over much of their body and a black cape in front of the dorsal fin.

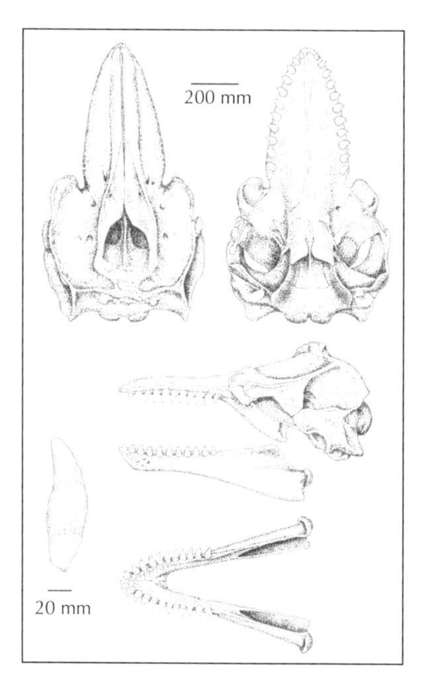

Measurements

length (metres):

Resident: male: 6.55 (6.08–7.25) n=13
female: 5.97 (4.70–6.44) n=20

Transient: male: 7.32 (6.81–7.95) n=4
female: 6.00 (5.60–6.71) n=5

Offshore: male: 6.52 (6.19–7.16) n=4
female: 5.49 (5.10–5.61) n=13

neonate (all populations): 2.37 (2.00–2.80) n=6

weight (kg):

male: 6600
female: 4700
neonate: 200

Dentition

10–14 pairs of large, conical teeth in both the upper and lower jaws (may be worn flat in Offshore Killer Whales).

Identification

With their distinctive black-and-white coloration and large dorsal fin (especially in adult males), Killer Whales are among the most recognizable of all cetaceans. In BC waters they are most likely to be confused at a distance with Risso's Dolphins, which have a relatively large dorsal fin for their body size, and Short-finned Pilot Whales, which are all black with a blunt head and generally found well offshore. False Killer Whales might also be confused with Killer Whales, but they also lack any white markings and are very rare in BC. Dall's Porpoises, with their striking white on black coloration, are often mistakenly reported as baby Killer Whales.

Distribution and Habitat

The Killer Whale is the most cosmopolitan of the cetaceans and one of the most widespread mammals on the planet. Killer Whales are

found in all oceans and most seas from the tropics to polar waters. They are sighted infrequently in most tropical areas and reach their highest densities in coastal, cold-temperate to subpolar waters with high productivity. There are notable concentrations along the northwestern coast of North America, around Iceland, along the coast of northern Norway and in the Southern Ocean around Antarctica. In the Antarctic, Killer Whales are commonly found up to the pack ice edge and may occur well into ice-covered waters. In the Canadian Arctic, Killer Whales are rarely seen in the vicinity of pack ice but do visit the region during the open water season in late summer. Information on the distribution of Killer Whales in most tropical and offshore waters is limited, but numerous scattered records attest to their widespread, if rare, occurrence.

In British Columbia, Killer Whales can be found in inshore channels and inlets, along the outer coast and in pelagic waters over the continental shelf and beyond. Three distinct, non-associating forms, or *ecotypes*, of Killer Whale are found in BC waters – these are known

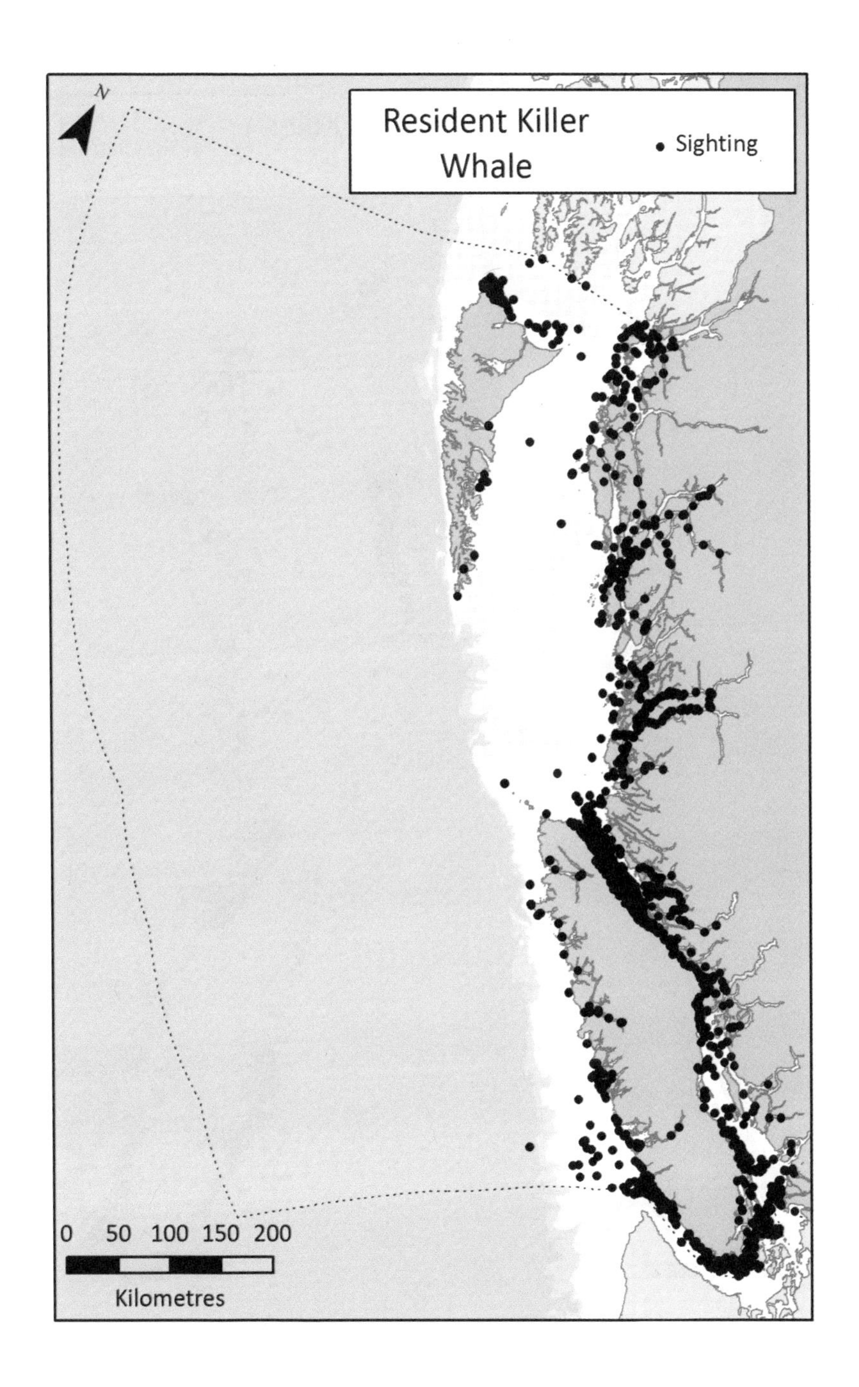
Resident Killer Whale
Sighting
N
0 50 100 150 200
Kilometres

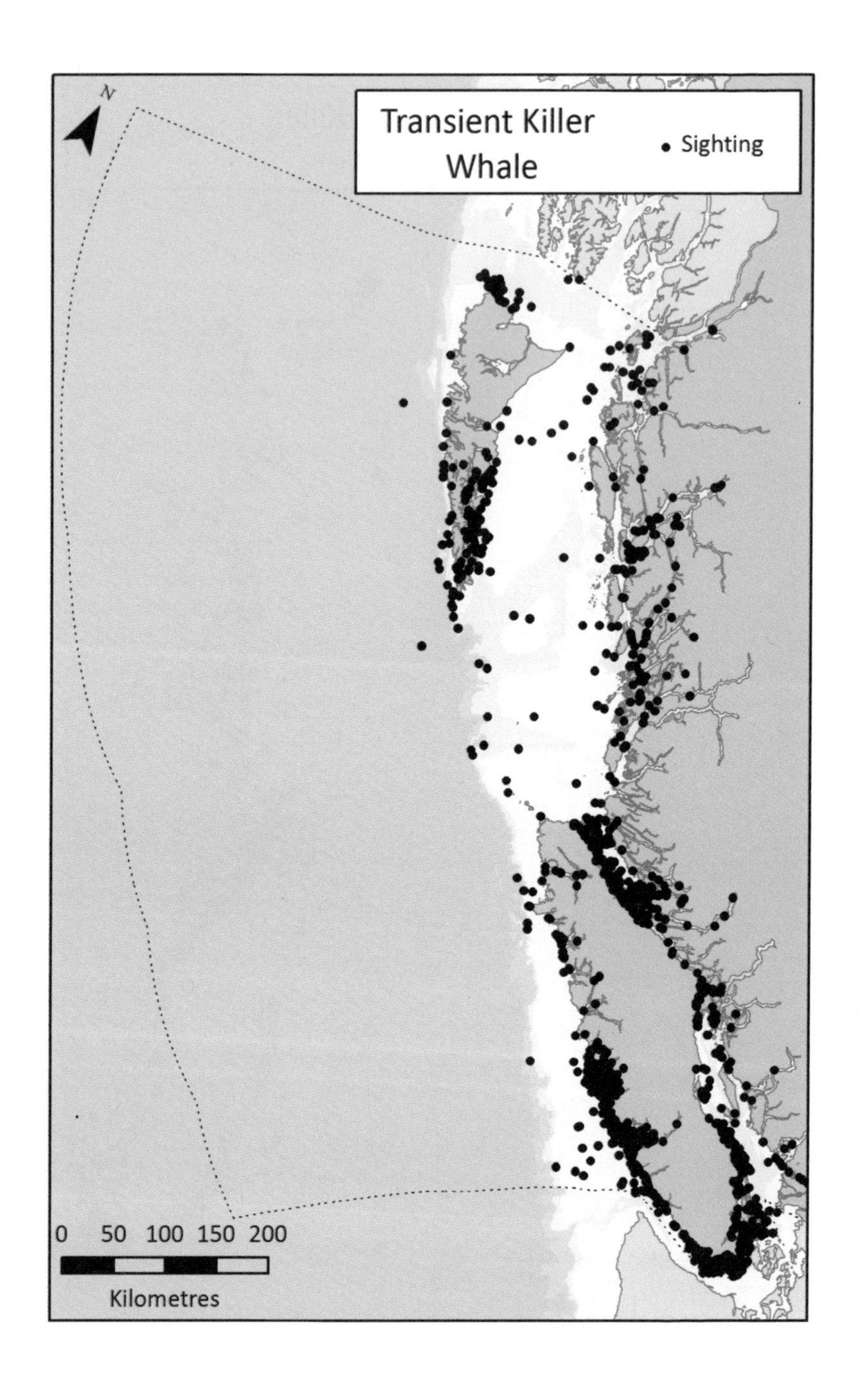
N
Transient Killer Whale
Sighting
0 50 100 150 200
Kilometres

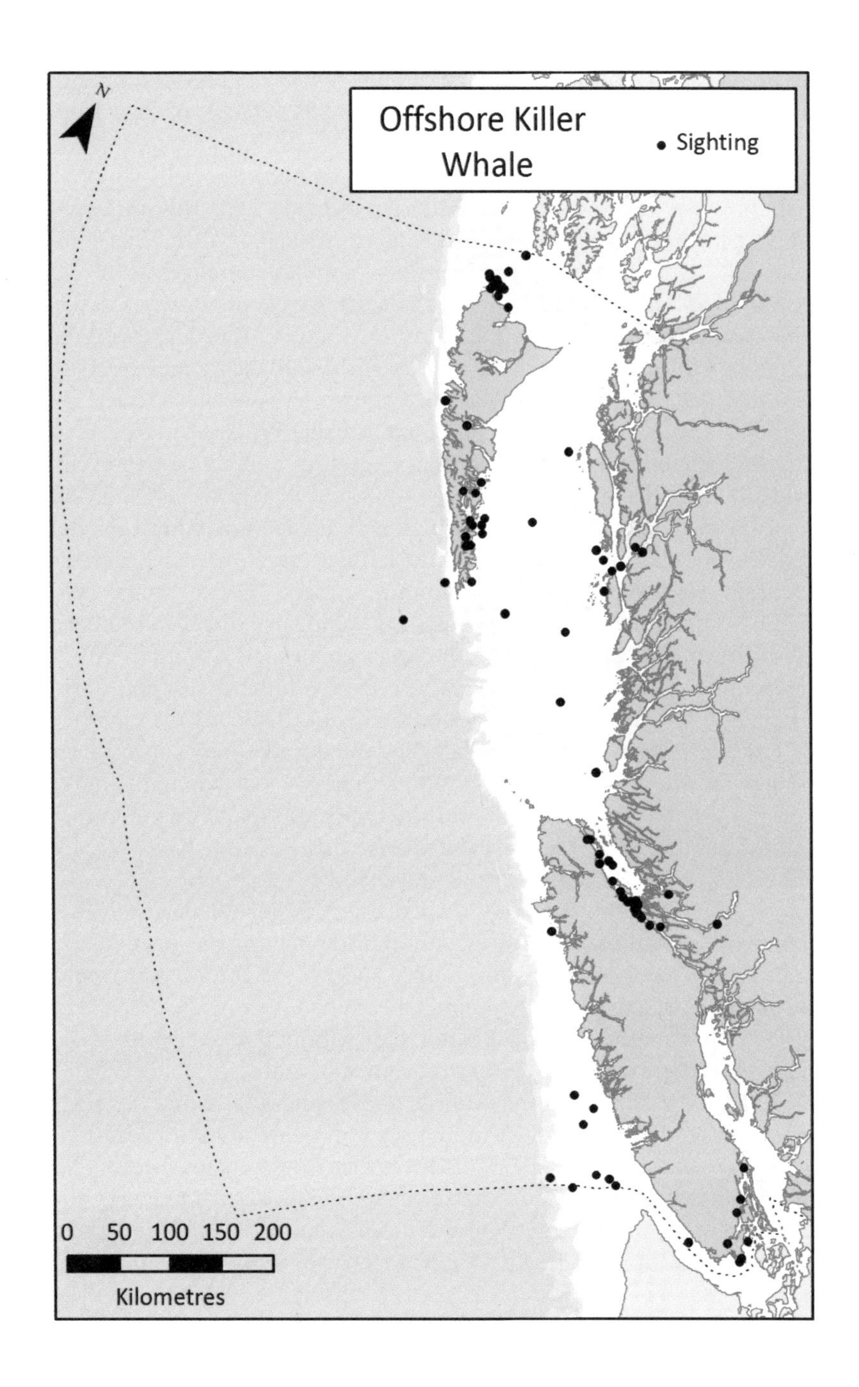
N
Offshore Killer Whale
Sighting
0 50 100 150 200
Kilometres

as *Residents*, *Transients* (or *Bigg's*) and *Offshores*. The distinctions among the three ecotypes are outlined here, and details of their feeding ecology are given in the next section.

Resident Killer Whales feed primarily on salmon, especially Chinook Salmon, and their movement patterns coincide with this preferred prey species. Two separate populations of Resident Killer Whales inhabit BC coastal waters. The *Northern* population ranges from the entrance to Juan de Fuca Strait to Glacier Bay, southeastern Alaska. The *Southern* population is found from Monterey Bay, California, to Chatham Strait, southeastern Alaska. Although the ranges of these two populations overlap, they are not known to associate or travel together. During summer and fall Resident Killer Whales congregate in coastal locations to intercept salmon migrating to their spawning rivers. Northern Residents are frequently found in Johnstone Strait, off northeastern Vancouver Island, which is designated as critical habitat for this population. Other important habitat for Northern Residents includes the waters of Dixon Entrance, eastern Hecate Strait, Caamaño Sound and Queen Charlotte Strait. Critical habitat for Southern Residents includes Juan de Fuca Strait, Haro Strait, the southern Strait of Georgia, Boundary Pass and Active Pass. Southern Residents also frequently forage during summer on Swiftsure Bank, off the entrance of Juan de Fuca Strait, and in fall they often make forays into Puget Sound. In winter Resident Killer Whales range widely along the outer coast and make only irregular and brief appearances in their summer critical-habitat areas. There is no evidence that Resident Killer Whales venture far offshore of the continental shelf.

Transient Killer Whales (also called Bigg's Killer Whales – see Remarks) feed almost exclusively on marine mammals, particularly seals and porpoises. Since these prey species are found year-round throughout BC waters, Transient Killer Whales are widespread and do not show marked seasonal shifts in distribution as seen with Residents. But there is a peak in the occurrence of Transient Killer Whales off Vancouver Island during August and September, which is when Harbour Seal pups are most numerous and vulnerable to predation. The Transient Killer Whale population that frequents inshore waters of British Columbia ranges from Glacier Bay, Alaska, to Oregon. Other populations of Transient Killer Whales occur to the west in Alaska and to the south in California. Individuals from those populations are occasionally seen in BC.

Offshore Killer Whales are found in continental shelf waters from southern California to the eastern Aleutian Islands, Alaska. They feed on bony fish and elasmobranchs, especially sharks. Offshore Killer

Whales are encountered far less often in BC waters than Residents and Transients – they make up only one per cent of all Killer Whale encounters in the photo-identification database at the Pacific Biological Station. In fact, there are just 103 sighting records for Offshore Killer Whales in BC waters, all since 1988. This is at least partly because these whales appear to prefer waters near the continental shelf slope off southwestern Vancouver Island and around Haida Gwaii, where there is relatively little sighting effort.

In June 1945 a group of 20 Killer Whales that stranded en masse at Estevan Point on the west coast of Vancouver Island was later determined from DNA analysis to have been Offshores. Although they rarely appear in waters off eastern Vancouver Island, Offshore Killer Whales tend to be identified and documented there whenever they do, so the number of encounters recorded for these inshore waters suggests more frequent use than is really the case. Prior to 1992, no Offshore Killer Whales had been seen in the waters of Johnstone Strait or the Strait of Georgia despite two decades of study effort by Killer Whale researchers. Since then, Offshore Killer Whales have ventured into those inside waters on at least 30 occasions. Offshore Killer Whales have been sighted in BC waters in all months of the year, but most sightings occur in summer. This may be due in part to increased sighting effort, but there does appear to be a seasonal shift northward. Researchers studying Killer Whales off California find Offshores mostly in the winter months and rarely in summer. The individuals identified off California are the same as those documented in BC waters, thus suggesting a seasonal movement along the coast.

Natural History

Feeding Ecology

The Killer Whale once had an almost mythical reputation as a ruthless predator with a prodigious appetite. The 19th-century whaler and naturalist Captain Charles Scammon wrote:

> *[Killer Whales] may be regarded as marine beasts, that roam over every ocean; entering bays and lagoons, where they spread terror and death among the mammoth balænas and the smaller species of dolphins, as well as pursuing the seal and walrus, devouring, in their marauding expeditions up swift rivers, numberless salmon or other large fishes that may come in their way.*

Although Scammon's observations are somewhat embellished, the Killer Whale is indeed the apex predator in marine ecosystems. As a species it can be considered a generalist predator, with an extremely diverse diet of almost 150 species of marine vertebrates and invertebrates. Globally, documented prey species include 31 cetaceans, 19 seals and sea lions, 44 bony fishes, 22 sharks and rays, 20 seabirds, 5 squid and octopuses, 1 sea turtle, and even 2 terrestrial mammals. But local populations of Killer Whales can have remarkably specialized diets and may forage selectively for only a very small subset of the prey species that Killer Whales overall are known to consume. This can be seen in the three genetically distinct and ecologically specialized forms of Killer Whales found in British Columbia waters. Residents feed primarily on fish, particularly salmon; Transients consume marine mammals; and Offshores eat sharks and other fish. These three ecotypes share the same waters but do not associate with each other and do not interbreed. Their dietary preferences, described in more detail below, are likely acquired cultural traits rather than genetically determined, and each ecotype is associated with a particular suite of features in its seasonal distribution pattern, social organization, behaviour and vocal repertoire.

Resident Killer Whales. Prey species identified from scales and bits of tissue collected from the water after a kill, as well as from stomach contents of stranded carcasses, indicate that the diet of Resident Killer Whales is predominately salmon. Although six species of Pacific salmon in BC waters are eaten by Resident Killer Whales, Chinook Salmon – one of the least abundant – is by far their preferred species (figure 76). This is likely due to the Chinook's large size, high fat content and year-round availability in the coastal range of Resident Killer Whales. Pink Salmon and Sockeye Salmon, considerably smaller in size but vastly more abundant during their spawning migrations through coastal waters, are surprisingly not significant prey items. Chum Salmon, second in size to Chinook, are important during fall. Resident Killer Whales also feed on bottom-dwelling fishes such as Pacific Halibut, Lingcod and Dover Sole, and beaks from Boreopacific Armhook Squid were found in the stomachs of two stranded Resident Killer Whales.

Transient Killer Whales. Eight marine mammal species have been documented as prey of Transient Killer Whales in BC waters. Of the 408 kills we have recorded to date, over half were of Harbour Seals and one quarter were of Harbour Porpoises or Dall's Porpoises. The remainder, in order of importance, involved Steller Sea Lions, Pacific White-sided Dolphins, Common Minke Whales, California Sea Lions and Northern Elephant Seals (figure 77). Larger whales found in BC, such as Humpback Whales, Sperm Whales and Fin Whales, are rarely

Figure 76. A Resident Killer Whale off southwestern Vancouver Island grasping a Chinook Salmon, the preferred prey of this ecotype. Photo: B. Gisborne, August 2008.

Figure 77. A mammal-hunting Transient Killer Whale in Johnstone Strait using its tail flukes to strike and kill a Pacific White-sided Dolphin. Photo: J. Towers, April 2012.

Figure 78. An Offshore Killer Whale near La Pérouse Bank, southwestern Vancouver Island, with a Blue Shark in its mouth.
Photo: B. Gisborne, August 2010.

attacked as they are either too dangerous or difficult to catch. But several attacks on Grey Whale calves have been observed in BC during their northward spring migration with their mothers. Predation on terrestrial mammals by Transients has also been witnessed, although this is rare. In June 1961, whales were seen feeding on a Sitka Black-tailed Deer carcass in Jackson Pass on the BC central coast, and in the 1990s a group of whales attacked a pair of Moose swimming in southeastern Alaska's Icy Strait; they killed and ate one of the pair. Transient Killer Whales also kill swimming seabirds, though they usually abandon the carcass rather than eat it. Species killed include the Common Murre, Marbled Murrelet, Rhinoceros Auklet and Surf Scoter. Predation on fish has not been observed, nor have fish remains been found in the stomach contents of stranded Transient Killer Whales. But we recently recovered numerous squid beaks from the stomachs of two Transients found stranded on Nootka Island and Price Island, indicating that the diet of Transient Killer Whales includes cephalopods. Squid species, identified from distinctive features of the beaks, included the Humboldt Squid, Boreopacific Armhook Squid, Boreal Clubhook Squid and Robust Clubhook Squid. These species are all much larger than the familiar Opalescent Inshore Squid (formerly Market Squid) – based on beak size, one of the Robust Clubhook Squid had a mantle length of more than one metre and a weight of more than 13 kg.

Offshore Killer Whales. The diet of Offshore Killer Whales is poorly known, but we suspect that sharks are important and possibly primary prey. On 26 occasions, my colleagues and I have collected samples from chunks of liver tissue that had floated to the surface from kills made at depth by Offshore Killer Whales. These samples, collected in Johnstone Strait, on Learmonth Bank in Dixon Entrance and in Prince William Sound, Alaska, were all identified by the Genetics Lab at the Pacific Biological Station to have been from Pacific Sleeper Sharks, large deepwater animals found in temperate waters of the North Pacific. Offshores have also been observed feeding on Blue Sharks and Pacific Spiny Dogfish off southwestern Vancouver Island (figure 78). From examination of stranded whales, it is evident that the teeth of Offshore Killer Whales typically suffer extreme wear, with the crowns being worn flat to the gum line and pulp cavities exposed, even in subadults. We believe this results from abrasion by the hardened denticles embedded in shark skin, which give it a sandpaper-like quality. Pacific Sleeper Shark skin is particularly coarse. We have never observed such tooth wear among Resident and Transient Killer Whales, even in very old individuals (figure 79). But Offshore Killer Whales do not restrict their predation to elasmobranchs. They have also been seen feeding on Pacific Halibut and Chinook Salmon in BC, and the remains of two Opah, large bony fish that occur in warm pelagic waters, were found in the stomach of an Offshore Killer Whale harpooned off California in the 1960s.

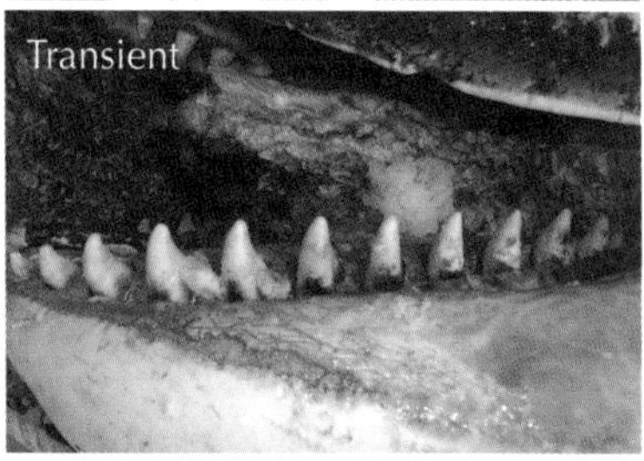

Figure 79. Extreme tooth wear is normal in Offshore Killer Whales, likely from eating sharks that have highly abrasive skin. Minimal tooth wear is typical in Resident and Transient Killer Whales. Photos: J. Ford (top two panels) and R. Palm (bottom panel).

Behaviour and Social Organization

Like most dolphins, Killer Whales are highly social, group-living animals. But the social structure and behaviour of Resident, Transient and Offshore Killer Whales differ considerably, and these differences are often clearly associated with the ecological specializations of the three types. Residents live in matrilines that are exceptionally stable in composition. A typical matriline is composed of an older female, her sons and daughters, and her daughters' offspring. Because the lifespan of females can reach 80 years and females have their first calf at 12–14 years, matrilines and pods may contain four and occasionally five generations of maternally related individuals. Over 35 years of demographic data based on individual photo-identification have shown that dispersal from the matriline is virtually absent in Resident Killer Whales – both males and females remain in their natal group for life. In no case has an individual been observed to leave its matriline and join another on a long-term basis, other than in a few rare cases involving orphans.

Members of Resident matrilines travel together, seldom separating by more than a few kilometres or for more than a few hours. Contact is maintained among matriline members using repertoires of distinct underwater calls, or dialects, that are unique to the group. Matrilines frequently travel in the company of certain other matrilines that are closely related, based on high degrees of call similarity, and likely shared a common maternal ancestor in the recent past. Matrilines that spend the majority of their time together are known as pods (figure 80). Pods are less stable than matrilines, and member matrilines may spend days or weeks apart. But matrilines still spend more time with others from their pod than with those from other pods. In British Columbia waters, Resident pods are composed on average of three matrilines (range: 1–11), with a mean total size of 18 whales (range: 2–49). Residents sometimes form large temporary aggregations, often called *superpods*, involving multiple matrilines and pods, especially at times when prey is abundant. A level of social structure above the pod is the *clan*, which is defined by the inter-pod similarity of vocal dialects. Clans are composed of pods that share a portion of their call repertoires; different clans have no calls in common. Pods belonging to a clan are likely descendants of an ancestral pod and their acoustic similarities reflect this common heritage. Call repertoires are passed on across generations by vocal learning, and calls actively or passively change in structure or use over time. The calls are retained in a lineage due to the lack of dispersal from matrilines.

Figure 80. Resident Killer Whales live in highly stable kin groups, known as pods, composed of one or more matriarchs and their offspring of several generations. Photo: B. Gisborne, southwestern Vancouver Island, September 2008.

Transient Killer Whale society lacks the closed, strictly matrilineal structure seen in Residents. Transients usually travel in groups of two to six, much smaller than the typical size of Resident matrilines and pods. In contrast to Residents, Transient Killer Whale offspring often disperse from the natal matriline for extended periods or permanently. Female offspring may leave their natal group around the time of sexual maturity and travel with other Transient groups. These young females usually give birth to their first calf shortly after dispersing. Once dispersed, these females may rejoin their natal matriline but generally only for brief periods after they have calves of their own. Male dispersal does take place, but less frequently than for females. The associations of Transient matrilines are very dynamic, and they do not form consistent groupings of related matrilines as do pods of Residents. In recent years, large temporary aggregations of 30 or more Transient Killer Whales have been observed around Vancouver Island. Also unlike Residents, Transient populations do not seem to be acoustically subdivided into clans. Instead, all whales in a population share a distinctive set of calls, although some additional calls or variants of shared calls may be specific to a subregion or portion of the population. The normally small size of Transient groups is likely a result of the foraging strategy of this ecotype. Transient Killer Whales mostly hunt other marine mammals with stealth – they swim quietly to prevent detection by their acoustically sensitive prey, then attack using the element of surprise. This strategy no doubt constrains group size, as larger groups would increase the probability of being detected by potential prey.

As with most details of their life history and behaviour, the social organization of Offshore Killer Whales is poorly understood. Their group sizes tend to be relatively large, certainly much larger than those of Transients and possibly larger on average than those of Residents. Groups of two to more than a hundred have been documented off the coast of British Columbia, with about half of these sightings involving twenty or more whales. These larger groups probably represent temporary gatherings of smaller social units, possibly related to prey density as in Residents. We have documented persistent bonds lasting over a decade between females and adult males, which likely represent mothers and their adult sons. But we have not observed long-term associations between reproductive females, as seen in the multi-generation matrilines of Residents. This suggests Offshores have a dynamic society with dispersal from the natal matriline as in Transients, but that they also often come together in fairly large aggregations as Residents do.

The distinct diets of Killer Whale ecotypes lead to corresponding contrasts in their foraging behaviour. When foraging, members of a Resident Killer Whale matriline or pod spread out, often over areas of several square kilometres, with individuals or small subgroups diving and surfacing independently while swimming generally in the same direction. They maintain contact and likely coordinate movements through the frequent exchange of loud underwater calls, which are effective to ranges of 10–25 km. When foraging in coastal inlets, channels and straits, individuals and small maternal groups usually forage along the shoreline, while other whales, particularly mature males, forage alone farther from shore and in deeper water. Foraging Residents dive for two or three minutes, usually to depths of less than 100 metres but occasionally to more than 250 metres.

Foraging Resident Killer Whales find their prey using echolocation, which may be effective for detecting Chinook Salmon at ranges of 100 metres or more. Residents do not appear to herd or capture fish cooperatively. Rather, prey capture is an individual effort with occasional cooperation from offspring, siblings or other close kin. Most salmon captured by adult females and subadults are brought to the surface where they are broken apart for sharing in the matriline or are fed to young. Adult males usually capture and consume salmon independently.

While in their core summer feeding areas, Resident Killer Whales divide their time between foraging, resting and socializing. They spend 50–65 per cent of their time foraging, and between foraging bouts group together to socialize or rest, which together represents about 30–40 per cent of their time. Resting whales line up abreast and make long, five- to six-minute synchronous dives separated by four or five surfacings and short, shallow dives. Socializing involves much physical interaction between individuals, including play behaviour on the part of juveniles. Spy-hopping, breaching, tail-slapping and flipper-slapping are common during socializing. Northern Resident Killer Whales also spend considerable time rubbing their bodies on certain sloped pebble beaches that have been used traditionally for many years. Although such beaches are found in several locations along the coast, the best known are in the Robson Bight (Michael Bigg) Ecological Reserve in Johnstone Strait. Rubbing appears to be a behavioural tradition unique to Northern Residents – Southern Residents, Transients and Offshores have not been observed to rub anywhere. It is unclear whether rubbing serves any physiological purpose or if it is simply a pleasurable social or recreational activity for the Northern Residents.

Unlike Residents, Transient Killer Whales forage in near silence, apparently to avoid detection by their marine mammal prey, all of which have good hearing. Since emitting calls or echolocation clicks would reduce the chance of hunting success, Transients seem to depend on passive listening to detect prey from a distance, likely cueing on the animals' vocalizations or swimming noises. There is little cost associated with the production of underwater sounds by Residents, as salmonids and most other fish have relatively low hearing sensitivity at the frequencies of sounds produced by Killer Whales, so they are unlikely to hear approaching whales at a distance.

Transient Killer Whales spend about 80 per cent of their time foraging, and they rest and socialize less often than Residents. Transients employ two fairly distinct modes of foraging – nearshore and open-water. When foraging near shore, the whales swim in relatively tight groups and follow the contour of the shoreline, rounding headlands and entering bays without hesitation. They often circle small islets and reefs, particularly those that serve as pinniped haulouts, particularly those of Harbour Seals. Residents, in contrast, forage along more direct routes, usually swimming from headland to headland. Foraging

Transients typically dive for twice the duration of Residents and may exceed 10 minutes. When foraging in open water, Transient groups may spread out over a larger area, with individuals swimming several hundred metres apart, often roughly abreast. Most prey species captured during open-water foraging are porpoises or dolphins, but seals or sea lions are also taken. Transients share the majority of their kills, likely to an even greater extent than Residents because of the larger body sizes of their marine mammal prey.

Transient Killer Whale group members use cooperative hunting tactics to catch and subdue their prey. Attacks on Steller Sea Lions, for example, are often prolonged events that can be risky for the attacking whales. These sea lions can be large (up to 1000 kg for males), with sizeable canine teeth that can inflict significant wounds when defending themselves. Groups of Transient Killer Whales attack single sea lions in open water by circling the animal to prevent it from reaching shore, while individual whales take turns ramming it, leaping on it, or striking it with their tail flukes. This can continue for one to two hours until the animal is sufficiently debilitated to be safely grasped, at which point it is drowned and shared among group members. The whales often hunt fast-swimming Dall's Porpoises using a cooperative "tag team" tactic where the whales take turns chasing the porpoise to exhaustion. Transients have been also been seen herding groups of 50 or more Pacific White-sided Dolphins into confined or shallow bays where individuals can be readily captured. A similar technique is used to hunt Common Minke Whales, which employ high-speed flight to avoid attack but are occasionally caught if they can be driven into an inlet or bay (figure 81). There is no evidence that particular individuals or groups of Transient Killer Whales specialize on particular types of prey, such as seals versus porpoises.

Like most dolphins, Killer Whales have a well-developed acoustic communication system, with the types and extent of vocalizations showing major differences among ecotypes. Resident Killer Whales frequently exchange loud, strident calls from stable repertoires of a dozen or more distinct call types. These learned call types or their variants form dialects specific to clans, pods and matrilines, and they therefore encode the matrilineal genealogy of individuals. This likely enhances the effectiveness of these calls as intra-group contact signals, especially when whales are dispersed and travelling in association with other matrilines or pods. These group-specific dialects may also play a role in preventing inbreeding. As there is no dispersal from the natal matriline, Resident Killer Whales would be at risk of inbreeding without a reliable means of distinguishing between kin and non-kin

Figure 81. Transient Killer Whales attacking a Minke Whale they have driven into the shallows at the head of Ganges Harbour, Salt Spring Island. Photo: D. Lundy, October 2002.

mating partners. Group-specific call repertoires appear to serve such a function, and genetic studies by Lance Barrett-Lennard as a UBC graduate student showed that Resident whales usually mate with individuals that are outside the pod or clan and are acoustically dissimilar.

Although Transient Killer Whales spend much of their time in silence while foraging, they are often very vocal while attacking and consuming their prey. Calling at such times likely carries little cost as stealth is no longer needed, and it may help coordinate cooperative attacks in the group or serve other social functions after the kill is made. Like Resident Killer Whales, Transients have repertoires of distinctive stereotyped calls. Unlike Residents, however, their repertoires generally do not differ among groups. Because there is dispersal from the natal matriline in Transients, group-specific calls would not be expected. Dispersal reduces the risk of inbreeding, so the need for an acoustic outbreeding mechanism may not be as great for Transients as it is for Residents.

The fish-eating Offshore Killer Whales are as vocal as Resident Killer Whales. Preliminary analyses indicate that Offshores produce calls that are distinct from any of those made by Residents or Transients, but it is not yet known whether any calls are specific to particular groups.

Life History and Population Dynamics

Long-term photo-identification studies of Killer Whales in British Columbia have provided 40 years of detailed demographic data, especially for Residents. This has led to a good understanding of Resident population dynamics and life history, and that of Transient and Offshore Killer Whales is likely similar. Females give birth to their first viable (defined as surviving its first year) calf at 12–14 years of age, following a gestation of 15–18 months. Calves are nursed for up to a year but may start taking solid food from the mother while still also being nursed. Females can produce a calf every three years, but because of relatively high calf mortality, intervals between viable calves average about five years. Females therefore produce, on average, about five surviving calves over their 25-year reproductive lifespan, which ends at about 40 years of age. Females then become post-reproductive, or reproductively senescent, for the rest of their lives. Female life expectancy is about 50 years but some may live to 80 or older.

Males reach sexual maturity at about 15 years of age, as indicated by a rapid growth of the dorsal fin. Such males are referred to informally as *sprouters*. Males continue to grow until they reach physical maturity at about 20 years. Average life expectancy for males is about 30 years, with a maximum longevity of about 50–60 years. Paternity tests have shown that males do not breed successfully until they are in their 20s, likely when they have attained social maturity and status.

Because of the length of time it takes for Killer Whales to attain sexual maturity, as well as the long interval between surviving calves and the prolonged post-reproductive phase of older females, the potential for population growth is limited to 3–4 per cent annually.

As the top predators in the oceans, Killer Whales themselves have no natural predators. Causes of death determined for stranded individuals include infectious diseases associated with the bacteria *Brucella*, *Staphlococcus*, *Aspergillus*, *Erysipelothrix* and *Edwardsiella*, as well as trauma from vessel strikes and an infection triggered by a fish hook perforating the esophagus.

Exploitation

At various times in the past, Killer Whales have been the targets of directed hunting, culling and persecution. Annual hunts that took place for many decades in Japan and Norway killed an average of about 40–50 whales per year in each country. In Japan whales were taken primarily for food, but in Norway the main motivation was to reduce competition with fisheries, and Killer Whale carcasses were also used for animal feed. An average of 26 Killer Whales were taken annually by Soviet whalers in the Antarctic from 1939 to 1975, with an exceptionally large kill of 916 animals in the 1979–80 season alone. This hunting ended in 1981, following an assessment by the Scientific Committee of the International Whaling Commission. Killer Whales are still hunted in small numbers in Japan, Greenland, Indonesia and some Caribbean Islands. Killer Whales were captured alive for display in aquariums in British Columbia and Washington during the 1960s and early 1970s (see below) and in Iceland during the late 1970s and 1980s. There have been two recent live-captures of Killer Whales in eastern Russia, also for aquarium display.

First Nations along the British Columbia coast have long revered the Killer Whale, and the species continues to play an important role in their art, mythology and social structure. Because of this stature, as well as the danger associated with hunting Killer Whales, they were seldom exploited by First Nations. However, young Nuu-chah-nulth hunters occasionally took *kakaw'in* – their name for Killer Whale – as proof of their whaling prowess. The meat and fat was considered good-tasting.

When Europeans arrived on the Northwest Coast, their attitudes shifted toward fear, malice and contempt of Killer Whales. Fishermen saw Killer Whales as threats to their lives and livelihood, and the animals were often shot intentionally. In response to concerns raised by sport fishing lodge guests in the Campbell River area in the late 1950s, the federal Department of Fisheries initiated a program to cull Killer Whales. In the summer of 1960, a 0.50-calibre machine gun was mounted on the shore of Seymour Narrows, north of Campbell River, with the intention of shooting Killer Whales to kill them or drive them from the area. But the whales failed to materialize in Seymour Narrows that summer and the program was abandoned without a shot being fired.

In the early 1960s a live-capture fishery developed in British Columbia to supply aquariums with Killer Whales for display. The first was live-captured unintentionally off Saturna Island in 1964. The Vancouver Aquarium had hired a harpooner to collect a Killer Whale to use in the creation of an accurate, life-sized model for display. A whale was harpooned, but the wound was non-lethal and the plan quickly changed to displaying the living animal. The whale, named

Moby Doll (later determined to be a young male), was moved to Vancouver where it was held in a floating net pen for three months before it succumbed to an infection. Public interest in Moby Doll was keen, and soon other aquariums were interested in acquiring live Killer Whales. Over the next 13 years, 62 additional whales were captured in the Strait of Georgia and adjacent US waters to supply aquariums in North America and Europe. An unknown number of whales were also killed in the capture process. Most captured whales were Residents, particularly Southern Residents, and this fishery further depleted these populations that were likely already depressed from years of intentional shootings. Public opinion against the live-capture hunt grew as it expanded, and the fishery in BC and Washington ended in 1977.

Taxonomy and Population Structure

Only a single species of Killer Whale, *Orcinus orca*, is currently recognized. But the existence of morphologically, ecologically and genetically distinct populations indicate that taxonomic revision may be warranted. Four discrete forms of Killer Whales – types A, B, C and D – have been described by biologist Robert Pitman and co-workers in waters around Antarctica, and they have suggested that these represent separate species. Types B and/or C may correspond to one or both of two putative species, *O. nanus* and *O. glacialis*, that were first proposed by Soviet researchers working in the Antarctic during the early 1980s. These proposed species have not been validated and accepted due to inadequate documentation and a lack of holotypes (a holotype is a specimen used as the basis for describing and naming a species).

Recently, geneticist Philip Morin and colleagues undertook an extensive study of genetic divergence among Killer Whales in the North Pacific, North Atlantic and Southern Ocean, based on variation in the complete mitochondrial genome. They found that Transients in the northeastern Pacific are by far the most genetically divergent and likely split from other Killer Whale lineages, including the sympatric Resident and Offshore ecotypes, about 700,000 years ago. They suggest that Transients, as well as the Antarctic types B and C, are sufficiently distinct that all should be considered separate species. The Society for Marine Mammalogy's Committee on Taxonomy considers Resident and Transient Killer Whales to be distinct subspecies, but neither these nor any potential new species have yet been formally described and named.

Conservation Status and Management

Despite the cosmopolitan distribution of the Killer Whale, it is not an abundant species. Over much of its range it is scarce, and Killer Whales are found in relatively large numbers only in cold, productive waters at high latitudes. Because of their wide distribution and generally low density, they are difficult animals to census. Photo-identification studies in coastal waters of the northeastern Pacific from the Aleutian Islands to California have yielded a total population count of about 3100 whales. Similar studies have identified about 450 Killer Whales off northern Norway, 115 around New Zealand, and approximately 900 in waters off the Russian Far East. Line-transect vessel surveys have yielded estimates of 8500 Killer Whales over an area of 19 million square km in the eastern tropical Pacific, and at least 25,000 in the Southern Ocean. After reviewing all available abundance estimates, biologists Karin Forney and Paul Wade suggested that there are at least 50,000 Killer Whales globally, but the total number is almost certainly higher because estimates are not available for large oceanic areas.

The abundance of Killer Whales in British Columbia is assessed using individual photo-identification, but methods vary with ecotype. Resident Killer Whale populations can be censused annually because most or all pods or matrilines can be found and their members photo-identified each year. Due to their tight social structure, any whale missing in more than a couple of good encounters with its natal matriline can be reliably considered dead. In 2013 the Northern and Southern Resident populations numbered 274 and 82 individuals, respectively. The Northern Residents have more than doubled their population size since the mid 1970s, growing at an average annual rate of 2.6 per cent. The Southern Resident population, however, is only about 20 per cent larger than it was in the mid 1970s, for an average annual growth rate of only 0.6 per cent.

For Transient and Offshore Killer Whales, their dynamic social structure and movement patterns mean that several years can pass between identifications of some individuals. As a result, the sizes of these two populations are less certain and must be estimated using statistical techniques. Transients seem to be composed of two subpopulations, inshore and offshore, that often mix. Whales belonging to the inshore subpopulation are encountered frequently and are thus fairly well known – 304 individuals have been photo-identified in this subpopulation. The outer coast subpopulation includes 217 individuals that have rarely been encountered, likely because they keep their distance from shore. Population modelling using mark-recapture techniques with photo-IDs indicate that the inner coast subpopulation of Transient Killer Whales has been increasing steadily at about three per cent a year since the mid 1970s. Similar modelling for Offshore Killer Whales resulted in an abundance estimate of about 300 whales and an apparently stable trend.

Killer Whales face a variety of threats from human activities. They have long been feared as dangerous predators or vilified as perceived or real threats to fisheries in many regions, and harassment or intentional shootings are still known to occur. Killer Whales began to remove fish from longlines in Sablefish fisheries in Prince William Sound, Alaska in the mid 1980s, which resulted in the shooting of several local Resident Killer Whales. Depredation has more recently spread to other fisheries in the Aleutian Islands and southern Bering Sea. Depredation of fisheries also occurs in the western Mediterranean, South Pacific, South Atlantic and the Southern Ocean, where it is a growing problem. Aside from being at risk of retaliation by fishermen, depredating Killer Whales can be injured by hooks ingested along with fish removed from longlines. In 2005 a Killer Whale stranded in Glacier Bay, Alaska, had died from an infection likely resulting from a perforation of its esophagus caused by a longline hook. Depredation of Pacific Halibut longline fisheries and salmon troll fisheries is occasionally reported in BC, but so far it has not become a serious management problem.

Other conservation concerns include direct effects of large-scale oil spills and other forms of toxic pollution. In Alaska 14 Killer Whales from a pod that had been seen swimming through oil slicks shortly after the *Exxon Valdez* oil spill disappeared. Although it is not possible to unequivocally attribute the deaths to oil exposure, the extreme mortality rate in this pod is difficult to explain with other known factors.

Persistent bio-accumulating contaminants are a potentially serious risk to some Killer Whale populations. Marine mammal toxicologist Peter Ross at the Institute of Ocean Sciences in Sidney, BC, analysed biopsy samples of blubber collected from Killer Whales in BC and found levels of polychlorinated biphenyls (PCBs) and other pollutants to be among the highest of any marine mammal in the world. As these lipid-bonding contaminants accumulate on their way up the food web, Transients, which are at a higher trophic level than Residents and Offshores, have the highest levels of PCBs. Although it is not yet known whether there is a direct effect of these PCBs on the health of Killer Whales, impairment of the immune system and reduced reproductive success are possible.

Other potential threats to Killer Whales are disturbance and risk of injury or death from vessels. This is of particular concern in areas of intense vessel activity, including whale-watching, such as Haro Strait and Johnstone Strait. The presence of boats moving in close proximity to Killer Whales has the potential to disrupt their normal activities. Noise from vessels may also interfere with the whales' echolocation and communication signals as well as the ability of Transient Killer Whales to detect prey through passive listening. The fact that Resident Killer Whales continue to return each summer to places like Haro Strait and Johnstone Strait indicates that they can adapt to a certain amount of disturbance, as these are busy waterways. But if chronic disturbance leads to reduced foraging efficiency and lower energy intake, the whales may be displaced from important feeding areas. Killer Whales show a particularly strong response to the intense sound from mid-frequency tactical sonar used on navy ships. Killer Whales are also vulnerable to being struck by fast-moving vessels. Although they seem to be adept at manoeuvring around boats and ships, several injuries and deaths due to vessel strikes have been documented in BC.

Because many Killer Whale populations are ecologically specialized, they are vulnerable to changes in their preferred food supply. Killer Whales are intelligent and adaptable animals, but their foraging strategies appear to be fixed behavioural traditions that are quite inflexible. Both the Northern and Southern Resident Killer Whale populations increased in size after the live-capture fishery ended in the mid 1970s, but they both experienced sharp declines due to greatly increased mortality rates and reduced calving rates in the late 1990s. These declines were strongly correlated with a coast-wide reduction in the abundance of Chinook Salmon, the Residents' primary prey species, during the same period. We have hypothesized that the reduced availability of Chinook caused nutritional stress, which led to the whales metabolizing their blubber reserves to make up the energy deficit. This in turn released PCBs and other stored toxins into the whales' systems, which led to impaired immune response to infections and disease, so increasing their mortality. The future well-being and recovery of Resident Killer Whales in BC depend on adequate supplies of Chinook Salmon and clean, quiet habitat in which to pursue their prey.

Mammal-hunting Transient Killer Whales have also experienced major fluctuations in the availability of their preferred prey. Both Harbour Seals and Steller Sea Lions were severely depleted in BC by culling during the late 19th and early 20th centuries, which aimed to reduce these animals' interactions with fisheries. This no doubt had a profound effect on Transient Killer Whales, which feed extensively on these species. Harbour Seals and Steller Sea Lions have recovered since control programs ended, and both are now at or above historical levels of abundance. This apparently has had a positive effect on Transient Killer Whales, which have been steadily increasing in numbers by about three per cent a year since monitoring began in the early 1970s. Transients were rarely seen in the Strait of Georgia in the 1970s, but now they're observed there on an almost daily basis. In fact, the number of individual Transient Killer Whales using the Strait of Georgia area each year is now greater than the total Southern Resident population.

The Killer Whale is listed on the IUCN Red List as Data Deficient, primarily as a result of the taxonomic uncertainty in what may be a species complex. COSEWIC has assessed the conservation status of each Killer Whale population, or Designatable Unit, separately. Northern and southern Resident populations are listed as Threatened and Endangered, respectively. The west-coast Transient and Offshore populations are both listed as Threatened. All four populations are on the BC Provincial Red list. DFO has developed recovery strategies for Resident, Transient and Offshore Killer Whale populations off Canada's Pacific coast.

Remarks

Those of us who study or simply enjoy viewing and learning about the lives of wild Killer Whales today owe much to the pioneering and visionary work of the late Dr Michael Bigg. In the early 1970s, Bigg was faced with the challenge of determining the status of Killer Whales in coastal waters of BC for Fisheries and Oceans Canada. At that time, almost nothing was known about the species, either in our area or elsewhere. Early in his study Bigg devised a novel field technique for studying the species – photographic identification of individuals using natural markings. This was a radical approach and some questioned whether it was even possible. But Bigg proved beyond any doubt that photo-ID is the key to understanding the lives of Killer Whales, and it is now the standard tool used in field studies of Killer Whales globally. For over 15 years Bigg, working closely with colleague Graeme Ellis, documented in meticulous detail the demography and population dynamics of Killer Whales in BC coastal waters. Bigg was driven by a passion to solve the mysteries of Killer Whale life history, and his enthusiasm was infectious. Of all the interesting facets of that life history, he was particularly fascinated by the relationship between the distinct populations of Killer Whales that he had dubbed Residents and Transients. Sadly, Bigg was never able to write up his studies on Transient Killer Whales – he died in 1990, at the age of 51.

In the more than two decades that have passed since Michael Bigg's death, the body of evidence that Transient Killer Whales represent a distinct species has become compelling. Although it may take some time before this is resolved and a new species is formally proposed, there is a growing consensus among researchers that Transient Killer Whales should be called "Bigg's Killer Whales". This would indeed be a fitting way of honouring the memory of this remarkable pioneer of Killer Whale science.

Selected References: Barrett-Lennard et al. 1996; Bigg et al. 1990; Carl 1946; Dahlheim and Heyning 1999; Deecke et al. 2005; Ellis et al. 2011; Ford 1989, 1991 and 2009; Ford et al. 1998, 2000, 2011, 2013; Ford, Abernethey et al. 2010; Ford and Ellis 1999 and 2006; Forney and Wade 2006; Monks et al. 2001; Morin et al. 2010; Morton 1990; Olesiuk et al. 1990a and 2005; Pike and MacAskie 1969; Pitman and Ensor 2003; Ross et al. 2000; Scammon 1874.

False Killer Whale *Pseudorca crassidens*

Other Common Names: Pseudorca.

Description

The False Killer Whale is a medium-sized toothed whale with a slender body and a relatively small, rounded or blunt head. The melon overhangs the lower jaw, especially in adult males, and there is no discernable beak. The dorsal fin is falcate, usually with a rounded tip, and is positioned at the midpoint of the back. The flippers have slightly rounded tips and a bulge or hump about midway along the leading edge, giving them a distinctive S-curved margin. The body colour is black or dark grey, with a faint grey blaze on the ventral surface between the flippers.

Measurements

total length (metres):

male: ~ 6
female: ~ 5
neonate: 1.6–1.9

weight (kg):

adult: 1500–2000
neonate: unavailable

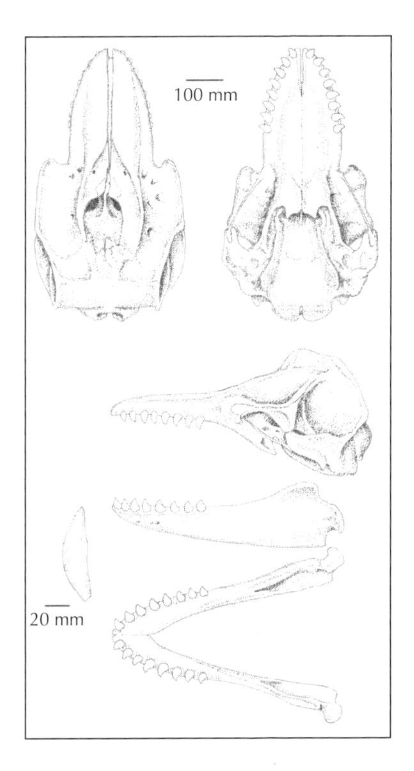

Dentition

7–12 pairs of large, conical teeth in both the upper and lower jaws.

Identification

False Killer Whales can be confused with other "blackfish", such as Short-finned Pilot Whales, Pygmy Killer Whales and Melon-headed Whales. Of these species, only the Short-finned Pilot Whale is known to occur in British Columbia waters. The head shape of False Killer Whales is more rounded and slender than the

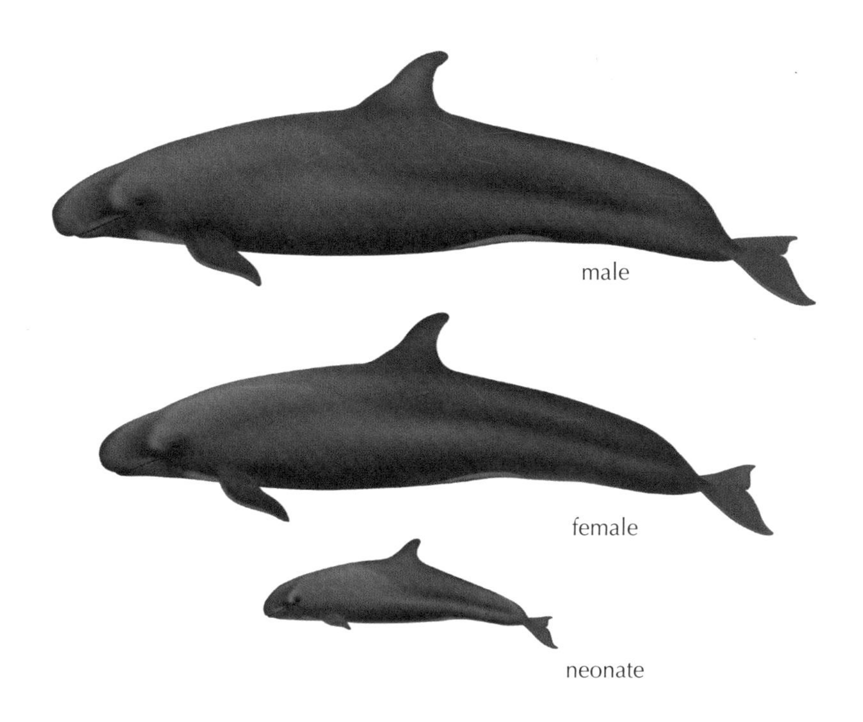

bulbous head of the Short-finned Pilot Whale, and the dorsal fin shape and position on the back differ substantially between the two species. But both False Killer Whales and Short-finned Pilot Whales are very rarely sighted in BC waters. At a distance, a small Killer Whale could be mistaken for a False Killer Whale, but closer inspection should reveal the grey saddle patch at the base of the Killer Whale's dorsal fin, as well as its distinctive white eye and flank patches.

Distribution and Habitat

False Killer Whales are found in tropical to warm-temperate waters throughout the world's oceans. Although they mostly occupy deep offshore waters, they are seen regularly around some oceanic islands, such as the Hawaiian Islands. Although generally confined to latitudes of 50° or less in both hemispheres, they occasionally are found in the colder waters of higher latitudes. In the eastern North Pacific they do not normally venture north of Mexico, and none was seen during extensive ship surveys by NOAA's Southwest Fisheries Science Center from 1991 to 2008, which covered waters out to 550 km (300 nautical miles) off the coasts of California, Oregon and Washington.

There are 115 published and unpublished records of False Killer Whales in British Columbia, but the species is far more rare in these waters than this tally would suggest. All sighting records are since 1987 and all involve single individuals. It is possible that most, if not all, of these records can be linked back to a single group of at least 12 False Killer Whales that was sighted repeatedly in Puget Sound, Washington, during May–July 1987. (This was a rare visit, as the only previous record of the species in Puget Sound was a lone individual shot and collected in 1937.)

The first Canadian record of the species was an adult male found stranded and dead on Denman Island on May 3, 1987. Another False Killer Whale, presumably a member of this group, was found dead in Puget Sound on May 5. Next, whale-watch operator Jim Borrowman photographed a single False Killer Whale westbound in Johnstone Strait on June 22. And then a 3.5 metre individual stranded alive in Ucluelet harbour on June 28 was successfully returned to deeper water by local residents (figure 82). During 1988 and 1989 there were multiple sightings of a single False Killer Whale in Barkley Sound. Then, from 1990 to 2003, repeated sightings were made of a lone individual in Vancouver harbour, Howe Sound, the Gulf Islands and, most frequently, near Roberts Bank shipping port and the Tsawwassen ferry terminal. In the summer of 2003, sightings of single False Killer

Figure 82. This male False Killer Whale survived after being stranded in Ucluelet harbour. Photo: M. Hobson, June 1987.

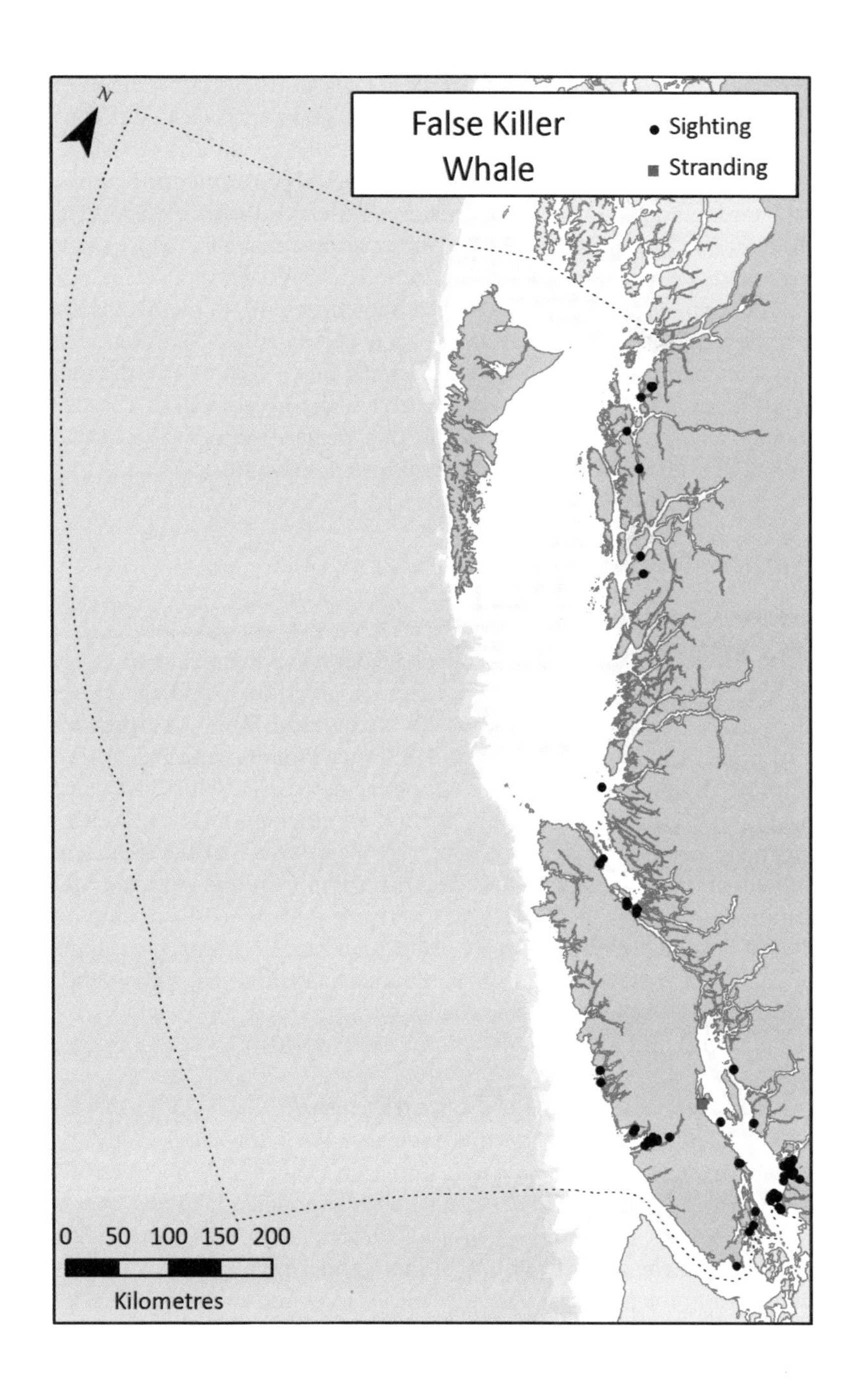
N
False Killer Whale
Sighting
Stranding
0 50 100 150 200
Kilometres

Whales were made at several locations off northeastern Vancouver Island and on the central and northern BC mainland coast, including Prince Rupert harbour. The most recent sighting record is of a single False Killer Whale that followed our research ship near Cape Caution on August 18, 2004, during a Cetacean Research Program survey. As discussed in Remarks, below, it is quite possible that all of these sightings since 1988 involved the same False Killer Whale.

On two occasions in August and September, 2010, the distinctive whistles of False Killer Whales were recorded by a Cetacean Research Program's autonomous acoustic recording buoy anchored to the seafloor about 5 km off Brooks Peninsula on the northwestern coast of Vancouver Island. It is possible that the species occurs at least occasionally in outer coast waters despite the lack of sightings.

Natural History

Feeding Ecology

False Killer Whales are opportunistic feeders, preying on a wide range of species and sizes of organisms, including various squid and fishes such as bonitos, Yellowfin Tuna and Mahi-mahi. They also prey on Pantropical Spotted Dolphins and Spinner Dolphins in the eastern tropical Pacific, but these incidents were associated with tuna seine fishing operations and it is not certain whether False Killer Whales normally eat dolphins. They have also been seen harassing Sperm Whales and Humpback Whales, and there is at least one report of predation of a Humpback Whale calf in Hawaii. The only information on diet of False Killer Whales in BC waters is from the stomach contents of the adult male stranded on Denman Island in 1987, which contained squid (either Magister Armhook Squid or Boreopacific Armhook Squid) and salmon (species not determined).

Behaviour and Social Organization

False Killer Whales are extremely social, usually travelling in groups of 20 to 100. Photo-identification studies in Hawaii have documented long-term associations and strong bonds among individuals over a 15-year period, but details of social structure are not yet known. It is thought that the high sociability of False Killer Whales is a contributing factor in the propensity for the species to strand en masse, often by the hundreds. The vocalizations of False Killer Whales are composed of a variety of high-pitched whistles and echolocation clicks.

Life History and Population Dynamics

Information on the life history of False Killer Whales comes from stranded individuals, animals taken in hunts in Japan and animals held in captivity. Females reach maturity at 8–11 years of age and 3.4–3.9 metres long, and males at about 18 years and 4.3 metres. Females give birth following a gestation of 14–16 months and nurse their calves for up to 2 years. Females older than about 45 years are post-reproductive. The reproductive rate may be quite low in this species, with one study estimating the average calving interval at nearly 7 years. The maximum known lifespan calculated from growth layers in teeth is 58 years for males and 63 years for females.

Exploitation

Conflicts with fisheries, especially a tendency to take fish from longline gear, have made this species a target of directed drive fisheries in Taiwan and Japan. Some are also hunted opportunistically by harpooning in the Caribbean, Indonesia, Japan and Taiwan. There is no record of False Killer Whales being taken historically or recently in British Columbia.

Taxonomy and Population Structure

Studies on genetic population structure have shown both broad-scale (between oceans) and smaller-scale (within oceans) restrictions on gene flow, but no subspecies of False Killer Whales have been proposed. A small, nearshore and insular population of False Killer Whales around the main Hawaiian Islands is genetically distinct and evidently reproductively isolated from an offshore population in the region.

Conservation Status and Management

Killing to reduce conflicts with fisheries (i.e., culling) may have reduced False Killer Whale numbers in Japan and Taiwan, and incidental catches of small numbers of False Killer Whales in gillnet fisheries have been reported from various regions such as Brazil, Australia and China. These whales are occasionally hooked in longline fisheries, presumably when they attempt to take fish off lines, and injuries from hooks can be serious, even lethal in some cases. False Killer Whales feed on fish species such as tuna and Mahi-mahi that are the focus of human fisheries and therefore may be depleted due to overfishing in some areas, but the potential effects of prey depletion on False Killer Whales are unknown.

The species is considered Data Deficient on the IUCN Red List and Not at Risk in Canada by COSEWIC. It is considered Accidental on the BC List.

Remarks

As described above, the appearance of False Killer Whales in British Columbia waters in recent decades may be ascribed to a single group that wandered into the region in 1987. Prior to that year there was no record of the species in BC and only one record of a single individual in 1937 from adjacent Washington waters. Starting in May 1987 there were two strandings and a series of sightings that are all probably linked to the arrival of that group. What became of the group is unknown, but it appears that at least one surviving individual remained in local waters for the next 17 years. This whale, a male, live stranded in Ucluelet harbour on the west coast of Vancouver Island in July 1987, and was kept moist by local residents until he refloated and swam off on the next tide (figure 82). He was sighted repeatedly during 1988 and 1989 in Barkley Sound, and became well known to local boaters due to his habit of approaching boats and surfing in their wake. The whale developed a particular attachment to a classic old Union Steamship vessel, the MV *Lady Rose*, which at the time operated a regularly scheduled ferry service in Barkley Sound. He would frequently follow the *Lady Rose* on its run between Port Alberni and Bamfield and came to be called Rufus by the ship's crew and other locals. This bond between the *Lady Rose* and Rufus strengthened during the remaining months of 1989 and early 1990, and the whale became a regular travelling companion.

On May 10, 1990, a group of students and educators from the Vancouver Aquarium were on an outing in English Bay when they observed and photographed a lone False Killer Whale. This was the first-ever sighting of this species in Vancouver harbour, but it became a regular occurrence from that day on. Over the remainder of 1990 the whale was reported dozens of times in various locations in and around Vancouver harbour. It often followed the SeaBuses on their trips back and forth across the inner harbour, or swam alongside freighters and water taxis (figure 83). It became known as Willy the Whale by waterfront workers, before that name became famous in the movie *Free Willy*.

Figure 83. Willy the Whale (likely the same whale shown in figure 82) frequented the waterfront of Vancouver harbour and Roberts Bank coal port near Tsawwassen from 1991 to 2003.
Photo: Cetacean Research Program, PBS.

Willy the Whale's attraction to ships and habit of energetically breaching and splashing around boats were very much like those of Rufus from Barkley Sound. Interestingly, sightings of Rufus in Barkley Sound ended abruptly in early April 1990, and no False Killer Whale has been seen there since. Suspecting that these two whales could actually be the same animal, I spoke with the owner of the *Lady Rose* and discovered that the ship had made a rare trip from Barkley Sound to Vancouver in April 1990 to attend a Union Steamship reunion at the Vancouver Maritime Museum. It seems very likely that, unbeknown to its crew, Rufus accompanied the ship on its 300-km journey to Vancouver. He must then have lost track of the *Lady Rose* after the ship tied up in the Museum's heritage harbour, and then shifted his interest to other vessels operating in Vancouver harbour.

After a year in Vancouver harbour, Willy evidently departed because sighting reports stopped in May 1991. But new sightings of a lone False Killer Whale soon began coming in from other areas. Although it can't be confirmed that these were all the same whale, the timing and locations of sightings are all consistent with that of a single individual, and descriptions of the whale's attraction to vessels, particularly ferries, are all quite similar. It appears that Willy spent most of the rest of 1991 and 1992 in Puget Sound, first at the Point Defiance ferry terminal and later near the ferry landing on Lummi Island. The whale became well known to ferry travellers at Lummi Island and was given yet another name, Foster. In the fall of 1993 the whale disappeared from Lummi Island and became a regular sight at Tsawwassen ferry terminal and adjacent Roberts Bank shipping terminal just south of Vancouver, where it interacted with tugs and other vessels. Except for occasional visits back to Vancouver harbour and the waters near Horseshoe Bay in West Vancouver, Willy made the Tsawwassen ferry terminal and Roberts Bank port his home for the next decade. Then, in April 2003, sightings ended at Tsawwassen, but a sequence of new sightings suggested that the whale had made a long excursion to the north between May and July, eventually reaching Juneau, Alaska, before returning to the Strait of Georgia in August. The last recorded sighting of a False Killer Whale in BC waters was a lone individual – possibly Willy – that followed our survey ship off Cape Caution, north of Vancouver Island, in August 2004.

The 15-year association of this lone False Killer Whale with boats and people is similar to other documented cases involving solitary whales and dolphins in other regions of the world. These individuals are typically members of highly social species but, for some reason, appear to have lost their group. In the absence of social bonds with others of their kind, they seem to transfer their behavioural affiliation to mobile objects in the vicinity – often a ferry or tug plying a regular course that becomes predictable and familiar to the whale. The best-known example of this was a Risso's Dolphin known as Pelorus Jack, which for several years regularly escorted a ferry between the north and south islands of New Zealand.

Selected References: Baird 2009, Baird et al. 1989, Barlow and Forney 2007, Odell and McClune 1999, Osborne et al. 1988, Stacey and Baird 1991a, Taylor et al. 2008.

Striped Dolphin *Stenella coeruleoalba*

Other Common Names: None.

Description

The Striped Dolphin has the basic body shape typical of small oceanic dolphins, though it is slightly more robust than many of them. It has a moderately long beak that is demarcated from the melon by a distinct crease. The dorsal fin is tall and falcate, and the flippers are curved and taper to a sharp point.

The colour pattern of the Striped Dolphin is striking. The beak, flippers, tail flukes and back are dark bluish-grey or black, the sides of the torso are light grey, and the throat and belly are white. There are two distinguishing thin black stripes on each side of the body: one starts at the eye and runs diagonally toward the genital region, dividing the light grey side from the white belly; the other starts below the eye and extends to the flipper. A thin accessory stripe usually branches downward from the side stripe behind the eye. There is normally also a light grey blaze below and anterior to the dorsal fin, which splits the overall dark grey or black background.

Measurements

length (metres):

male: ~ 2.5
female: 2.1–2.2
neonate: 0.9–0.95

weight (kg):

adult: ~ 150
neonate: 7–11

Dentition

40–55 pairs of slender, pointed teeth in each jaw.

Identification

Although similar in shape and size to other small oceanic dolphins, Striped Dolphins can be easily distinguished at sea by the unique striped pattern sweeping backwards from the eyes, as described above (figure 84). In BC waters they are most likely to be confused with the common Pacific White-sided Dolphin and the rarely seen Short-beaked and Long-beaked Common Dolphins. But the colour pattern on all of these species is distinctive and different. No sighting of a living Striped Dolphin has been reported in BC waters to date.

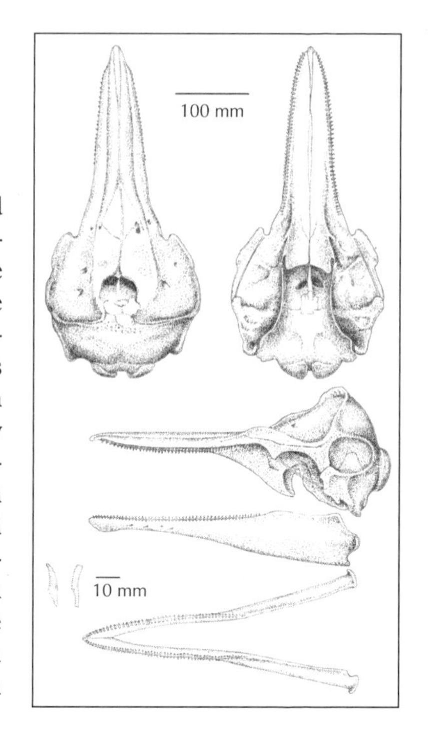

Distribution and Habitat

Striped Dolphins are widely distributed in tropical to warm-temperate waters of the Pacific, Atlantic and Indian oceans, as well as in most adjacent seas. Sightings have been reported in waters of 10–26°C, but most are in areas with surface temperatures ranging from about 18–22°C. Striped Dolphins generally occur in deep-ocean regions and are found inshore only where deep waters extend close to a coast. In the North Pacific the Striped Dolphin is the most northerly representative of the genus *Stenella*, which includes the Spinner Dolphin and Pantropical Spotted Dolphin. Striped Dolphins are associated with the central North Pacific gyre and with productive regions that have upwelling areas in the eastern tropical Pacific and along the edges of the California and Kuroshio current systems. In the eastern North Pacific they range from the equator to British Columbia.

Striped Dolphins are rare in BC waters. The 14 confirmed records for the species between 1948 and 2011 are all of stranded animals or skeletal remains from Vancouver Island (see Appendix 2 and figure 85). One was just a skull and several vertebrae snagged in a trawl net

Figure 84. A Striped Dolphin leaping in the wake of a ship off the California coast. Photo: C. Oedekoven, September 2005.

about 12 km off Kyuquot Sound on the west coast of Vancouver Island. A dolphin identified as "*Stenella* sp." was recovered entangled in a drift net during an experimental fishery for flying squid about 400 km west of Ucluelet. This was most likely a Striped Dolphin, as other species in the genus are usually found much farther south in tropical or subtropical waters. To date there has not been a single verified sighting of a living Striped Dolphin in BC waters, and therefore the species is considered extralimital off Canada's west coast. All sightings of Striped Dolphins during extensive cetacean surveys off the US west coast in 1991–2008 were south of 42°N, which is the latitude of the northern border of California. However, stranded individuals have been recorded in both Washington and Oregon.

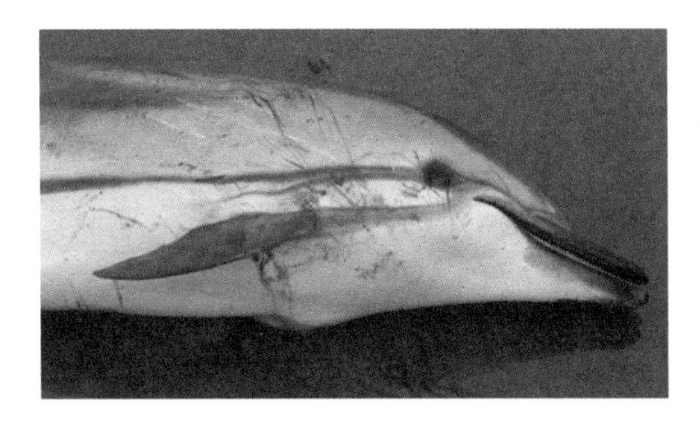

Figure 85. A male Striped Dolphin stranded on Long Beach, Vancouver Island, on March 13, 1972. Photo: Pacific Rim National Park Reserve.

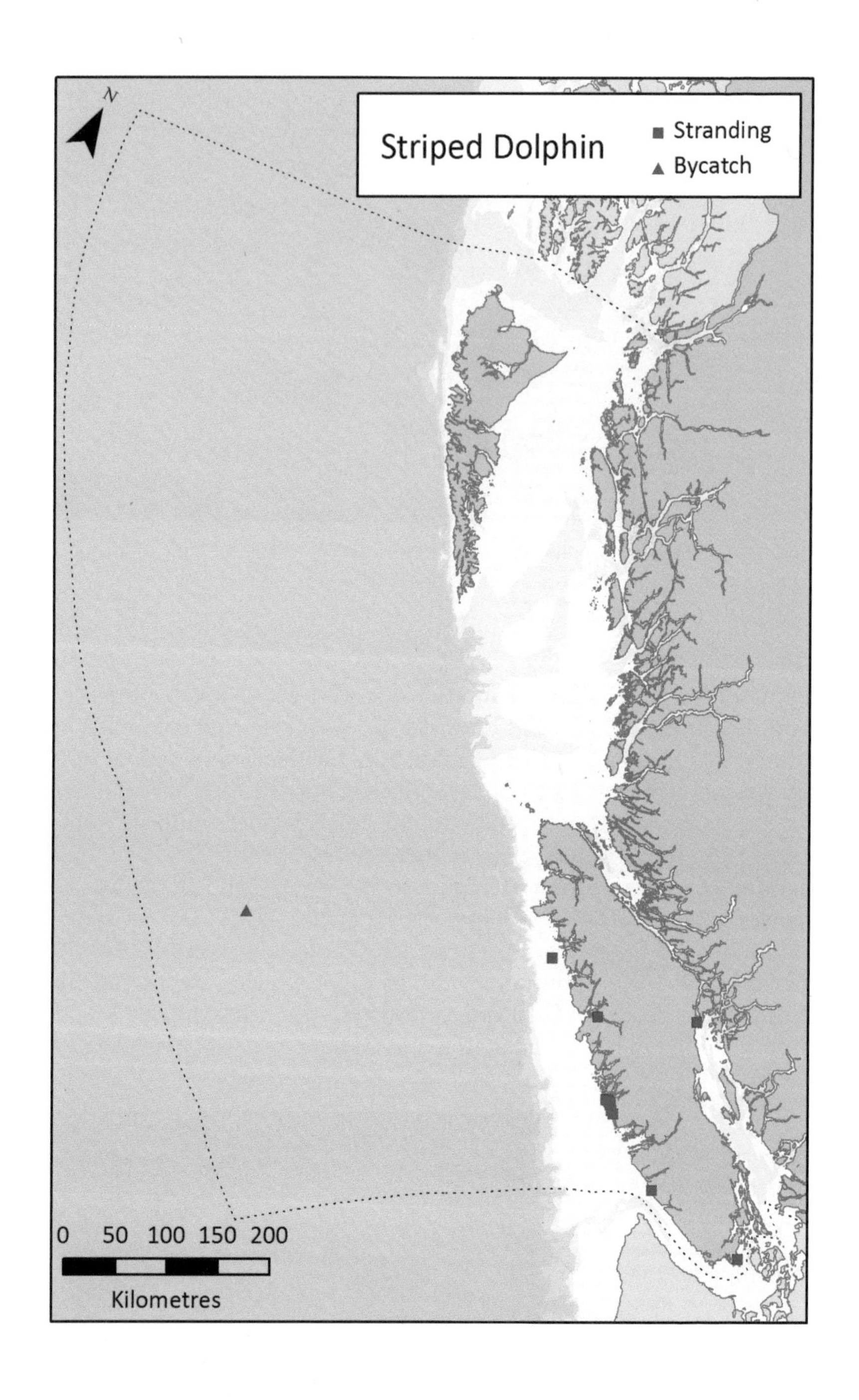
N
Striped Dolphin
Stranding
Bycatch
0 50 100 150 200
Kilometres

Natural History

Feeding Ecology

Striped Dolphins feed on a variety of small (less than 13 cm) pelagic fishes and squid. Prey fish species include lanternfish (family Myctophidae), anchovies and cod. The dolphins are reported to feed anywhere in the water column to depths of 700 metres.

Behaviour and Social Organization

Striped Dolphins are energetic, fast-swimming animals that often travel in dense schools of 100 or more. The composition of schools is variable, with some containing only adults and others both adults and juveniles. They are acrobatic – displaying many types of leaps, breaches and other aerial manoeuvres – and will occasionally ride on the bow waves and stern wakes of vessels. Striped Dolphins produce a variety of high-frequency whistles and pulsed calls for communication as well as clicks for echolocation.

Life History and Population Dynamics

Females become sexually mature at 5 to 13 years of age, and males at 7 to 15 years. The average length at sexual maturity is 2.1–2.2 metres for both sexes. Calving is seasonal, following a gestation of 12–13 months. Maximum longevity is estimated to be about 60 years.

Exploitation

Striped Dolphins have been the targets of drive and hand-harpoon hunting in Japan for over 150 years, and annual catches have exceeded 15,000–20,000 in some years since 1978 (when catches started being recorded). The annual take in recent years has averaged about 500 animals. This long history of exploitation has no doubt had a significant effect on dolphin numbers around Japan. Striped Dolphins are taken occasionally in local hunts for human consumption in Taiwan, Sri Lanka, the Solomon Islands and St Vincent. In the past, some were also taken for food in the Mediterranean and northeastern Atlantic by fishermen from Spain, Italy and France. There is no history of exploitation of Striped Dolphins in the northeastern Pacific.

Taxonomy and Population Structure

Variations in the Striped Dolphin's skull morphology and body size have been documented in several regions, and genetic evidence suggests that population subdivision may be warranted. The genus *Stenella* is in need of taxonomic revision. The obsolete species names *Stenella styx* and *S. euphrosyne* are considered synonyms of *S. coeruleoalba*.

Conservation Status and Management

The Striped Dolphin is very abundant in several oceanic regions. Over 500,000 have been estimated for the western North Pacific and about 1.5 million for the eastern tropical Pacific, with surveys off California yielding estimates of about 19,000 in recent years. Striped Dolphins are vulnerable to incidental mortality in net fisheries. Bycatch in drift-net, seine and trawl fisheries is widespread, including in the western Mediterranean, northeastern Indian Ocean, eastern tropical Pacific and northeastern Atlantic. In some areas, Striped Dolphins prey on fishes and squid considered commercially valuable in human fisheries. In the western Mediterranean, several important prey species of Striped Dolphins have been depleted and food limitation may be a growing threat to this population.

The Striped Dolphin is designated Least Concern on the IUCN Red List and is considered Not at Risk in Canada by COSEWIC. It is on the BC Yellow List.

Remarks

The specific name *coeruleoalba* refers to this dolphin's striking coloration. It is derived from the Latin *caeruleus* for "sky blue" and *albus* for "white". The common name Striped Dolphin was often used in the past for the Pacific White-sided Dolphin. And in *The Mammals of British Columbia* Cowan and Guiguet refer to the only BC specimen of *Stenella coeruleoalba* that existed at the time as a Long-beaked Dolphin.

Selected References: Archer and Perrin 1999, Baird et al. 1993, Carretta et al. 2011, Cowan and Guiguet 1965, Hammond et al. 2008b.

Harbour Porpoise *Phocoena phocoena*

Other Common Names: Common Porpoise.

Description

The Harbour Porpoise is the smallest cetacean in British Columbia waters, with a maximum length of about two metres. The body is stocky, with a girth of about two-thirds of its length in adults. Females are slightly larger than males at maturity. The Harbour Porpoise has a blunt, short-beaked head and a broad-based, triangular to slightly falcate dorsal fin situated midway along the back. Several rows of small bumps, or tubercles, line the leading edge of the dorsal fin. The flippers are small with slightly rounded tips, and the flukes have a prominent median notch and concave trailing edge.

The Harbour Porpoise has a counter-shaded colour pattern that is generally medium to dark grey or brownish-grey on the back, lighter grey on the sides and white on the belly. There are variable amounts of grey streaks, flecks or blotches on the sides. The flippers and flukes are dark grey, and at least one, sometimes several, grey stripes of variable width extend from the corner of the mouth to the flipper. There is no consistent difference in colour or body shape between the sexes.

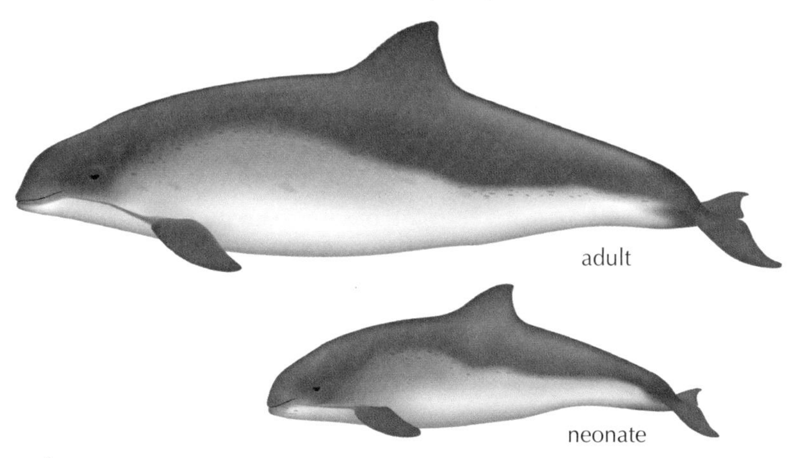

Measurements

length (metres):

male: 1.43 (1.27–1.86) n=23
female: 1.60 (1.24–1.98) n=50
neonate: 0.70–0.75

weight (kg):

male: ~ 50
female: ~ 60
neonate: 5–6

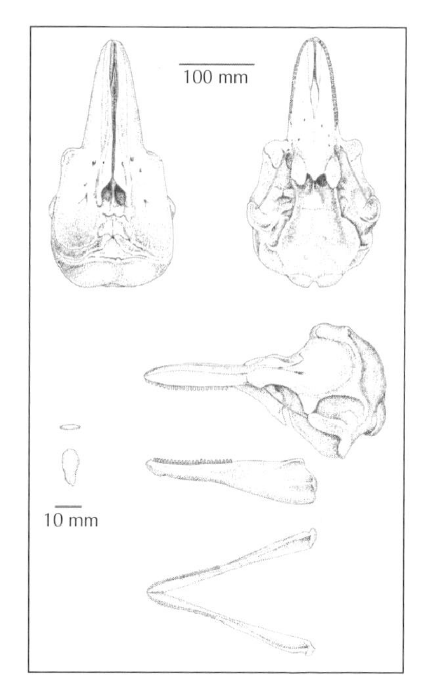

Dentition

Small, spade-shaped teeth, 22–28 pairs in the upper jaw, 21–25 pairs in the lower jaw.

Identification

Harbour Porpoises are fairly easy to identify at sea, although they can be hard to spot in anything but calm conditions due to their diminutive size and low-profile behaviour. They are usually found alone or in small groups of two to five in areas of shallow water (less than 150 m). Only the triangular dorsal fin and dark grey back are exposed when the animal "slow rolls" at the surface to breathe (figure 86). Harbour Porpoises are timid and wary of moving boats, and they rarely exhibit any aerial or percussive behaviour such as breaching. Dall's Porpoises are larger, have a black-and-white dorsal fin and resemble a square rolling at the surface (due to their robust and deep tail stock), unlike the smooth roll typical of Harbour Porpoises. And Dall's Porpoises often streak through the water and surface with a V-shaped rooster tail (see figure 89), whereas Harbour Porpoises seldom swim at high speed and usually do not make splashes when doing so. Hybrids between Dall's Porpoises and Harbour Porpoises are seen regularly off southern Vancouver Island. These hybrids lack the white side patches of Dall's Porpoises and could be mistaken for Harbour Porpoises. Pacific White-sided Dolphins are occasionally seen alone or in small groups as well, but they are noticeably larger than Harbour Porpoises and have a falcate, two-tone black and grey dorsal fin. There are other differences between the two species in body shape, markings and behaviour.

Figure 86. Harbour Porpoises typically show little more than their dorsal fin and back when surfacing. Photo: C. McMillan, Weynton Passage, May 2011.

Distribution and Habitat

Harbour Porpoises are found in cool-temperate to subarctic coastal waters of the North Pacific (including Okhotsk and Bering seas), the North Atlantic (including North and Baltic seas) and the Black Sea. They usually inhabit bays, harbours, estuaries and shallow offshore banks with depths of less than 150 metres, but they can occasionally be found in deeper waters over the continental shelf. Harbour Porpoises prefer areas with surface temperatures of less than 16° C. In the North Pacific, they are distributed from Point Conception, California, and northern Honshu, Japan (about 34°N), to the southern Beaufort and Chukchi seas (72°N). In British Columbia, Harbour Porpoises may be seen year-round in shallow inshore waters, as well as along the outer coast throughout the region. They are occasionally found in shallow areas of Queen Charlotte Sound, Hecate Strait and Dixon Entrance, and on Swiftsure and La Pérouse banks off southwestern Vancouver Island. When water conditions are calm, these porpoises can often be seen from ferries travelling between the BC lower mainland and Vancouver Island. There is one record of a Harbour Porpoise entanglement in a salmon gillnet about 55 km up the Fraser River, but generally the species does not venture past the river's estuary.

Aerial surveys off southeastern Vancouver Island in the mid 1990s showed that Harbour Porpoise densities were considerably lower in the relatively deep waters of the Strait of Georgia than in the shallower waters of the Gulf Islands or Juan de Fuca Strait. UBC graduate student Anna Hall, studying Harbour Porpoises off southern Vancouver Island, found that 90 per cent or more of sightings were in waters of less than 150 metres. Sighting densities tended to be greatest in areas of tidal eddies, such as off Discovery Island in southeastern Haro Strait. These concentrations are likely due to greater prey availability in these highly active waters.

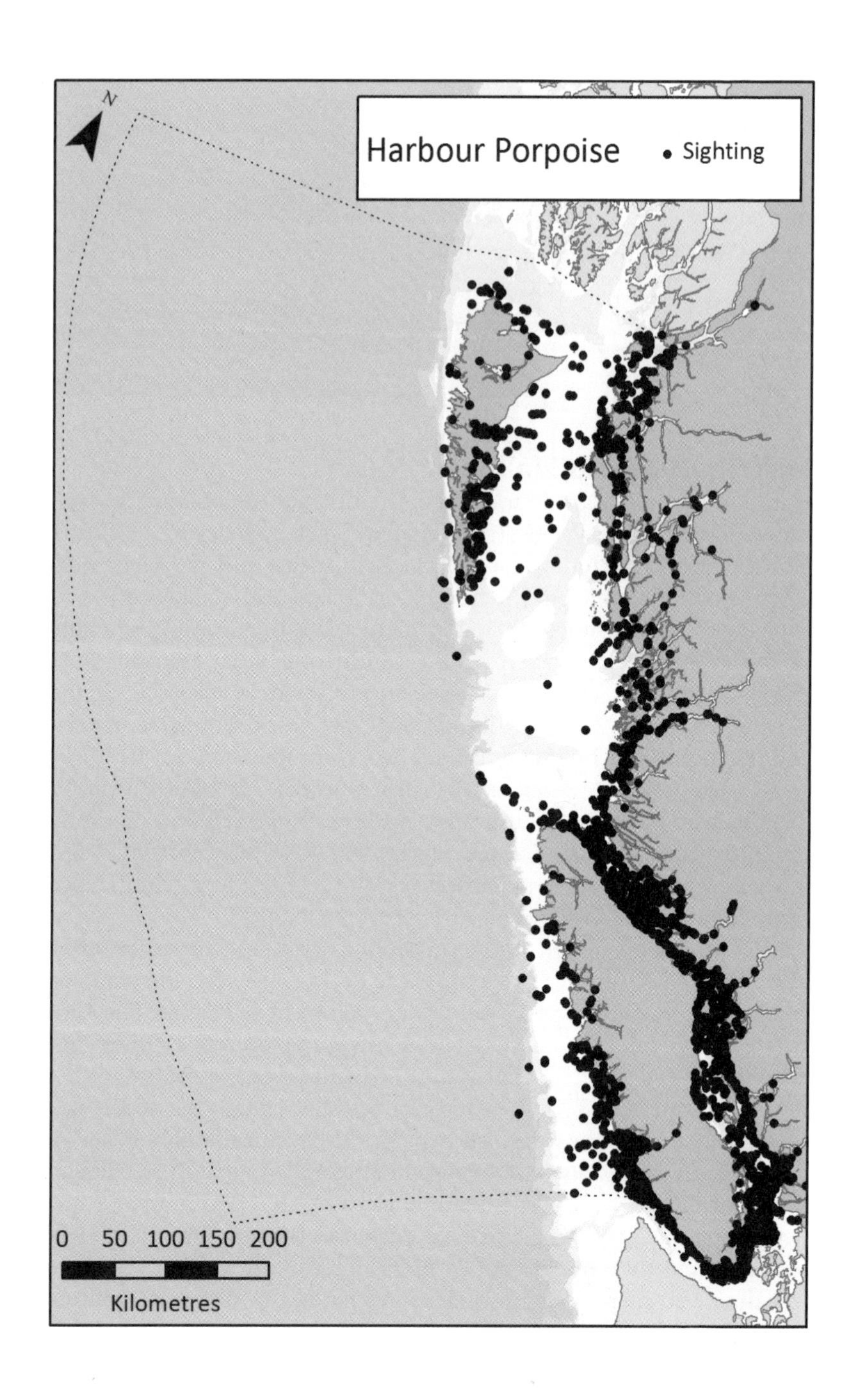
N
Harbour Porpoise
• Sighting
0 50 100 150 200
Kilometres

Natural History

Feeding Ecology

Harbour Porpoises feed on a wide variety of small fishes and squid, and their diet varies regionally. Studies of the stomach contents of over 50 stranded Harbour Porpoises recovered from the shores of the Strait of Georgia and Juan de Fuca Strait between 1990 and 2010 revealed a diet of 15 different species of fishes and several species of invertebrates. Important prey species included Pacific Herring, Blackbelly Eelpout, Walleye Pollock, Eulachon, Pacific Sand Lance, Pacific Hake, Northern Anchovy and Opalescent Inshore Squid. Prey of lesser importance were Plainfin Midshipman, Shiner Perch, Northern Sculpin, Pacific Sanddab and High Cockscomb. Also consumed was the Mussel Worm (a free-swimming polychaete annelid that reaches lengths of 15 cm), the Boreal Clubhook Squid and Berry Armhook Squid. The diets of Harbour Porpoises and Dall's Porpoises are generally quite similar in the Strait of Georgia region, although Dall's tend to take greater numbers of deepwater species, such as lanternfish (family Myctophidae). The diet of Harbour Porpoises in other areas of British Columbia has not been studied.

Behaviour and Social Organization

Harbour Porpoises are normally seen in small groups of two to five, which often contain a mother and calf pair, but solitary individuals are also commonly found. They occasionally form large, loosely structured gatherings of fifty or more that are likely attracted to high concentrations of prey. Harbour Porpoises are active but not very splashy swimmers – they seldom display aerial behaviour such as breaching or tail slapping and do not exhibit the rooster-tailing behaviour typical of Dall's Porpoises travelling at high speed (see figure 89).

Harbour Porpoises spend as much as 75 per cent of their time foraging for food and swim at speeds of about 5 km per hour, which is efficient for their body size. They usually dive for periods of less than two minutes, but some dives can exceed five minutes, and they normally surface to breathe three or four times before sounding. Studies using digital data loggers have shown that Harbour Porpoises spend most of their time in the top 10–20 metres of the water column but can dive to more than 200 metres. Harbour Porpoises will sometimes rest at the surface for extended periods, lying motionless with the body at an angle so that only the blowhole and part of the back is exposed.

Sometimes individuals restrict their movements to preferred foraging areas. One tagged female in northern Washington, for example, remained within a 65 square km area for more than six months.

Harbour Porpoise vocalizations consist exclusively of high-frequency clicks, with most energy between 100 and 180 kHz – far higher than the range of human hearing. Series, or *trains*, of clicks are repeated at rates of several hundred per second and appear to be used primarily for echolocation. Research on echolocation in wild Harbour Porpoises in Denmark indicates that prey fish the size of herring should be detectable at ranges of 40 metres. Intense bursts of clicks up to three seconds long are used in aggressive interactions among Harbour Porpoises held in captivity, and they likely have a similar context in the wild.

Life History and Population Dynamics

The life history of Harbour Porpoises features rapid growth, a high calving rate and a short lifespan compared to most other cetaceans. They reach sexual maturity at three to four years of age; males become physically mature at five years and females at seven, and both sexes have an average life span of only eight to ten years. Breeding is seasonal, with calves born in summer following a 10.5-month gestation, and mating taking place about 1.5 months later. The mating system is promiscuous, or *polygynandrous*, where a female pairs with several males, each of which also pairs with several different females. This reproductive strategy involves sperm competition, where males with the most sperm volume and most frequent matings have an advantage over other males in fathering the most offspring. The testes of male Harbour Porpoises increase markedly in size during the breeding season, reaching three to six per cent of the total body mass – 13 times greater than the typical mammalian testes/body mass ratio.

One source of natural mortality for Harbour Porpoises in British Columbia is predation by Transient (Bigg's) Killer Whales. Harbour Porpoises are particularly vulnerable because they lack the speed to escape once detected and pursued. Of the more than 400 kills by Transients that we have documented, 16 per cent have been of Harbour Porpoises. Some predation on Harbour Porpoises by large sharks, such as the Pacific Sleeper Shark, is possible but has not yet been confirmed. The Harbour Porpoise is the most commonly stranded cetacean in BC – most drift ashore already dead, but some become stranded alive. The Vancouver Aquarium has successfully rescued two live-stranded Harbour Porpoises (figure 87). Necropsies of dead stranded Harbour Porpoises by Stephen Raverty of the BC Animal

Figure 87. This live-stranded Harbour Porpoise, named Levi, was rescued from the shore in Patricia Bay, north of Victoria, in March 2013. The Vancouver Aquarium rehabilitated Levi and released him five months later. Photo: L. Thorpe.

Health Centre have revealed a variety of pathogens, including the fungus *Cryptococcus*.

Exploitation

Harbour Porpoises have been exploited for food in many parts of their range, but the only substantial hunt still remaining is in Greenland waters, where First Nations hunters kill several hundred animals annually. In British Columbia, Harbour Porpoises were widely hunted and consumed by First Nations people until about 1900. Hunters speared or harpooned them from canoes, sometimes waiting quietly at night until a porpoise approached, its outline revealed by bioluminescence in the water. There has been no other directed killing of Harbour Porpoises in BC.

Taxonomy and Population Structure

Early morphological studies suggested that the Harbour Porpoise is closely related to Burmeister's Porpoise, found in cool coastal waters of South America, and the Vaquita, found only in the upper Gulf of California. These species were all placed in the genus *Phocoena*. More recent DNA evidence suggests a close relationship to the Dall's Porpoise, which is consistent with the frequent hybridization observed between the two species in the Strait of Georgia region.

Significant differences in the skulls of Harbour Porpoises from different regions have led to their separation into subspecies, with three subspecies being distinguished to date. These are *Phocoena phocoena phocoena* in the North Atlantic, *P. p. vomerina* in the eastern North Pacific and an as yet unnamed subspecies in the western North Pacific. Some early references to eastern North Pacific Harbour Porpoises use *P. vomerina* as the species name. Despite the species' continuous distribution along the west coast of North America from central California northward, the Harbour Porpoise population appears to be divided into genetically distinct genetic subunits. But recent studies by UBC graduate student Carla Crossman have found few genetic differences in Harbour Porpoises from Haida Gwaii to southern Vancouver Island.

Conservation Status and Management

Although there is no estimate of the total number of Harbour Porpoises throughout their range, estimates for selected portions suggest a range-wide total of more than 700,000. Surveys conducted off the west coast of the US mainland during 2002–07 estimated about 55,000 from California to Washington. Aerial surveys by the US National Marine Fisheries Service in 2002–03 yielded estimates of approximately 700 Harbour Porpoises in the Canadian portion of Juan de Fuca Strait and about 6200 in the Gulf Islands and Strait of Georgia. Boat-based surveys undertaken by Raincoast Conservation Foundation in 2004–05 resulted in an estimate of around 9000 for nearshore waters of British Columbia, excluding the outer coasts of Vancouver Island and Haida Gwaii which were not surveyed.

The most significant threat currently facing Harbour Porpoises is mortality from entanglement in fishing gear, particularly in gillnets. Incidental catches have been a serious problem throughout much of the Harbour Porpoise's range, but levels have declined to some extent thanks to mitigation efforts and reduced use of gillnets in a few areas. Acoustic deterrent devices (also known as *pingers*) that alert porpoises to the presence of nets have reduced bycatch substantially, and monofilament nets impregnated with substances such as barium sulfate to increase acoustic reflectivity have also shown promise. It is likely that fewer than 100 Harbour Porpoises are now killed annually in salmon gillnet fisheries off the south coast of BC. This is much reduced from past decades, when gillnet fisheries were far more extensive than today.

Because they occur predominantly in shallow coastal habitat, Harbour Porpoises are particularly vulnerable to adverse effects from human activity and development. Environmental contaminants and habitat degradation in heavily industrialized areas may pose a threat. They also appear to be susceptible to disturbance by vessels and the noise they generate. Noise from other activities can also affect Harbour Porpoises. For example, loud underwater acoustic deterrent devices that were widely used during the 1980s and 1990s to keep seals and sea lions away from salmon aquaculture pens were found to displace Harbour Porpoises from habitat in the Broughton Archipelago. The use of these devices is no longer permitted in BC. Displacement from important habitat is also a possible threat for this species.

Although little survey data are available, there are encouraging signs that Harbour Porpoise numbers are increasing in southern British Columbia and the adjacent waters of Washington. Aerial surveys around the San Juan Islands and in the US portions of Juan de Fuca Strait and the Strait of Georgia in 2002–03 yielded estimates that were three times higher than similar surveys in the mid 1990s. Harbour Porpoises were common in southern Puget Sound in the 1940s but had been essentially absent there until recently, when sightings have been frequent enough to suggest they are making a comeback. Harbour Porpoises are now seen regularly in Juan de Fuca Strait near Victoria, in Boundary Pass near Saturna Island and in Haro Strait. This resurgence of Harbour Porpoises may be at least partly due to decreased mortality in salmon gillnet fisheries.

Globally the Harbour Porpoise is designated as Least Concern by IUCN. In the Canadian Pacific the species is listed as Special Concern by COSEWIC, and it is on the BC Blue List.

Remarks

This species was the first to be called "porpoise", which is derived from the Old French word *porpeis,* which in turn is a corruption of the Latin *porcus piscus*, meaning "pig fish". The word has been spelled at least 40 different ways since it entered English in about 1300. Harbour Porpoises are commonly referred to as "puffing pigs" off Canada's east coast.

Selected References: Bjørge and Tolley 2009, Carretta et al. 2011, Chivers et al. 2002, Crossman 2012, Fisheries and Oceans Canada 2009, Hall 2004 and 2011, Nichol et al. 2013, Scheffer and Slipp 1948, Walker et al. 1998, Williams and Thomas 2007.

Dall's Porpoise *Phocoenoides dalli*

Other Common Names: None.

Description

The Dall's Porpoise has a powerful, stocky body and a relatively small head with a very short beak. It has a short, wide-based, triangular dorsal fin with a slightly falcate tip. The flippers are small and positioned well forward on the body, and the tail stock is robust, with pronounced dorsal and ventral keels. Adult males have very pronounced keels, a hump in front of the dorsal fin, which is canted forward, and a convex trailing edge of their tail flukes – these features become more pronounced with age (figure 88).

The colour pattern of Dall's Porpoises is striking. The body is almost entirely black except for bright white patches on the belly and flanks. These patches extend midway up the sides and end in line with the front of the dorsal fin. As individuals age they develop grey to white "frosting" on the tip of the dorsal fin and the trailing edges of the flukes, as if they had been dipped in paint. Calves are slate grey with rather muted, greyish-white flank patches.

Figure 88. The dorsal fin of an adult male Dall's Porpoise is canted forward, and this, together with a hump forward of the fin, gives it a distinctive appearance at the surface. This animal is swimming from right to left. Photo: J. Towers, Queen Charlotte Strait, July 2010.

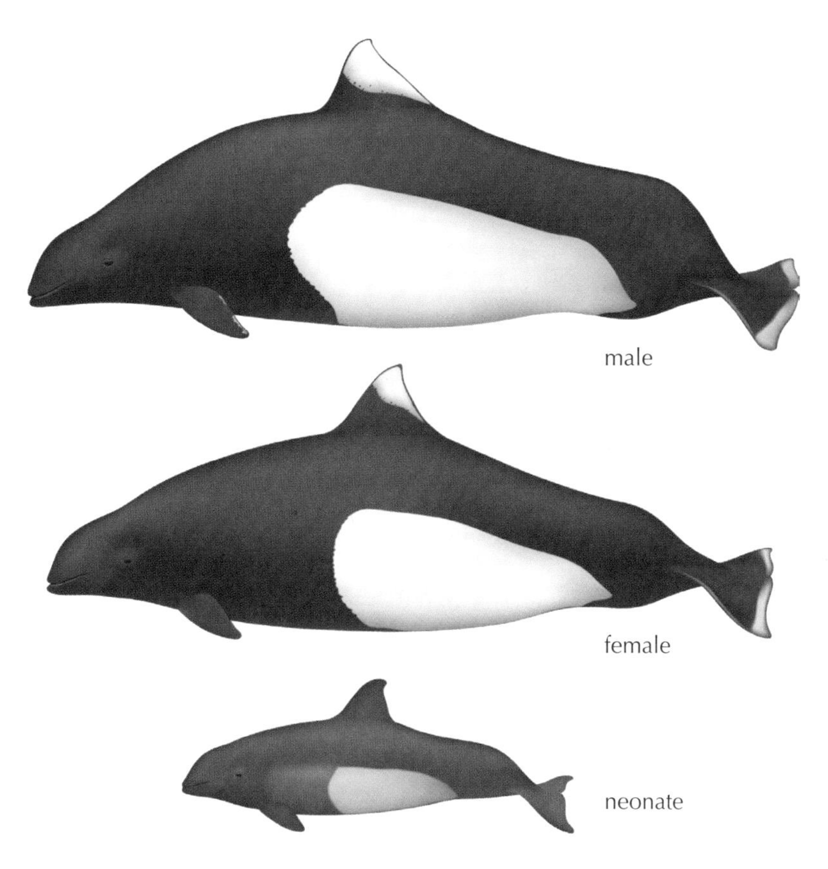

Measurements

length (metres):

male: 1.97 (1.60–2.35) n=13
female: 1.78 (1.52–2.06) n=14
neonate: ~1.0

weight (kg):

adult: 200
neonate: 5.5

Dentition

23–28 pairs of extremely small teeth in both the lower and upper jaws.

Identification

If a Dall's Porpoise's distinctive black-and-white markings cannot be seen, it can be confused with a Harbour Porpoise or perhaps a dolphin at a distance. The grey or white-tipped dorsal fin is a sure sign of a Dall's, but this is not fully reliable since not all individuals have this feature. Dall's Porpoises are usually seen in small groups of 10 or fewer – 80 per cent of 947 sightings during our Cetacean Research Program surveys were of groups of five individuals or fewer. Occasionally, as many as a few dozen Dall's Porpoises gather in a temporary aggregation. Dall's Porpoises are fast swimmers, slicing through the surface explosively to breathe every 5–10 seconds or so and creating a V-shaped burst of spray, known as a rooster tail, that can obscure them from view (figure 89). These porpoises are avid bow-riders and regularly approach moving vessels, darting back and forth to get a push from the bow wave. They do not leap clear of the surface when breathing, as most small dolphins do when travelling at high speed, and they very rarely display aerial behaviour such as breaching or tail slapping. When swimming at a slow pace they surface quietly to breathe, and the hump of the thick tail gives them the appearance of a black-and-white cube rotating in the water when viewed from the side. In contrast, Harbour Porpoises generally inhabit shallower water than Dall's Porpoises, swim more slowly without rooster-tailing, do not bow ride and tend to shy away from boats. Inexperienced observers may also sometimes mistake Dall's Porpoises for baby Killer Whales.

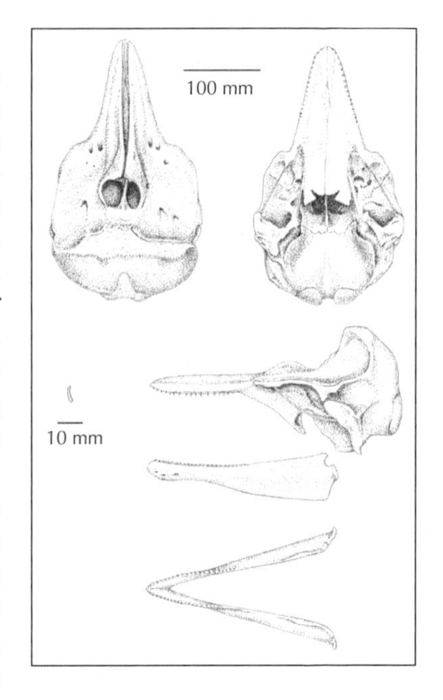

Distribution and Habitat

Dall's Porpoises are endemic to the cool-temperate North Pacific Ocean and adjacent Bering, Okhotsk and Japan seas, preferring water temperatures of less than 17°C. They are widely distributed in oceanic and coastal waters from 32–35°N (southern California and southern Japan) to about 63°N (central Bering Sea). Off California the distribution of Dall's Porpoises has been observed to shift inshore and southward during winter and offshore and northward during summer,

Figure 89. Three Dall's Porpoises swimming rapidly from left to right show the explosive spray they produce when breaking the surface to breathe.
Photo: J. Towers, Queen Charlotte Strait, October 2010.

possibly in response to changes in prey densities. Dall's Porpoises can be found year-round throughout British Columbia waters, including inshore channels, passes and straits, continental shelf waters, and offshore waters beyond the shelf break. They appear to be particularly abundant in areas of strong tidal mixing and over continental shelf and slope waters. There is some evidence of a shift inshore during summer and offshore in winter, but this has yet to be clearly demonstrated. University of BC graduate student Alison Keple conducted a year-round survey of cetaceans in the Strait of Georgia from BC Ferries ships and found that the density of Dall's Porpoises was lowest in fall (3.5 animals per 100 square km) and highest in spring (17.5 animals per 100 square km).

Natural History

Feeding Ecology

Dall's Porpoises prey on a wide variety of small fishes and cephalopods. Most common are schooling fishes (such as Pacific Herring, Northern Anchovy, Pacific Saury and Pacific Sardine), mesopelagic fishes (such as lanternfish) and squid. Stomach contents of Dall's Porpoises found stranded along the shores of the Strait of Georgia and Juan de Fuca Strait have provided insight into the species' diet in this region. Pacific Herring, Walleye Pollock and squid were the most

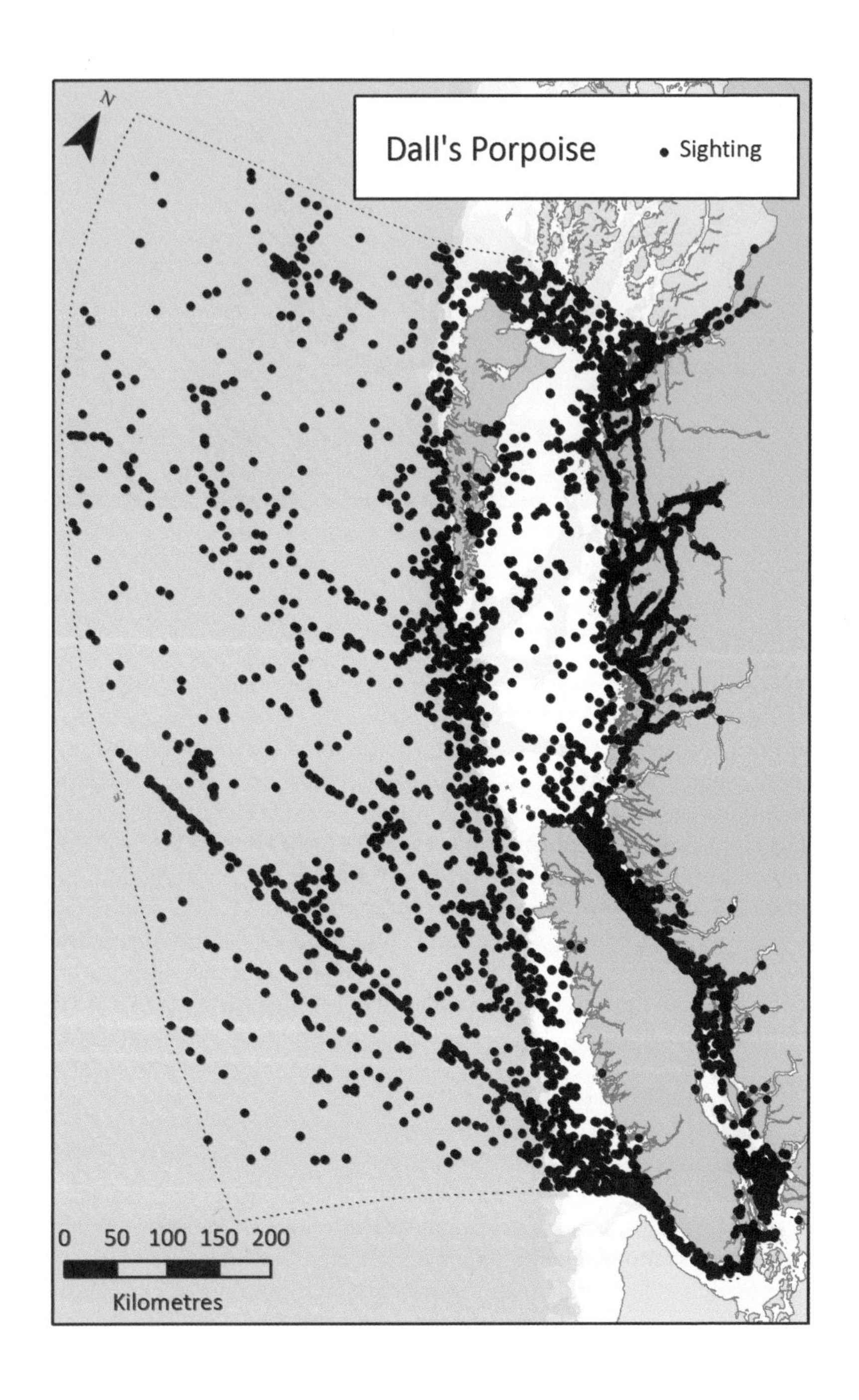
N
Dall's Porpoise
Sighting
0 50 100 150 200
Kilometres

common prey, but appearing in smaller numbers were Blackfin Sculpin, Pacific Hake, Eulachon, Northern Smoothtongue, Blackbelly Eelpout, Pacific Sand Lance and various lanternfish (myctophids). So far there is no evidence from stomach contents that Dall's Porpoises feed on salmonids. Although there is considerable dietary overlap with Harbour Porpoises in inshore waters, Dall's feed more extensively on deeper-water species such as Pacific Hake and lanternfishes. The apparent influx of Dall's Porpoises into the Strait of Georgia during spring may coincide with spawning aggregations of Pacific Herring and Pacific Hake.

Behaviour and Social Organization

Dall's Porpoises are social but not particularly gregarious animals. They are usually found in groups of 2–10. While larger aggregations are occasionally observed, these are temporary associations of animals likely attracted to prey concentrations and lack the cohesion of dolphin schools. Cetacean biologist Tom Jefferson studied Dall's Porpoise behaviour in Johnstone Strait off northeastern Vancouver Island, and observed that group size averaged 2.6 individuals (range: 1–5). Simon Fraser University graduate student Pam Willis, studying Dall's Porpoises off southeastern Vancouver Island, found that single adult males typically maintained close associations with adult females that had recently given birth, and often they reacted aggressively toward other adult males that approached. Willis interpreted this as "mate guarding", when an adult male defends his escort position to get priority mating access when the female comes into estrus. These associations are often observed during summer, following the peak in calving around early to mid July.

Groups of Dall's Porpoises seen in Johnstone Strait usually moved at a slow pace, averaging about 5 km per hour, and made feeding dives two to four minutes long, interrupted by 10–15 short, shallow dives of less than 15 seconds. The animals exhibited "slow rolling" surfacings most of the time and seldom broke into high-speed rooster-tailing behaviour. They seldom approached boats to bow ride, perhaps because vessels are so frequent in the Strait that they are no longer a novelty.

Dall's Porpoises recognize the difference between Resident and Transient Killer Whales and react very differently to the two ecotypes. They usually show little reaction to Residents and at times approach and swim among them, clearly not considering the fish-eating Killer Whales a threat. But if Dall's Porpoises detect mammal-hunting Transients approaching, they flee at high speed. Transients attack Dall's Porpoises frequently – constituting about ten per cent of their 581 predation events (chases, attacks and kills) on marine mammals documented in BC. But Dall's Porpoises are among the fastest cetaceans – at least for short-duration sprints – and have out-paced Transients in almost half of the chases we've observed.

The vocalizations of Dall's Porpoises appear to be restricted to echolocation clicks in frequencies many times higher than the range of human hearing, peaking in a narrow band of 120 to 140 kHz, with some reaching almost 200 kHz. It has been suggested that the use of such high frequencies may have evolved as a tactic to avoid predators, since these are above the hearing limit of Killer Whales.

Life History and Population Dynamics

Dall's Porpoises reach sexual maturity at 4–7 years in females and 3.5–8 years in males. Females gestate for 10–12 months before giving birth to a single calf, generally during early summer, and are ready to breed again within a month. Nursing continues for less than a year. Individuals may live to 20 years, but animals older than 10 are rare.

Natural hybridization of Dall's Porpoises and Harbour Porpoises appears to occur regularly in BC waters. The Harbour Porpoise is usually the father. Hybrids most often resemble the Dall's Porpoise in appearance and behaviour, except they lack the distinctive white flank patches. Recent genetic studies of stranded porpoises by UBC graduate student Carla Crossman have found that some hybrids do not differ in appearance from normal Dall's or Harbour Porpoises and that Dall's Porpoises may also be fathers to hybrids. It has been estimated that hybrids represent 1–2 per cent of the Dall's Porpoise population off southeastern Vancouver Island.

Exploitation

Dall's Porpoises have long been the focus of directed hunts for human consumption in Japan. Porpoises are harpooned by hand as they ride the bow waves of catcher boats, and annual catches in recent years have exceeded 15,000 animals in the Sea of Okhotsk. There is no hunting of the species elsewhere in the North Pacific, and no evidence that Dall's Porpoises were taken in significant numbers in the past by First Nations in BC.

Taxonomy and Population Structure

There are two major colour forms of Dall's Porpoises that are considered to be separate subspecies. *Phocoenoides dalli dalli* has flank patches that extend forward to a point in line with the leading edge of the dorsal fin, while in *P. d. truei* the flank patches extend further forward to the flippers. The *truei* type is found mostly in western North Pacific waters and is not seen in BC. Genetic studies have revealed distinctions between the two forms but not to the extent that separation into species is warranted. Forms intermediate in appearance between the *dalli* and *truei* types have also been observed.

Conservation Status and Management

Although Dall's Porpoises are common in many parts of their range, counts can lead to over-estimates because of the animals' tendency to approach vessels, including those used in surveys. Total abundance in the North Pacific has been estimated at more than 1.2 million, but it is not clear to what extent this is biased. Recent surveys in Alaska estimated 83,400 animals, which includes a correction for the positive bias due to vessel attraction. Surveys off the US west coast have yielded highly variable estimates, apparently due to changes in ocean conditions over the years. The most recent abundance estimate of 42,000 is based on the average of surveys in 2005 and 2008. Surveys of the inshore waters of BC by Raincoast Conservation Foundation in the mid 2000s resulted in an abundance estimate of about 5000 individuals. UBC graduate student Anna Hall, who studied porpoises near Victoria for many years, noted a decline in Dall's Porpoises in Haro Strait and eastern Juan de Fuca Strait after 2002, and a corresponding increase in the densities of Harbour Porpoises. Naturalists aboard whale-watch boats operating out of Victoria have made similar observations.

Between 1981 and 1994 more than 45,000 Dall's Porpoises were killed in Japanese drift-net fisheries and almost 250,000 in the Japanese harpoon hunt, which may have depleted populations in the western North Pacific. The magnitude of such takes has diminished in recent years, but there is still concern about the conservation status of Dall's Porpoises in that region. In BC waters Dall's Porpoises are occasionally killed by entanglement in fishing gear, mostly salmon gillnets but occasionally bottom or midwater trawl nets. This mortality is not considered significant at a population level.

The Dall's Porpoise is considered Least Concern internationally by IUCN and Not at Risk by COSEWIC in Canadian waters. It is on the BC Yellow List.

Remarks

The Dall's Porpoise was named after the noted American naturalist and paleontologist William Healey Dall (1845–1927). In 1865–1867 Dall explored British Columbia, Alaska and the Russian Far East aboard the USS *Nightingale*, under the command of renowned whaler Charles M. Scammon. On this and a subsequent expedition, Dall observed and took notes on this distinctively marked black-and-white porpoise, then collected a specimen on a later expedition in 1873. A new species was officially described from this specimen in 1885 by Fredrick W. True, mammal curator at the US National Museum (now part of the Smithsonian Institution), who dedicated the species to his friend Dall "not alone on account of his prominence as a zoologist, but also because the specimen and notes from which the description has been drawn are the fruits of his labour".

Selected References: Allen and Angliss 2011; Barlow and Forney 2007; Cowan 1944; Crossman 2012; Hall 2011; Jefferson 1987, 1988 and 1990; Jefferson et al. 2008; Keple 2002; Nichol et al. 2013; Pike and MacAskie 1969; True 1885; Walker et al. 1998; Williams and Thomas 2007; Willis et al. 2004.

Northern Fur Seal *Callorhinus ursinus*

Other Common Names: North Pacific Fur Seal.

Description

The Northern Fur Seal is a small to medium-sized otariid (eared seal) with a relatively small head and a very short, downcurved snout. Adults are highly sexually dimorphic, with mature males being up to 3.4 times larger than females and 5.4 times heavier during the breeding season. Males and females are difficult to distinguish until about four or five years old, when males develop a more stocky build, with a much wider and thicker neck. Adult males also develop a prominent sagittal crest on the top of their skulls, giving them a prominent forehead compared to females. The Northern Fur Seal has long, prominent external earflaps (pinnae) that have no fur on their tips, and the eyes are large and conspicuous, especially in young animals. This seal has the largest hind flippers of any otariid, with extremely long digits that extend well beyond the small claws. The broad fore flippers have a fur covering on the dorsal surface that ends with a distinct line at the wrist, beyond which the flippers are naked.

As their name suggests, Northern Fur Seals have a dense, luxuriant pelage. The thick underfur is water-repellant and provides thermal insulation; it is overlaid with longer, coarser guard hairs. The hair and fur fibres of Northern Fur Seals average about 57,000 per square

centimetre (about 200 times the density of human hair). Guard hairs are particularly coarse and long in the mane of adult males, which covers the head, neck and shoulders. Fur colour varies with age and sex: newborn pups are dark blackish-brown with a lighter belly; adult females and subadults are dark silver-grey to brown, with lighter undersides. Adult males are uniformly dark, with shades ranging from black to brown to reddish; their guard hairs lack pigmentation, giving them a silvery appearance.

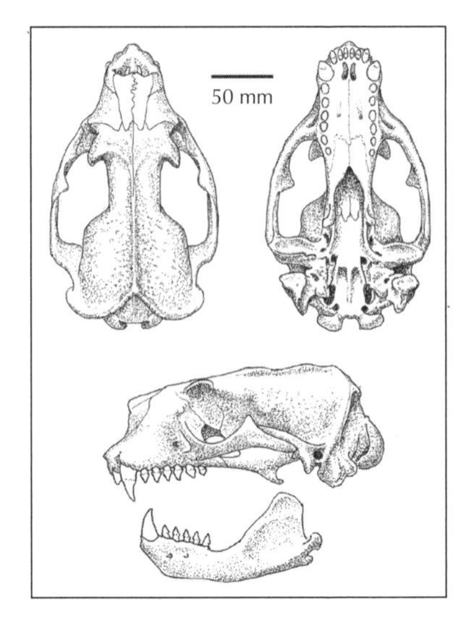

Measurements

length (metres):

male: 1.50–2.1
female: 1.2–1.3
neonate: 60–65

weight (kg):

male: 150–270
female: 35–50
neonate: 5–6

Dental formula

incisors: 3/2
canines: 1/1
post-canines: 6/5

Identification

Although the Northern Fur Seal is British Columbia's most abundant pinniped, it is seldom seen within 20 km or so of the coast. In BC waters Northern Fur Seals are most likely to be confused with California Sea Lions, which has similarly dark pelage. But the two species are easily distinguished upon closer observation – Northern Fur Seals have a shorter, pointed muzzle, longer, more obvious earflaps (figure 90), and much longer hind flippers with extended digits. Northern Fur Seals off the BC coast are mostly females and subadults. They are considerably smaller than male California Sea Lions, which are the

Figure 90. Northern Fur Seals have more prominent eyes and external earflaps than sea lions. Photo: C. McMillan, southwestern Vancouver Island, May 2012.

only members of that species normally occurring in BC. Northern Fur Seals rarely haul out on the BC coast, whereas California Sea Lions can be found in substantial concentrations along both coasts of Vancouver Island during winter and spring. Northern Fur Seals are found mostly off the outer coast of BC but are occasionally seen in inner coastal waters. While resting at sea, Northern Fur Seals routinely adopt a distinctive posture known as the "tea pot" or "jug handle" position (figure 91), in which the seal floats on its back with the underside of one fore flipper lying over the hind flippers, which are rotated forward to meet the fore flipper. The other fore flipper remains

Figure 91. A Northern Fur Seal in Beauchemin Channel, resting in "tea pot" or "jug handle" position. Photo: G. Ellis, June 2012.

along the animal's side or in the water, presumably to stabilize the body at the surface. Northern Fur Seals can often be observed sleeping in mats of drifting kelp and other flotsam.

Distribution and Habitat

Northern Fur Seals inhabit the North Pacific Ocean and the Bering and Okhotsk seas. They are among the most pelagic pinnipeds in the North Pacific, spending about 90 per cent of their lives at sea. They are found in offshore waters from latitudes of central Japan (36°N) and southern California north to the Gulf of Alaska in the east and Okhotsk Sea in the west, and throughout the Bering Sea. During summer they migrate north to breeding rookeries located mainly on Alaska's Pribilof Islands and Bogoslof Island as well as the Commander Islands of Russia. Smaller breeding rookeries are also found on Russia's Kurile Islands and Robben Island in the west and on the Farallon Islands and San Miguel Island off California in the east.

Migration patterns of Northern Fur Seals are complex, varying with age and sex. After the breeding season from June through October, seals begin migrating southward away from their rookeries. Adult males leave the rookeries first, followed by adult females and juveniles, and finally pups. On average, females travel further south than males, and pregnant females travel the furthest south. Adult males tend to stay in northern waters around the Aleutian Islands and Gulf of Alaska, while females and subadults disperse widely on the high seas of the North Pacific, with some reaching California by December. Sealing records and field surveys indicate that densities of Fur Seals off the west coast of North America (within 200 km of shore) increase rapidly in January and remain fairly stable from February to May. But there is a northward shift of peak density over the wintering period, from California in February, to Oregon and Washington in April and then BC and southeastern Alaska in May.

Off British Columbia, Northern Fur Seals begin to appear in outer coastal waters during December; they reach peak numbers in May, but most have left the area by July. They are concentrated mostly over the outer continental shelf and slope, particularly on La Pérouse Bank off the west coast of Vancouver Island, and to a lesser extent in Queen Charlotte Sound and Hecate Strait. Some young seals can occasionally be seen in protected passes and inlets along the inner BC coast. About one-third of the population found off the west coast of North America – or roughly 125,000 seals – use BC waters during winter and early spring. Northern Fur Seals rarely haul out except

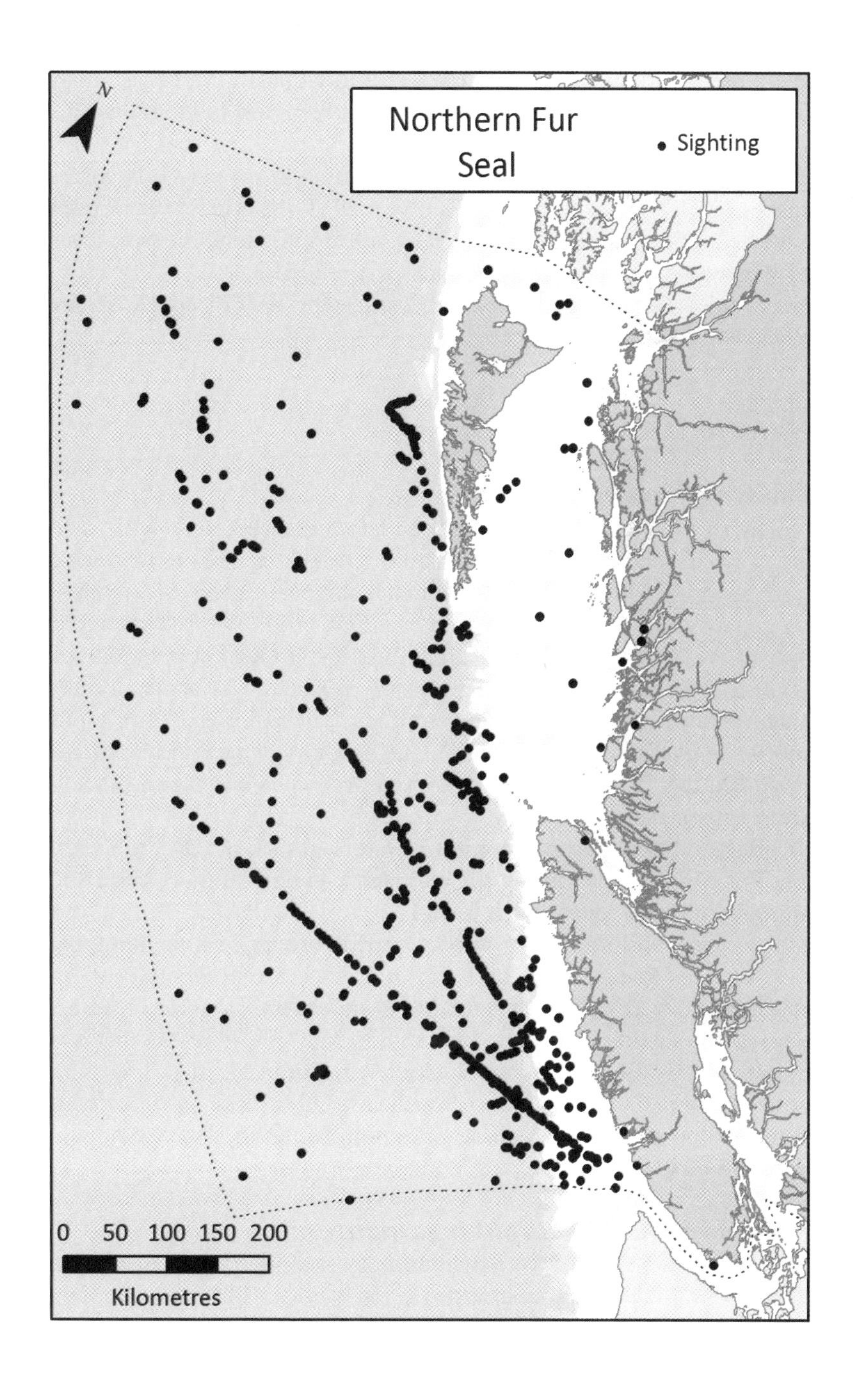
Northern Fur Seal
Sighting
N
0 50 100 150 200
Kilometres

at their traditional breeding rookeries. Individuals have occasionally been seen among Steller or California sea lions on haulouts in BC, such as at Race Rocks near Victoria, but no breeding rookeries currently exist in the province. This apparently has not always been the case, however, since remains of unweaned Northern Fur Seal pups have been found in excavated archaeological sites along the west coast of Vancouver Island. The locations of prehistorical breeding rookeries for Northern Fur Seals in BC, and why they disappeared, are unknown.

Natural History

Feeding Ecology

Northern Fur Seals are generalist predators that feed on a wide variety of small schooling fishes and squid. Over 70 species of prey have been identified from scats and stomach samples collected over their range, but diets in any given region or month tend to be dominated by only a few key prey species. Overall, the most important prey include Walleye Pollock, Pacific Herring, Northern Anchovy, Capelin, Pacific Hake, Eulachon, rockfishes, myctophids, salmonids and numerous species of squid. It is likely that Pacific Sardines, which almost disappeared in the 1940s but have recovered dramatically since the 1980s, are now also important prey.

Researchers in British Columbia waters studied the diet of Northern Fur Seals by examining the stomach contents of more than 300 animals collected for research from 1958 to 1974. Pacific Herring was the primary food in inshore waters from February to June along the west coast of Vancouver Island and in Hecate Strait; Opalescent Inshore Squid was also important in coastal inlets. In oceanic waters beyond the shelf break off the BC coast in May and June, the diet was dominated by Boreal Clubhook Squid and salmon, including Chinook, Chum, Coho, Pink and Steelhead. Other prey species documented off the BC coast include Eulachon, Sablefish, Pacific Cod and Pacific Saury.

Behaviour and Social Organization

Northern Fur Seals are strongly migratory, and their social behaviour depends on the phase of migration. In the waters of British Columbia, Fur Seals are usually found at sea, mostly 20–150 km from shore, where their primary activity is feeding. They are mostly seen alone, though pairs and groups of three or four are fairly common. Larger aggregations can sometimes be observed in areas of prey concentration. Northern Fur Seals feed mostly at night, since this is when their prey rises close to the surface and is more accessible. Foraging dives last about five minutes and may be as deep as 200 metres. During the day Fur Seals can often be seen sleeping at the surface in their distinctive "jug handle" or "tea pot" position. They also spend considerable time grooming their dense fur to maintain its insulative properties. Fur Seals can be inquisitive and, if not alarmed, approach boats for a better look or to surf in the wake.

Northern Fur Seals have a polygynous breeding system, with adult males competing for territories on traditional breeding rookeries and mating with females that haul out in their territory. Adult males arrive on the breeding rookeries in May or June, about a month before females start arriving, and form territories averaging about 100 square metres; individuals use the same territory from year to year. They stay on shore and fast during this time, maintaining territory boundaries by repelling rival males with vocal and postural threats. Fights between neighbouring males are rare, but can be violent and bloody. When females arrive at the breeding rookery, many are pregnant from mating in the previous season. They show strong site fidelity to past birth sites, often giving birth within 10 metres of where they delivered the previous year. Mating takes place within a week of pups being born. Females are gregarious but not sociable – they are aggressive toward other females and will threaten and bite them if they come too close.

Males leave their breeding territory and return to sea in late July or early August. Females go on foraging excursions for a week or so at a time, returning for one or two days to nurse their pups. While their mothers are away, pups withdraw to unoccupied parts of the rookery to avoid being bitten by other adult females. They often gather in dense "pods", and re-establish contact with their mothers when they return through the exchange of individually distinctive calls. Pups are weaned at about four months of age, at which point they head to sea for up to two years. The main predators of Northern Fur Seals are probably mammal-hunting Killer Whales and Great White Sharks, although Steller Sea Lions have been observed preying upon neonates.

Life History and Population Dynamics

Both male and female Northern Fur Seals reach sexual maturity at 3–7 years of age. Although females reproduce soon after reaching sexual maturity, males are not usually large enough to obtain and defend a breeding territory until they are 6–8 years old. Breeding males have a reproductive lifespan of only 1.5 years on average, although one male was observed to hold the same territory for 10 years. Males holding prime territories may mate with more than 100 females in a season. Females can produce a single pup annually until they are about 22 years old, although few survive to that age. Females give birth within a day of arriving at the breeding rookery. They typically mate only once a season, on average 5.3 days after giving birth, and the fertilized ovum does not implant and begin to develop until the newborn pup is weaned at four months of age. Mammal-hunting Killer Whales are important predators of Northern Fur Seals.

Exploitation

First Nations along the British Columbia coast hunted Northern Fur Seals for food and skins for at least 5000 years. Fur-seal bones dominate mammal remains at almost all of the excavated archaeological sites along the west coast of Vancouver Island and also occur in middens along the central mainland coast and on Haida Gwaii. At the ancient village site of Ts'ishaa in Barkley Sound, Northern Fur Seals accounted for almost two-thirds of marine mammal remains, six times more than any other marine mammal species. Fur Seals were hunted from dugout canoes using harpoons tipped with mussel shell or, after the arrival of Europeans to the coast, iron heads.

The Northern Fur Seal has a long and complex history of commercial exploitation for pelts, which made valuable and luxurious fur coats. Large-scale killing began with the discovery of the main breeding colonies in the Bering Sea in the mid to late 1700s. At that time the largest Fur Seal rookeries were on the Pribilof Islands, with an estimated population of 2–3 million. While under Russian management, the Pribolof seals were rapidly depleted, with hunters taking an average of about 100,000 animals a year, mostly pups. The Russian government imposed catch restrictions in the early 1800s, permitting the hunting of immature males only. The population had recovered to near historical levels when the US took ownership of Alaska in 1867. The Americans allowed unregulated hunting once again, and almost 250,000 Fur Seals were killed in 1868 alone.

Around the same time, a pelagic sealing industry began, peaking in the late 1800s and ending in 1911. During this period, sealers killed at least 800,000 Northern Fur Seals, mostly adult females. Sailing schooners made trips lasting many months, following the seals' migration from California to the Bering Sea. Because of restrictions on pelagic hunting by US-registered vessels, Victoria became the centre of the sealing industry by the late 1800s. The sealing companies employed mostly First Nations men, particularly the highly experienced hunters from Nuu-chah-nulth communities on the west coast of Vancouver Island. The hunters caught seals from dugout canoes or small skiffs deployed from schooners, using either harpoons or rifles. In 1880 alone, they took over 12,000 Fur Seals off the coast of Vancouver Island. Overall, the Victoria Sealing Company processed about 255,000 seal pelts from 1886 to 1911. Declining Fur Seal numbers resulting from this hunting on the high seas as well as a large unregulated kill on the Pribilof Islands led to the signing of the North Pacific Fur Seal Treaty in 1911, one of the first international agreements protecting marine wildlife. The treaty prohibited pelagic sealing and ended the commercial hunting of Northern Fur Seals in Canadian waters.

Taxonomy and Population Structure

The Northern Fur Seal was first described in 1751 by Georg Wilhelm Steller, who had been a naturalist on Vitus Bering's voyage that discovered the Commander Islands a decade earlier. Steller initially named the species *Ursus marinus*, the sea bear, which was changed to its current name by Linnaeus in 1758. The fossil record and genetic studies both indicate that *Callorhinus*, of which the Northern Fur Seal is the only living representative, is the oldest extant genus in the family Otariidae, having branched off some 3–6 million years ago from the line leading to other modern fur seals and sea lions. Although Northern Fur Seals breeding at different North Pacific sites were once thought to represent three distinct species or subspecies, only the one species is currently recognized, with no subspecies.

Conservation Status and Management

The abundance of Northern Fur Seals has fluctuated dramatically over the past two centuries during periods of unregulated commercial killing for pelts interspersed with periods of restricted or managed hunting. After the biggest decline, the North Pacific Fur Seal Treaty was enacted in 1911 to protect the remaining population, which was down to 300,000 on the Pribilof Islands. This treaty, signed by Japan, Russia, the United States and Britain (for Canada), implemented an immediate moratorium on all hunting until 1917, followed by a limited land-based take of immature males. These regulations allowed abundance on the Pribilofs to increase to 2.2 million animals by the 1950s. In that decade, a new management framework was implemented that included the taking of females, and the population declined once again. Commercial hunting of all Fur Seals on the Pribilof Islands ended in 1984, but the population has continued to decline as a result of poor pup production and survival. The decline is most pronounced on the Pribilof Islands, while numbers on other smaller rookeries are stable or increasing. But the overall abundance of Northern Fur Seals from all US rookeries combined declined by about 28 per cent from 1978 to 2008. An annual kill for subsistence on the Pribilof Islands has declined to a few hundred Fur Seals in recent years. There has been no hunting of Fur Seals in BC since 1974, when a field research program by DFO came to an end.

The causes of the continuing population decline of Northern Fur Seals in the Pribilof Islands are unknown. Several factors are likely involved, and these may be acting synergistically. Among such factors are mortality from entanglement in discarded fishing gear, prey limitation and ecosystem changes. Other threats include contaminants and oil spills.

As a result of the ongoing and unexplained decline, IUCN has assessed the species as Vulnerable, and it was listed as Threatened in Canada by COSEWIC in 2010. The species is currently on the BC Red list.

Remarks

The genus name of the Northern Fur Seal, *Callorhinus*, refers to its fine-quality pelage, which led to the population's depletion several times in the past from overhunting. The word is based on the Greek *kallos*, for "beauty", and *rhinos*, for "skin" or "hide". The species name *ursinus* means "bear-like".

Selected References: Bigg 1990, COSEWIC 2010, Crockford et al. 2002, Gentry 2009, Insley 2001, McKechnie and Wigen 2011, McMillan et al. 2008, Moss et al. 2006, Murie 1981, Olesiuk 2008, Perez and Bigg 1986, Trites and Bigg 1996.

Steller Sea Lion *Eumetopias jubatus*

Other Common Names: Northern Sea Lion.

Description

The Steller Sea Lion is the largest otariid (eared seal) in the world. It has a powerful, robust body with marked sexual dimorphism – adult males grow to two to three times the size of females and develop a thick neck and mane with age. The head and muzzle are relatively large and wide, with obvious earflaps (pinnae). The top of the head slopes evenly to the muzzle except in adult males, where an enlarged sagittal crest creates a clearly defined forehead. The fore flippers are large and broad, while the hind flippers are short and narrow.

Steller Sea Lions are mostly light tan to blonde in colour, darkening to a rusty brown on their ventral side. Males are often darker than females. Flippers are very dark brown to black, with only a thin stubble of hair covering part of their dorsal surface. Pups are born with thick blackish-brown fur that lightens after their first moult at three to five months of age and continues to get lighter with successive annual moults until they reach adult coloration at about three years.

Measurements

length (metres):
male: 2.7–3.1
female: 2.1–2.4
neonate: 1.0

weight (kg):
male: 400–1100
female: 200–300
neonate: 16–23

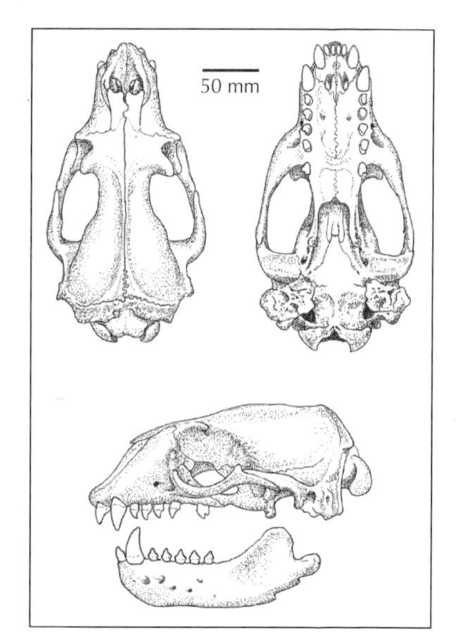

Dental formula

incisors: 3/2
canines: 1/1
post-canines: 5/5

Identification

Steller Sea Lions are often seen in association with California Sea Lions, particularly around Vancouver Island, and distinguishing between the two species can be tricky for the inexperienced observer. Stellers are larger than California Sea Lions, but because only males of the latter species normally occur in BC waters, the difference in size between them and adult female or subadult male Stellers may be difficult to discern. More reliable is the fur colour, which is much lighter in Steller Sea Lions, and the shape of the head and muzzle, which is smaller and sharper in California Sea Lions. When in the water, Stellers may appear greyish white, while California Sea Lions appear dark, almost black. Adult male Steller Sea Lions also have a sagittal crest that is smaller and more posterior than the pronounced crest on adult male California Sea Lions. Steller Sea Lions vocalize with a deep roar or growl, which is very different from the repetitive honking bark of California Sea Lions.

Distribution and Habitat

Steller Sea Lions inhabit cool-temperate to subarctic coastal waters of the North Pacific, from southern California to Hokkaido, Japan, including the Bering and Okhotsk seas. They breed on 55–60 rookeries and rest on more than 300 haulout sites in their range. Breeding rookeries are located in the Kurile Islands, Kamchatka and the Aleutian Islands, and off the coasts of Alaska, British Columbia, Oregon and

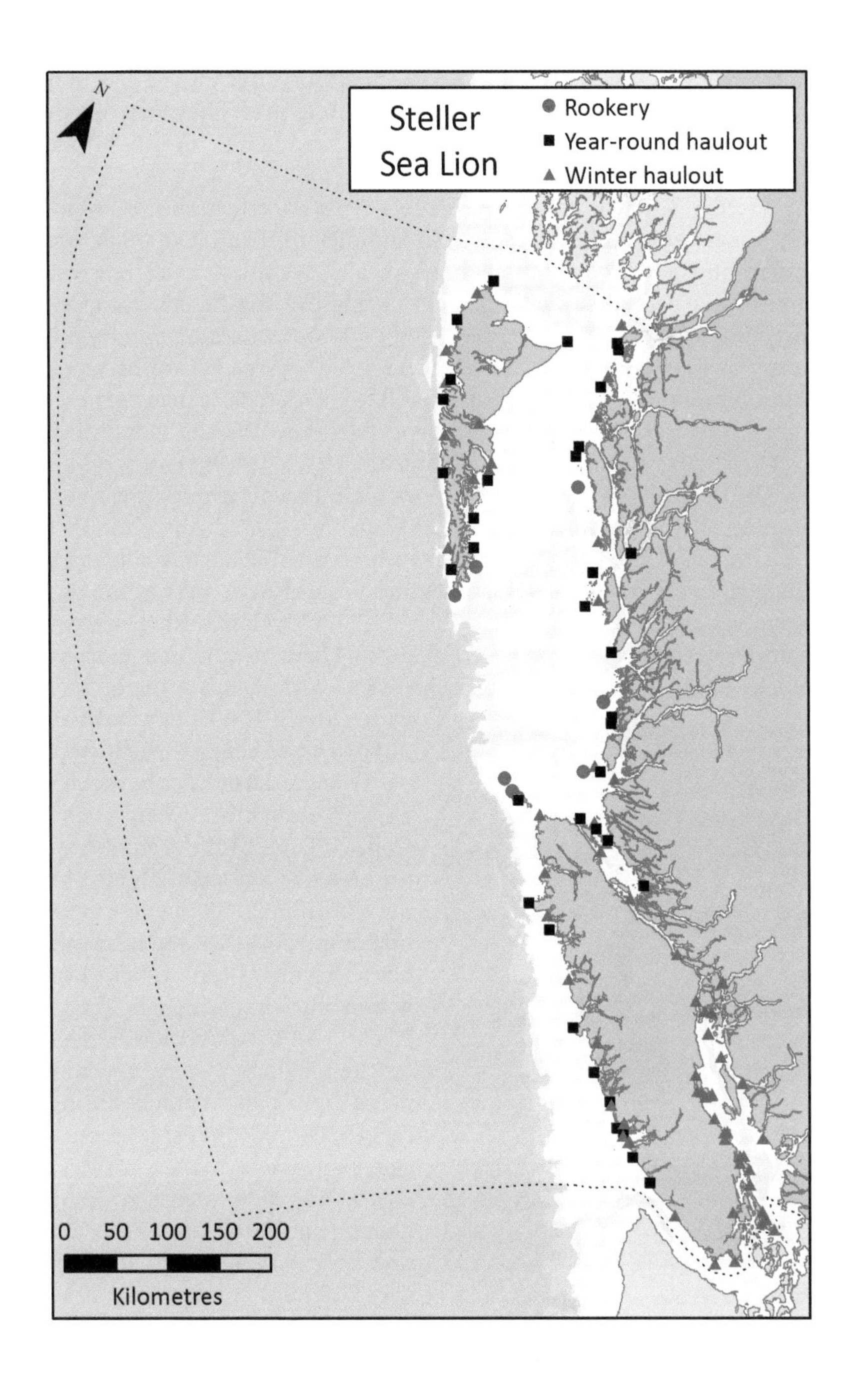
N
Steller
Sea Lion
Rookery
Year-round haulout
Winter haulout
0 50 100 150 200
Kilometres

California. Steller Sea Lions are non-migratory, but individuals may disperse a considerable distance from breeding sites. They feed mostly within 60 km of shore during summer but may range up to 200 km from shore in winter.

Steller Sea Lions use three distinct types of terrestrial sites: rookeries, where animals congregate in summer to mate, give birth and nurse pups; year-round haulouts typically occupied in all seasons; and winter haulouts, used primarily during the non-breeding season. Breeding rookeries tend to be located on remote, isolated and barren rocks or islands in areas of high ocean productivity. Use of breeding sites appears to reflect traditions that have persisted for many generations. Both sexes tend to have strong fidelity to the site where they were born, returning as adults to breed.

In British Columbia there are six main breeding areas: the Scott Islands off the northwestern tip of Vancouver Island; the Kerouard Islands, near Cape St James at the southern tip of Haida Gwaii; North Danger Rocks in eastern Hecate Strait; Virgin Rocks, part of the Sea Otter Group in eastern Queen Charlotte Sound; Garcin Rocks off the southeastern coast of Moresby Island in Haida Gwaii; and Gosling Rocks off the central mainland coast. The Scott Islands is the largest, with almost 4000 pups born in 2010, 96 per cent of which were on Triangle Island at the northwestern end of the island chain. Cape St James is the second largest breeding site, with about 850 pups born in 2010. The last three breeding rookeries are relatively new, having been established within the last decade. Virgin Rocks has been recolonized after the sea lions there were extirpated by intense culling during the 1920s and 1930s; it is once again an active rookery. Garcin Rocks was first identified as a haulout site in the late 1990s, but now hosts breeding animals, with over 200 pups counted in a 2010 survey. Gosling Rocks was found to have over 100 pups in a 2013 survey.

Figure 92. A Steller Sea Lion in Johnstone Strait swallowing a Pacific Spiny Dogfish.
Photo: J. Hildering, May 2013.

There are currently about 30 year-round haulout sites in BC, more than twice the number observed in the early 1970s. These are distributed among rocks and islands off the west coast of Vancouver Island, off the central and northern mainland coast from Queen Charlotte Strait to Hecate Strait, and around Haida Gwaii. The use of winter haulouts is more variable than year-round sites. Most winter haulouts are in protected waters such as the Strait of Georgia, Juan de Fuca Strait and Queen Charlotte Strait. Sea lions may also haul out on log booms, floats, jetties and docks during winter. If no suitable haulout locations are available in a good feeding area, animals may rest at the surface in a tightly packed group called a raft, extending their flippers into the air to conserve heat and periodically exposing their snouts to breathe (see figure 4).

Natural History

Feeding Ecology

Steller Sea Lions feed on over 50 species of fishes and invertebrates, with their diet varying regionally and seasonally depending on prey availability. In general, they prefer small to medium-sized schooling fish; in BC important prey species include Pacific Herring, Pacific Hake, Pacific Sand Lance, Spotted Spiny Dogfish, Eulachon, Pacific Sardine and salmon (figure 92). They also feed on squid, octopuses, and bottom fish such as various species of rockfish, Arrowtooth Flounder and skates. Steller Sea Lions are also known to prey on Harbour Seal and Northern Fur Seal pups, and on gulls. Overall, salmonids appear to represent about 10 per cent of the diet of Steller Sea Lions in British Columbia waters.

Behaviour and Social Organization

Steller Sea Lions are highly polygynous, with males competing for territorial space on rookeries and mating with multiple females in their territory. Adult males start arriving at rookeries in early May and establish territories of about 200 square metres, usually with definable boundaries along cracks, faults or ridges of rock. They defend these territories against other males with a variety of threat displays, hisses and roars at the boundaries. Boundary challenges may escalate to include bouts of rapid neck fencing, biting and occasionally all-out fighting. Successful breeding bulls may not leave their territory for as long as two months, fasting and meeting requirements for energy and hydration by metabolizing their reserves of body fat. Pregnant females begin arriving on rookeries in late May, and by the peak of the breeding season there will be 10–15 times as many females as there are breeding bulls. Females give birth to a single pup within a few days of their arrival and stay with their pup for 7–10 days before leaving the rookeries on regular feeding trips that each last a day. Females mate once with the territorial bull, usually just before the female's first feeding trip. Pups are quite precocious and are able to swim soon after birth. They call to their mothers using a bleating or braying sound that is higher in pitch than the adults' roar. At about four weeks of age, they start travelling with their mother to nearby haulouts. Most animals have left the rookeries by the end of August.

Outside of the breeding season, all ages and both sexes aggregate on year-round or winter haulouts (figure 93). Females with dependent pups may stay at a single haulout or move among several. Feeding trips by lactating females lengthen to about two days, followed by one day on shore. Animals without dependent young may range widely and stay at sea for extended periods between visits to haulouts. Compared to many other pinnipeds, Steller Sea Lions tend to make relatively shallow dives, with few being recorded at depths of more than 250 metres. Most dives are to depths of between 15 and 50 metres and last only 1.5–2.5 minutes.

Life History and Population Dynamics

Males reach sexual maturity at 3–7 years of age, but only territory-holding bulls are successful at mating. These are typically 9–13 years old and may hold territories for up to 7 years in succession. Females become sexually mature at 3–6 years, and most conceive annually into their early 20s. After mating, development of the fertilized ovum is suspended for about three months until implantation in September. This delay is followed by a gestation of 8–9 months, ensuring that females give birth at the same time the following summer. Most females conceive each year, but the rate of reproductive failure can be relatively high in some regions and years. The lactation period in Steller Sea Lions is extremely long compared to most pinnipeds. Most pups are weaned just before their first birthday, but some continue nursing into their second or even third year.

Although Steller Sea Lion pups are precocious, only 60–65 per cent typically survive their first year. Sudden storms can drown them in substantial numbers, not due to an inability to swim but because once swept off the rocks, they cannot clamber back from the rough sea. They are also subject to being trampled and bitten by older animals, and may get separated from their mothers.

Figure 93. Steller Sea Lions on a year-round haulout at Hope Island. The two adult males show the great size they can attain compared to the smaller females and subadult males. Photo: B. Hill, August 2010.

It is generally believed that Steller Sea Lion populations are limited by bottom-up ecosystem processes – that is, they are ultimately limited by food availability. But some researchers have proposed that mammal-hunting Killer Whales could be responsible for declines in sea lion abundance in the Gulf of Alaska due to a shift in the diet of these predators. Although this theory is controversial, Killer Whales could retard the recovery of sea lion populations that have declined for other reasons – this is known as a "predator pit". About 13 per cent of kills that we have observed by Transient (Bigg's) Killer Whales in BC waters involved Steller Sea Lions, juveniles and adults.

Exploitation

Steller Sea Lions were a food source for First Nations in this region for millennia. Remains of this species have been found in Nuu-chah-nulth archaeological middens along the west coast of Vancouver Island, although they are less common than the remains of Northern Fur Seals and Harbour Seals. Steller Sea Lion remains are also found in Coast Salish archaeological sites throughout the Strait of Georgia, although in this area the evidence of marine mammals in middens is generally less prominent than that of terrestrial mammals. Certain First Nations villages in the Strait of Georgia specialized in Steller Sea Lion hunting, especially the Penelekut of Kuper Island who hunted sea lions cooperatively from multiple canoes in the Porlier Pass area. More typically, however, sea lions were hunted by harpooning or clubbing on haulouts.

With the expansion of commercial fisheries along the BC coast in the early 1900s, Steller Sea Lions came to be viewed as serious nuisances that pilfered catches, damaged fishing gear and generally competed with humans for valuable fish, especially salmon. A predator control program was initiated in 1912, and by 1968 an estimated 49,100 Steller Sea Lions had been culled, and an additional 5700 animals killed in commercial hunts. The most intensive culling took place on the breeding rookeries at Virgin Rocks and Pearl Rocks, in the Sea Otter Group off the central BC coast, ostensibly to protect the Rivers Inlet salmon fishery. Federal fishery officers visited these rookeries near the end of the pupping season each year from 1923 to 1939, and used machine guns to shoot animals from boats, then went ashore and clubbed pups that were too young to escape into the water. In total, about 20,000 sea lions, including 7000 pups, were destroyed, effectively eradicating the populations at these two rookeries. Although the sites continued to be used as haulouts, it was not until 2006 that significant numbers of pups were born annually at Virgin Rocks, and this site is now considered a rookery once again.

Following the extirpation of rookeries in the Sea Otter Group, the focus of culling shifted to the Scott Islands off the northern tip of Vancouver Island, with about 7500 animals being killed during 1936–39. An attempt was made to market their hides, but this venture proved uneconomical. Culling was suspended during World War II, but significant numbers of animals may have been killed by the Canadian navy and air force during bombing practices. The kills resumed in 1956 and expanded to rookeries at Cape St James and North Danger Rocks both for predator control and in a commercially unsuccessful venture to sell the carcasses as mink food. A further 12,000 animals were slaughtered during this phase, which ended with the protection of the species under the Fisheries Act in 1970. By this time the population was only about one quarter to one third the size that existed at the turn of the century. Currently, there is minimal deliberate killing of Steller Sea Lions in BC. A few are killed annually by First Nations hunters for food and ceremonial purposes.

Taxonomy and Population Structure

Two separate subspecies have recently been recognized based on differences in morphology, demography and genetics. *E. j. jubatus* (Schreber 1776), also known as the Western Steller Sea Lion, is distributed from Asia through the Aleutian Islands to the Gulf of Alaska. *E. j. monteriensis* (Gray 1859), also known as Loughlin's Northern Sea Lion, is distributed from southeastern Alaska through British Columbia and south to California. These have often been referred to in the past as the western and eastern populations or stocks of Steller Sea Lions, respectively, with a division at 144°W longitude.

Conservation Status and Management

The two subspecies of Steller Sea Lions have had divergent population trends in recent decades and, as a result, have different conservation designations. The western subspecies experienced a dramatic and unexplained decline in abundance of about 70 per cent between the late 1970s and 1990, with the steepest declines of about 15 per cent per year occurring in the late 1980s. Factors identified as potentially contributing to this decline included reduced food supply due to competition with fisheries or to ecosystem regime shifts, deliberate killing by fishermen, disease, predation by Killer Whales, and entanglement in fishing gear and debris. This decline ended around 2000 and there has been a slight overall increase of 1.5–2 per cent annually since then. In contrast, the eastern subspecies has increased at an average annual rate of 3 per cent since the late 1970s.

The abundance of Steller Sea Lions in British Columbia has increased four- to five-fold since control programs and commercial hunting ended in 1970. From surveys in 2010, about 32,000 animals were estimated to inhabit BC coastal waters during the breeding season, increasing to about 48,000 in winter due to an influx of individuals from Oregon and southeastern Alaska. So far, population growth in BC has shown no sign of slowing and abundance today likely surpasses historical levels.

Currently, potential threats to Steller Sea Lions in BC include illegal shooting, incidental take in fishing gear, competition with commercial fisheries, entanglement in synthetic debris (net fragments, packing bands, etc.), and displacement from critical habitat due to disturbance. So far, none of these appear to have been significant enough to inhibit population growth. Between 1990 and 2003, 363 Steller Sea Lions were killed under 'Nuisance Seal Licences' at aquaculture sites, but the species has been protected under the Species at Risk Act since 2004. Several deaths from illegal shooting and entanglement are reported each year in BC.

The Steller Sea Lion is listed by IUCN as Near Threatened, but the eastern subspecies *E. j. monteriensis* is classified as Least Concern. In Canada, COSEWIC listed Steller Sea Lions in 2003 as Special Concern due to the unexplained declines in western Alaska and because there were only three breeding locations in BC. An updated COSEWIC assessment is currently under way. Steller Sea Lions are on the BC Blue List.

Remarks

Steller Sea Lions are named after Georg Wilhelm Steller, a German naturalist who described the species from his observations during a voyage to the Bering Sea with Vitus Bering in 1741. The genus name *Eumetopias* is based on the Greek words *eu* and *metopion*, and means "having a well-developed forehead". The specific name *jubatus* is a Latin word meaning "having a mane".

Selected References: Bigg 1985 and 1988, COSEWIC in press, Gentry and Withrow 1986, Loughlin 2009, McKechnie and Wigen 2011, Olesiuk 2011, Reeves et al. 2002, Suttles 1987.

California Sea Lion *Zalophus californianus*

Other Common Names: None.

Description

The California Sea Lion is a medium-sized to large otariid (eared pinniped). It is sexually dimorphic when mature, with males reaching 3–4 times the weight and 1.2 times the length of females. While females are slender-bodied, males become very robust in the neck, chest and shoulders as they age. Mature males develop a pronounced sagittal crest on the top of the head, which creates a prominent, clearly defined forehead. The muzzle is long and narrow in both sexes. The fore flippers are broad with thinly dispersed hair on the upper surface that extends beyond the wrist and tiny, vestigial claws that are rarely exposed. The hind flippers are relatively short with small claws set back from the fleshy tips at the end of the digits.

The pelage colour of California Sea Lions varies with age and sex. Females, juveniles and young subadult males are tan on top with variable shades of tan to tawny brown underneath. The coat of males begins to darken as they approach maturity, and physically mature males are typically dark brown except for a patch of tan hair on the sagittal crest and a pale face and muzzle (figure 94). Pups are born with thick, brownish-black fur that is shed by the end of their first month. After another moult at about six months they show typical juvenile coloration.

Measurements

length (metres):
- male: 2.0–2.5
- female: 1.4–1.7
- neonate: 0.80

weight (kg):
- male: 200–390
- female: 70–110
- neonate: 6–9

Dental formula

- incisors: 3/2
- canines: 1/1
- post-canines: 5/5 (some 6/5 or 7/5)

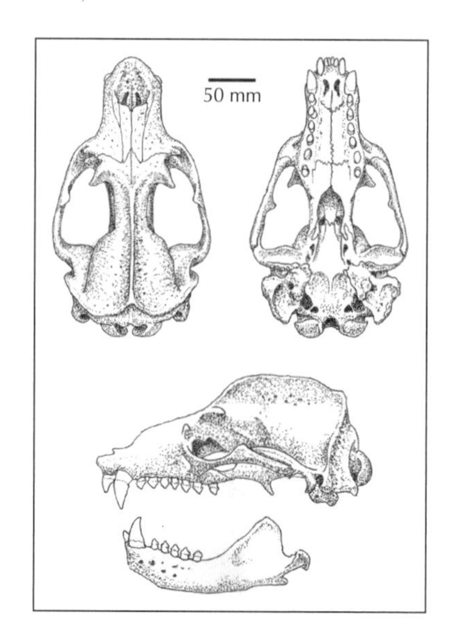

Identification

In BC waters California Sea Lions are most likely to be confused with Steller Sea Lions. The two species are often found together on haulouts or in mixed groups in the water. Only male California Sea Lions typically migrate to BC during fall through spring – they are noticeably darker and smaller than adult male Steller Sea Lions, though similar in size to female Stellers. The difference in coloration is especially obvious when animals are in the water – California Sea Lions are dark brown, almost black in appearance, whereas Stellers are very pale. Other distinguishing features are the pronounced light-coloured sagittal crest on the head of mature male California Sea Lions and their narrow, pointed muzzle (figure 94). Male California Sea Lions have a very distinctive, loud bark that is often given in a long repetitive series – a striking contrast to the deep, guttural growl of Steller Sea Lions. Another species that may be confused with the California Sea Lion in BC waters is the Northern Fur Seal. Although both are dark in colour, Northern Fur Seals have a short, pointed muzzle and are considerably smaller in size, since only females and juvenile males typically occur here. And Northern Fur Seals are usually found 20 km or more from shore, rarely hauling out in BC.

Distribution and Habitat

California Sea Lions are found in coastal waters of the eastern North Pacific from the northern mainland of Mexico near Puerto Vallarta

Figure 94. An adult male California Sea Lion showing his pronounced sagittal crest with lighter colouration than the subadult male behind him.
Photo: R. Mirza, Race Rocks, September 2010.

to southern Alaska. They breed at rookeries on islands off the coast of southern California and Baja California, Mexico, preferring sandy beaches to the exposed rocky ledges used by Steller Sea Lions. They also haul out on rocky islets, coastal headlands, beaches under cliffs that limit access by terrestrial predators, and on man-made structures such as wharfs, docks, channel buoys, log booms and moored boats. As the abundance of California Sea Lions has increased over the last few decades, so too has their range. Females were rarely found north of Point Conception, California, in the early 1980s but are now commonly seen in northern California, and some pups have been born at the Farallon Islands near San Francisco. During the breeding season the majority of California Sea Lions are found south of Oregon, but in late summer males begin a seasonal migration northward, with many reaching British Columbia and occasionally southeastern Alaska. Most return to California by late spring, at the start of their breeding season.

The California Sea Lion population was severely depleted due to overhunting in southern California and Mexico during the 19th and

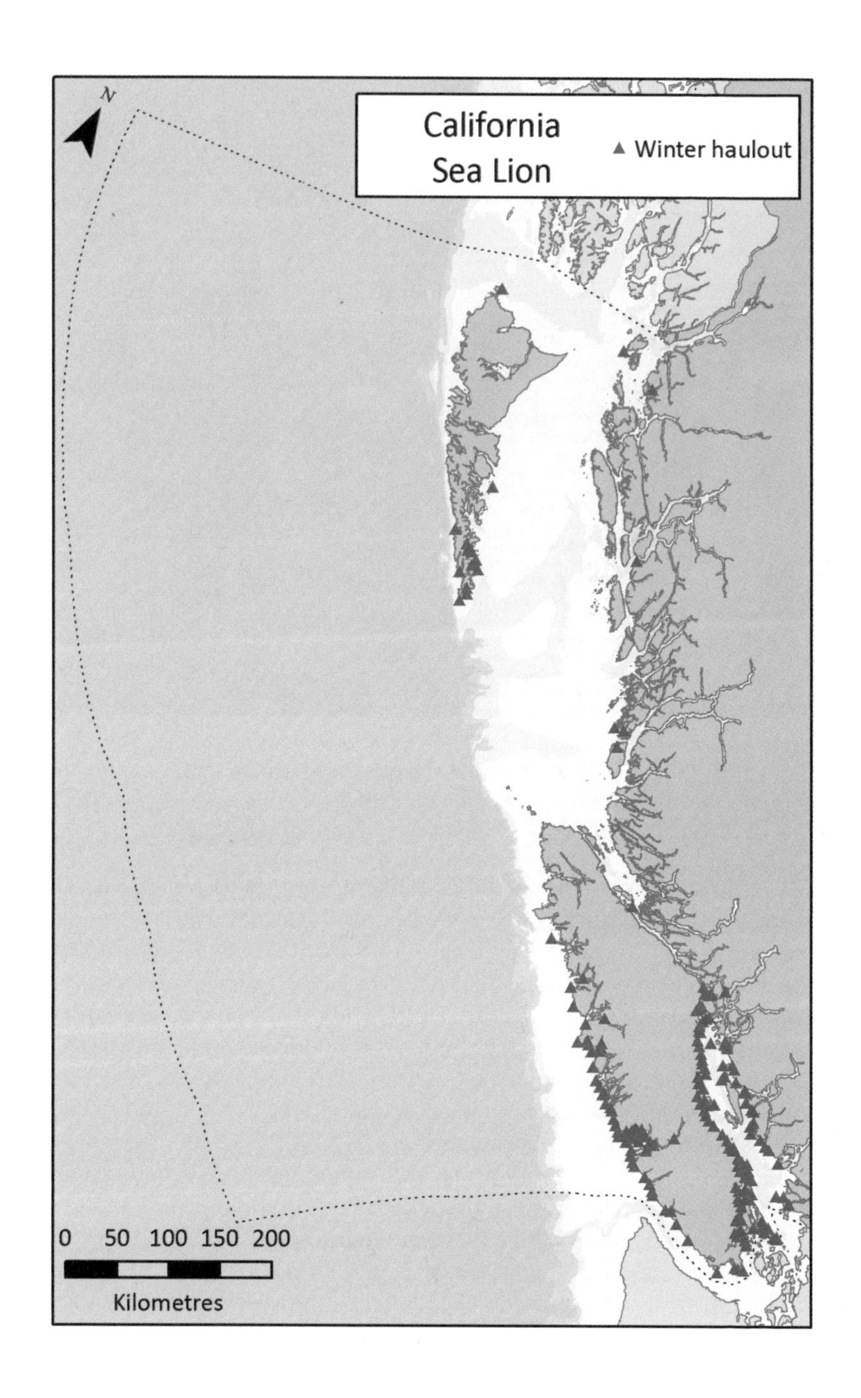
N
California
Sea Lion
Winter haulout
0 50 100 150 200
Kilometres

early 20th centuries, which contracted their range. Prior to the 1960s California Sea Lions were rare in British Columbia (at least during historical times), with only occasional reports of individuals mixed in with Steller Sea Lions at haulouts off southwestern Vancouver Island. More animals were seen in the 1960s, and numbers increased dramatically during the 1970s and 1980s. In recent years several thousand are typically present along the BC coast from late fall to spring. They are seen mostly along the west coast of Vancouver Island and in the Strait of Georgia, generally mixed with Steller Sea Lions on haulouts or resting in rafts at the surface. In some years, substantial numbers of California Sea Lions have been reported as far north as Haida Gwaii and the central mainland coast.

Natural History

Feeding Ecology

California Sea Lions feed opportunistically on a wide spectrum of prey, with their diet varying according to seasonal and regional availability of different species. Important prey in their overall range include Pacific Hake, Pacific Herring, Northern Anchovy, Pacific Mackerel, Jack Mackerel, Pacific Sardine, Opalescent Inshore Squid, octopuses and various rockfishes and salmonids. The diet of California Sea Lions in BC is described from prey remains recovered from scats collected by Pacific Biological Station researchers in the mid 1980s off southern Vancouver Island. Prey species identified were, in descending order of occurrence in scats: Pacific Herring, Pacific Hake, Pacific Spiny Dogfish, salmon (not identified to species), squid or octopus, Eulachon, Sand Lance, skates, and Lingcod. Pacific Sardines are likely important prey in BC today – these fish were uncommon off the BC coast in the 1980s but have increased dramatically in abundance since then, primarily off the west coast of Vancouver Island. In some recent years, however, Pacific Sardines were abundant off southern Haida Gwaii and along the central mainland coast, and they appeared to attract substantial numbers of California Sea Lions to those areas. Spring spawning aggregations of Pacific Herring are a major draw for both California and Steller Sea Lions in winter and early spring.

Behaviour and Social Organization

Like all otariid pinnipeds, California Sea Lions have a polygynous mating system. Adult males compete for and defend territories on traditional breeding rookeries, and they mate with females who choose to haul out and give birth in the territory. Boundaries between territories are maintained with ritualized displays and virtually incessant barking, escalating occasionally to physical combat. Males fast for the duration of their time on the territory, which may last up to 45 days. In May and June females arrive on the rookeries and give birth within a few days, then mate with the territory-holding male about a month later. During the first week postpartum, the mother stays with her pup, nursing frequently and establishing mutual recognition by scent and individually distinctive contact calls. After this period, the mother begins alternating feeding trips of two or three days at sea with bouts of suckling her pup for one or two days, a pattern that continues until the pup is weaned before the next breeding season. Older pups may join their mothers on foraging excursions prior to weaning.

Tagging studies in California have revealed details about the foraging and diving behaviour of California Sea Lions. Adult females travelled as far as 66 km from rookeries during feeding trips that lasted 48 hours on average. Depths of dives were highly variable, with those by some individuals averaging less than 30 metres whereas others averaged more than 100 metres. Adult males outside of the breeding season foraged mostly over the continental shelf but during years of low productivity due to anomalously warm water temperatures, they ventured as far as 650 km offshore. On average, males dove to 32 metres for periods of 1.9 minutes, although many dives were deeper (maximum 575 metres) and longer (maximum 20 min). Diving behaviour appeared to vary according to the types of prey being sought – short, shallow dives targeted Northern Anchovies, Pacific Sardines and Opalescent Inshore Squid, while longer, deeper dives were associated with Pacific Hake, flatfishes and rockfishes. California Sea Lions are intelligent, adaptable predators that can exploit novel feeding opportunities. For example, in Santa Monica Bay, California, foraging California Sea Lions frequently follow schools of dolphins, apparently to take advantage of the dolphins' superior ability to locate concentrations of prey by echolocation.

Life History and Population Dynamics

California Sea Lions reach sexual maturity at 4–5 years old, although males are not large enough to hold a breeding territory for several more years. Females give birth to a single pup within a few days of arriving on breeding rookeries in late May or June, then come into estrus and mate about 27 days later. Implantation of the fertilized ovum is delayed for about two months, followed by a nine-month gestation leading to parturition the following year. Pups are generally weaned at 10–12 months, but some continue to be suckled as yearlings or even two-year-olds. Longevity is estimated to be 15–24 years. Natural predators of California Sea Lions include Great White Sharks and mammal-hunting Transient (Bigg's) Killer Whales. California Sea Lions are vulnerable to increased mortality and low pup survival during years with strong El Niño conditions, which reduce prey availability.

Exploitation

California Sea Lions were exploited by First Nations hunters for thousands of years. They were especially important to aboriginal groups living in the vicinity of breeding rookeries in coastal southern California and the Channel Islands. In British Columbia, California Sea Lion remains have been found in archaeological sites dated to 300–2400 years ago along the west coast of Vancouver Island from Hesquiat to Barkley Sound. The remains are uncommon, however, and the species appears to have been less important to First Nations in this area than most other pinnipeds, particularly Northern Fur Seals.

Exploitation of California Sea Lions increased dramatically in the 19th century following the arrival of Europeans in California and Mexico. Uncontrolled killing for various purposes, including hides, oil, pet food and predator control, drove the population down to only about 1500 animals by the 1920s. Low market value of sea lion products and new protective legislation in the US allowed the population to begin increasing by the 1940s, and the US Marine Mammal Protection Act of 1972 has led to the population's continued strong recovery. The species has never been hunted commercially or culled in British Columbia. Limited hunting of California Sea Lions by Nuu-chah-nulth groups on the west coast of Vancouver Island and Kwakwa̱ka̱'wakw communities at Port Hardy and Campbell River is currently permitted for food and ceremonial purposes, but altogether they take fewer than 20 animals annually.

Taxonomy and Population Structure

Until recently *Zalophus californianus* was considered to comprise three subspecies – the California Sea Lion *Z. c. californianus*, the Galápagos Sea Lion *Z. c. wollebaeki* and the Japanese Sea Lion *Z. c. japonicus*. However, based on new evidence showing significant genetic and morphological differences, the three forms have been elevated to full species status. The Japanese Sea Lion is believed to have become extinct in the late 1950s.

Conservation Status and Management

Once reduced to only 1500 or so, the California Sea Lion is thriving today. Protection in the early 1970s has allowed the population in California to increase at an average annual rate of about 5.6 per cent, excluding El Niño years with poor survival. Current abundance is estimated to be about 238,000 in California and 105,000–117,000 in Mexico.

Sightings of California Sea Lions in British Columbia were rare from the late 1800s to around 1960, likely due to serious depletion of the population from over-exploitation in California and Mexico. Individuals were occasionally seen mixed in with Steller Sea Lions at haulouts on Sea Lion Rocks off Long Beach, in Barkley Sound and at Race Rocks west of Victoria during the 1950s and 1960s. Numbers of California Sea Lions in BC began to increase in the 1970s, coinciding with the strong recovery of the population generally. During a province-wide winter sea lion survey flown by Michael Bigg and co-workers from the Pacific Biological Station in 1971, about 260 California Sea Lions were counted between Barkley Sound and Race Rocks. Abundance continued to rise rapidly through the 1970s and 1980s, and California Sea Lions began to be sighted in the Strait of Georgia. By 1982 the total mid winter count was 1500 animals, and by 1984 it had reached 4500. Numbers thereafter declined to a mid winter average of 2600 during the late 1980s and to about 1500 during the 2000s. Since the mid 1980s there has been a shift in the winter distribution of California Sea Lions, with relative abundance declining off the west coast of Vancouver Island and increasing in the Strait of Georgia. This may correspond to the weakening of Pacific Herring stocks off the west coast of Vancouver Island. There has also been a shift northward of California Sea Lions in the Strait of Georgia, associated with a northward shift in the location of spawning aggregations of herring. Most sea lions are now found from Nanaimo north to Baynes Sound and Lambert Channel, near Denman and Hornby islands. Aerial surveys by the Pacific Biological Station in autumn along the southwestern coast of Vancouver Island found about 3000 California Sea Lions in 2008 and a surprising 10,000 in October 2012. These were located mostly on haulouts in Barkley Sound and at Perez Rocks off Hesquiat Peninsula. What caused this recent influx of animals is unknown.

Currently fisheries are the main source of human-caused California Sea Lion mortality. Accidental entanglement and drowning in drift or set gillnets kills about 75 each year off the US west coast. In BC, over the past decade, an annual average of 18 California Sea Lions have drowned accidentally in anti-predator nets at salmon aquaculture facilities, mostly in Clayoquot and Nootka sounds. Anti-predator nets are strong, large-mesh nets designed to keep pinnipeds away from vulnerable inner nets containing salmon, but faulty deployment or maintenance can allow sea lions to enter, become trapped between the nets and drown. Intentional shooting of pinnipeds at aquaculture sites is also permitted if other anti-predator measures have failed and the animals are threatening the farm's fish stock. Between 2000 and 2011, 815 California Sea Lions (average of 68 per year) were reported killed under licence, with 224 killed in 2011 alone. Improved management practices reduced California Sea Lion kills to 4 in 2012 and none in 2013.

The California Sea Lion is listed as Least Concern on IUCN's Red List, and its COSEWIC status is Not at Risk in Canada. It is on the BC Yellow List.

Remarks

As the well-known performing "seal" in aquatic parks, zoos and circuses, the California Sea Lion is one of the most widely recognized marine mammals. The genus name *Zalophus* is derived from the Greek prefix *za* and *lophos*, meaning "crest", referring to the enlarged sagittal crest on the head of adult males.

Selected References: Bearzi 2006, Guiguet 1953, Heath and Perrin 2009, Mate and DeMaster 1986, McKechnie and Wigen 2011, Olesiuk and Bigg 1988, Pike and MacAskie 1969.

Northern Elephant Seal

Mirounga angustirostris

Other Common Names: Sea Elephant.

Description

The Northern Elephant Seal is the largest pinniped inhabiting the northern hemisphere and is only surpassed in size by the Southern Elephant Seal. This species has pronounced sexual dimorphism: as males reach puberty, they grow much larger than females, especially in the neck and chest area; adult males are 3–4 times heavier and almost 1.5 times longer than females. The body of both sexes is long and robust, especially in the upper body, tapering to narrow hips. Typical of phocids (true or earless seals), the fore flippers are short and used mostly for propelling the animal on land, while the hind flippers are broad and are used mostly for propulsion in water. The eyes are relatively large and widely spaced. and they develop an elongated and inflatable fleshy snout, or *proboscis*, which extends 15–25 cm beyond the lower lip when fully mature. When relaxed, the pendulous nose droops over the mouth, looking somewhat like a shortened elephant trunk. When males rear their head to vocalize during breeding displays, the nose is inflated and projects into the open mouth, acting as a resonating chamber for their distinctive calls. The noses of females and juveniles are not so pronounced, extending only slightly beyond the mouth. Adult males also develop much larger canine teeth than females, which may protrude 10 cm from the gum.

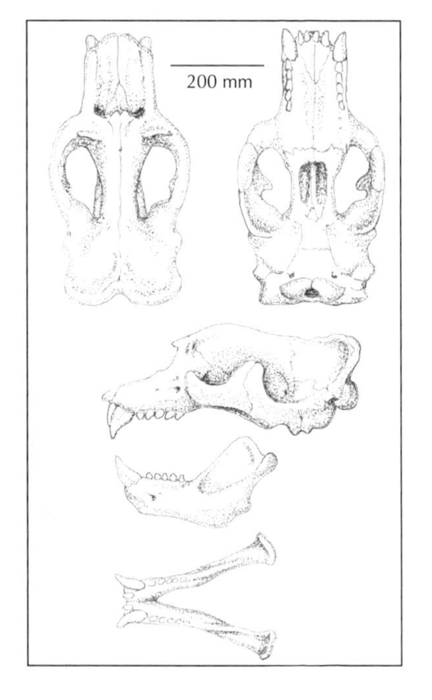

The pelage of Northern Elephant Seals varies in colour from light tan to dark brown or grey depending on age, sex and stage

in the annual moult, when they shed their fur and epidermis together in patches. After moulting, the new coat of both adults and juveniles is silver to dark grey, but this fades over the next few months to light brown or tan. Adult males have medium to dark brown fur during most of the year, except on the chest and neck, which are mostly hairless. This area is made up of thickened, heavily calloused and creased skin that serves as a shield in fights during the breeding season. Breeding bulls accumulate wounds and scars on the chest, neck and head from such fights as they age, and these scarred areas become light grey or tan, mottled with pink splotches. Pups are black at birth, silver-grey after their first moult at three to five weeks old, and light brown or tan at two to three months.

Measurements

length (metres):

male: 3.7–4.2
female: 2.6–2.8
neonate: 1.2–1.3

weight (kg):

male: 600–2700
female: 450–600 (max 710)
neonate: 30–40

Dental formula

incisors: 2/1
canines: 1/1
post-canines: 5/5

Identification

Although both male and female Northern Elephant Seals are observed in BC waters, adult males, which are unmistakable due to their large head and prominent proboscis, are more commonly seen. Sightings at sea typically involve single animals, and normally all that is seen is the head held almost motionless above the water's surface for up to three minutes or so as the seal replenishes oxygen stores following a long dive (figure 95). The seal then sinks vertically under the water to start its next dive. Mariners approaching what appears to be a deadhead in the distance have often been surprised to have it disappear before their eyes.

Both juvenile and adult Northern Elephant Seals may be found hauled out on beaches in BC while they moult. They are scruffy in appearance as they shed patches of epidermis and fur, often leading to the mistaken belief that they are sick or injured since they are reluctant to enter the water even when approached (figure 96). Adult Elephant Seals, particularly males, dwarf any other seal or sea lion that might be nearby. Subadults can be confused with Harbour Seals, which are similar in size, but Northern Elephant Seals lack the spotted pelage of Harbour Seals and are a uniform greyish-brown to tan colour, sometimes with patches of sloughed fur and outer skin if moulting.

Figure 95. An adult male Northern Elephant Seal near Race Rocks, showing the characteristic nose-up posture wihile breathing at the surface after a dive. Photo: V. Shore, April 2007.

Figure 96. A juvenile Northern Elephant Seal hauled out to moult on Perez Rocks on the west coast of Vancouver Island. Elephant Seals shed the epidermis as well as the fur, often resulting in superficial bleeding and infections. They are often lethargic when moulting and mistakenly thought to be sick or injured. Photo: J. Towers, May 2013.

Distribution and Habitat

Northern Elephant Seals range throughout the eastern and central North Pacific. They breed primarily on islands lying off the west coast of Baja California, Mexico, and off California. The largest breeding colonies are found off southern California at San Miguel and San Nicolas islands in the Channel Islands. Significant breeding sites are also located on mainland beaches, including Point Reyes, Año Nuevo Point and Piedras Blancas, California. Northern Elephant Seals are strongly migratory, returning to their rookeries twice each year, once to breed during December to March and then again a few months later to moult. Some individuals moult at beaches from Oregon north to British Columbia. Between the two visits to land, these seals range widely in deep waters on and beyond the continental shelf. Overall, about 8–10 months of the year are spent foraging at sea. Studies using instruments attached to and recovered from seals on their haulouts have shown that males tend to travel further north on their foraging excursions than the females, with some males going as far as the western Aleutian Islands and northern Gulf of Alaska, a round trip of up to 10,000 km from their breeding rookeries. Females spread out over a wide expanse of the open North Pacific, especially in a latitudinal

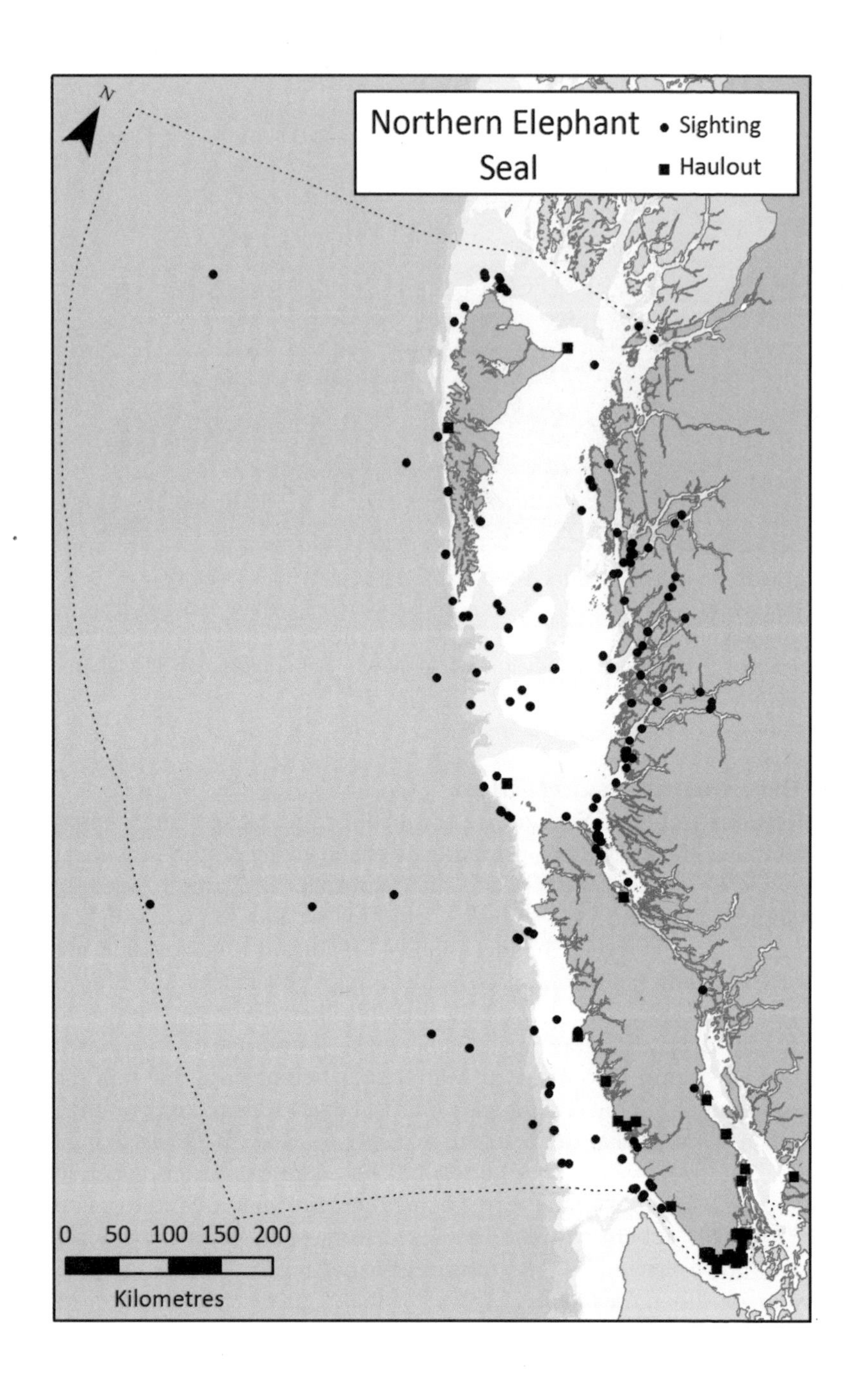
Northern Elephant Seal
Sighting
Haulout
N
0 50 100 150 200
Kilometres

band between 40°N and 50°N in a highly productive area known as the Transition Zone.

In British Columbia, Northern Elephant Seals may be observed in deep waters off the outer coast beyond the shelf break, over the continental shelf or in protected channels and fjords along the central and northern mainland coast. Being deep-diving animals, they are seldom seen in shallow waters unless they are approaching land to haul out. Solitary individuals, especially juveniles, have been observed with increasing regularity hauled out on beaches to moult, mostly from December to May but sometimes in summer and fall. This has been seen most frequently on beaches in the southern Vancouver Island area but also at Triangle Island off the northern tip of Vancouver Island and Rose Spit at the northwestern corner of Haida Gwaii.

In the late 1980s Northern Elephant Seals began hauling out occasionally at Race Rocks, about 15 km southwest of Victoria, and by the mid 1990s they had become regular visitors. Thanks to the efforts of Garry Fletcher and colleagues with the Friends of Ecological Reserves, detailed records have documented the establishment of a new, albeit small, breeding colony at this site. Over the past decade Northern Elephant Seals have been present year-round, with as many as 20–25 animals at a time. Mating behaviour has been observed and several pups have been born there in recent years, mostly in December and January (figure 97). Some of the individuals observed at Race Rocks have flipper tags that originate from the Año Nuevo rookery in central California, suggesting that this may be the main source of colonizing seals for this nascent northern breeding site.

Natural History

Feeding Ecology

Knowledge of Northern Elephant Seal diet is based on prey remains recovered from dead animals on shore or from lavaged (or pumped) stomachs of hauled-out seals sedated by researchers. These remains likely reflect predation in the days just prior to the seal's reaching land and therefore may not provide a detailed picture of diet on the high seas. Nonetheless, over 50 species of prey have been identified, including: a wide variety of deepwater, oceanic squid; bony fishes such as Pacific Hake and rockfishes; small sharks and rays; and cyclostomes (Pacific Hagfish and Pacific Lamprey).

Figure 97. A Northern Elephant Seal female, male and newborn pup at Race Rocks Ecological Reserve. Photo: R. Mirza, January 2011.

Differences in the diving patterns and foraging areas of males and females suggest broad dietary differences at sea, with females preying mostly on pelagic squid and fishes in the water column between 400 and 600 metres deep, and males targeting prey closer to shore but nearer to the bottom, possibly sharks and cyclostomes, along the continental slope. Most prey are consumed at depth, but on several occasions adult male Northern Elephant Seals in BC coastal waters have been observed swallowing whole Spotted Spiny Dogfish at the surface. In January 2013, a female Northern Elephant Seal eating a Pacific Hagfish was captured on live video at the Barkley Canyon node of the NEPTUNE Canada cabled observatory network. As the hagfish moved slowly on the sea floor at a depth of about 900 metres, the seal nosed up to it and appeared to suck it off the bottom simply by using the suction created as it opened its mouth.

Behaviour and Social Organization

Northern Elephant Seals are phenomenal divers that spend most of their life alone on the high seas, far under the surface. Overall, these seals spend 67–83 per cent of the year at sea, and for about 90 per cent of this time they are submerged on deep dives. Females generally dive deeper than males, averaging 450–520 metres, with a maximum recorded depth of 1735 metres. Males, on the other hand, dive to about 300 metres on average, rarely going deeper than 1000 metres. Dives by both sexes average about 22–23 minutes, followed by an average of

2.9 minutes at the surface. One recorded dive by a female lasted 109 minutes.

In contrast to their solitary life at sea, Northern Elephant Seals gather in dense colonies when they return to shore to breed and moult. They are polygynous breeders, with adult males establishing dominance hierarchies and defending their access to groups of females from rival males. But they do not defend territories with clearly defined boundaries, as do sea lions. Dominance among male Northern Elephant Seals is determined through ritualized physical and vocal threats and displays, which sometimes escalate into violent battles. When sparring, males face each other and rear up on their hindquarters, lifting the forward half to two-thirds of their body up off the beach in the process. Then they throw their head back and emit a loud, resonant "clap-threat" bellow. If neither male backs off, intense fighting can result, with each animal slashing at the other's head, neck and chest with its large canine teeth. The thick, calloused skin of older males provides some protection, but serious wounds can result from such fights. Once the hierarchy is established, conflicts are usually settled by the exchange of vocalizations and threat postures.

Females arrive on the rookeries a few weeks after males, from late December to March, and give birth to a single pup within a few days. A strong bond is formed between mother and pup soon after birth through the mutual recognition of individually distinct contact calls. Females emit a guttural growl when threatened and respond aggressively to any other pup that approaches. Just before weaning (about four weeks), the mother comes into estrus and mates with one or more socially dominant males. She then heads to sea on a foraging excursion that lasts about 75 days and when she returns to land she spends several weeks moulting. She follows this with a longer pelagic migration of about seven months before returning to the rookery to give birth. Males leave the rookery in late February or March, return to moult in July and August, then head to sea again until the start of the next breeding season in December.

Life History and Population Dynamics

Female Northern Elephant Seals are ready to breed at just three years old, but males do not breed until they are seven to nine years of age. The annual pregnancy rate of mature females is about 95 per cent. Gestation lasts 11 months, including a 4-month period of delayed implantation, and females give birth within a week of hauling out on the breeding rookery. They nurse their pups for an average of 27 days, fasting all the while and relying on their blubber reserves to sustain them and to supply the pup with many litres of milk. Between birth and weaning, pups increase their weight by four or five times. At the same time, mothers may lose almost half their body mass. After mating and weaning their pup, mothers head to sea to replenish their reserves. The pups, known as *weaners* at this stage, stay on the rookery beach for about a month, making short excursions to sea to gain swimming and diving experience. Once they leave the rookery, pups are particularly vulnerable to predators, such as Great White Sharks and mammal-hunting Transient (Bigg's) Killer Whales, and only about half of them survive their first year. Longevity is about 14 years for males and 21 years for females.

Exploitation

Northern Elephant Seals, as well as other pinnipeds, were important to the subsistence economy of First Nations living on the southern California mainland and adjacent Channel Islands. In BC, elephant seal bones have been recovered in some Nuu-chah-nulth archaeological sites on the west coast of Vancouver Island, but they are much less common than Northern Fur Seal, Harbour Seal and sea lion remains.

Northern Elephant Seals were the focus of intensive commercial exploitation following the arrival of Europeans to the west coasts of Mexico and California. Beginning around 1818 these seals were slaughtered relentlessly on breeding rookeries for the valuable oil that could be extracted from their blubber. By 1850, seals were scarce but the sealers continued to search for new rookery sites and killed any Northern Elephant Seals they found.

The species was considered extinct by the late 1880s, but in 1892 a group of nine animals was found by an expedition from the Smithsonian Institution at the site of the former large rookery on Isla Guadelupe off Baja California. Thinking that the species' survival was beyond hope, they killed seven of the nine seals for their collection. But a remnant population thought to number less than 100 individuals, possibly as few as 20, somehow survived and slowly fueled recovery. The species was given protection by the Mexican government in 1922 and by the US soon thereafter, and the population began to increase rapidly. By 1960 all of the significant former breeding colonies had been re-colonized and the population numbered about 15,000. Growth has continued at about 6–8 per cent per year, and the population was estimated at 171,000 when last surveyed in 2005.

Taxonomy and Population Structure

The Northern Elephant Seal is closely related to the Southern Elephant Seal, found exclusively in the southern hemisphere. No subspecies are recognized for either species. Northern Elephant Seals went through a severe genetic bottleneck when driven to the brink of extinction in the late 19th century, and all animals today have descended from very few breeding individuals. As a result, there is minimal genetic diversity in the population.

Conservation Status and Management

Northern Elephant Seal abundance is stable or slowly decreasing in Mexico, but it continues to increase on California rookeries. Numbering more than 171,000 and with relatively few threats from human activities, the species is ranked as Least Concern on the IUCN Red List and Not at Risk in Canada by COSEWIC, and it is on the BC Yellow List. A few deaths are recorded each year from entanglement in marine debris and vessel strikes, but these are insignificant at the population level.

Remarks

We tend to think of seals and sea lions as land-based mammals that must go to sea to feed, and that once in the water they must dive to forage but return to the surface as soon as they can. But Northern Elephant Seals are so proficient under water that a different perspective is warranted – some say that they should be thought of as "surfacers" rather than "divers". They spend the great majority of their lives in the deep ocean, on average at 300–500 metres but often far deeper under the surface. These seals are at sea for up to 300 days per year, and more than 20 out of every 24 hours is spent at depth. This is equivalent to spending seven to nine months of the year under water. Every 20–25 minutes they must come back to the surface to breathe, but they waste little time doing so. They ascend rapidly, ventilate for less than three minutes, then descend back to depth just as quickly.

Selected References: Cowan and Carl 1945, Le Boeuf 1974, Le Boeuf et al. 2000, Le Boeuf and Laws 1994, McGinnis and Schusterman 1981, McKechnie and Wigen 2011, Pike and MacAskie 1969, Robinson et al. 2012.

Harbour Seal *Phoca vitulina*

Other Common Names: Common Seal, Hair Seal.

Description

The Harbour Seal is a small phocid pinniped with a short body, a large head and short limbs. The head is rounded with large eyes and V-shaped nostrils on the end of the snout. The prominent whiskers are flat with bead-like expansions that give them a spiral appearance. The ear openings are relatively large and prominent but, of course, lack the pinnae of otariids. The fore flippers are short with robust claws. Males are slightly larger than females when mature, but otherwise there is minimal sexual dimorphism.

The pelage of Harbour Seals varies considerably in pattern and colour, both regionally and among individuals. Small markings – spots, rings and blotches – range from greyish-white to dark brown to black, and these are set against a background of similar colour variability. Harbour Seals tend to be mostly dark in southern California and Baja California, and are typically lighter from central California to the eastern Aleutian Islands. Both light and dark morphs, as well as individuals with intermediate coloration, can be found along the BC coast (figure 98), but higher proportions of dark Harbour Seals are seen in Haida Gwaii and in mainland inlets than in the Strait of

Figure 98. The pelage of Harbour Seals varies in colour from mostly blonde to brown and black. Photo: B. Hill, Mussel Inlet, September 2008.

Georgia. The soft silvery-grey fur of neonates, called the *lanugo*, is usually shed before birth, so the pelage of newborn Harbour Seals resembles that of adults.

Measurements

length (metres):

male: 1.49 (1.31–1.69) n=11
female: 1.45 (1.28–1.62) n=9
neonate: 0.83 (0.68–0.93) n=25

weight (kg):

male: 80 (43–136) n=9
female: 63 (43–81) n=7
neonate: 12 (10–15) n=13

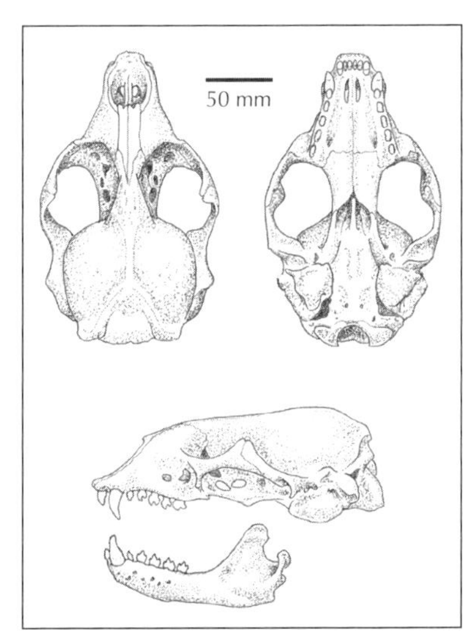

Dental formula

incisors: 3/2
canines: 1/1
post-canines: 5/5

Identification

Harbour Seals are ubiquitous in nearshore waters of British Columbia, and most people who have spent time on or near the ocean should be familiar with this species. It is the only small true (or "earless") seal in BC. It is possible to confuse Harbour Seals with juvenile Northern Elephant Seals, which sometimes haul out on beaches, especially on southern Vancouver Island, to moult. But Northern Elephant Seals are substantially larger (even as juveniles), are uniformly grey to tan in colour and are comparatively uncommon in BC. Steller Sea Lions, California Sea Lions and Northern Fur Seals all have conspicuous pinnae (external earflaps) and are usually larger than Harbour Seals (although juvenile Fur Seals may be similar in size). These species also tend to dive with a forward swimming motion whereas Harbour Seals quietly sink vertically to swim beneath the surface.

Distribution and Habitat

The Harbour Seal lives in a wide variety of habitats in temperate, subarctic and some arctic marine waters of the Northern Hemisphere. In the North Pacific, Harbour Seals are found from the central west coast of Baja California northward to the Gulf of Alaska and the southeastern Bering Sea, westward to the Kamchatka Peninsula and south to Hokkaido Island, Japan. Harbour Seals usually occur within 20 km of the coast but are sometimes seen up to 100 km offshore. They favour nearshore areas with tidal mud flats, sand bars and reefs, where they can haul out with minimal threat from terrestrial predators. Though primarily marine, they also occur in some large rivers and accessible lake systems. Harbour Seals are non-migratory but have local movements associated with tides, food, reproduction and season.

Harbour Seals are found in all coastal areas of British Columbia, including the exposed outer coast, protected bays, straits and sounds, and narrow coastal fjords. They are found in at least 20 rivers in BC and range at least 250 km up the Skeena River and as far as Hell's Gate in the Fraser River, about 500 km from the ocean. They can also be found year-round in several BC lakes, including Pitt Lake and Harrison Lake in the lower Fraser River valley, Owikeno Lake on the central coast, and Kitlope Lake on the north coast. The highest concentration of Harbour Seals in BC is in the Strait of Georgia, which has an average density of 13.1 seals per kilometre of shoreline. In other areas of the BC coast, densities average about 2.6 seals per kilometre. Nearly 1400 Harbour Seal haulout sites have been identified in the province, with over 500 in the Strait of Georgia region alone. Major haulout sites with more than 200 seals in the Strait of Georgia region include the Belle Chain Islets, Java Islets, Boiling Reef, Boundary Bay, Roberts Bank, Chain Islets, Snake Island, Norris Rocks, Flora Islet, Mitlenatch Island and Pam Rocks. These seals also often haul out on

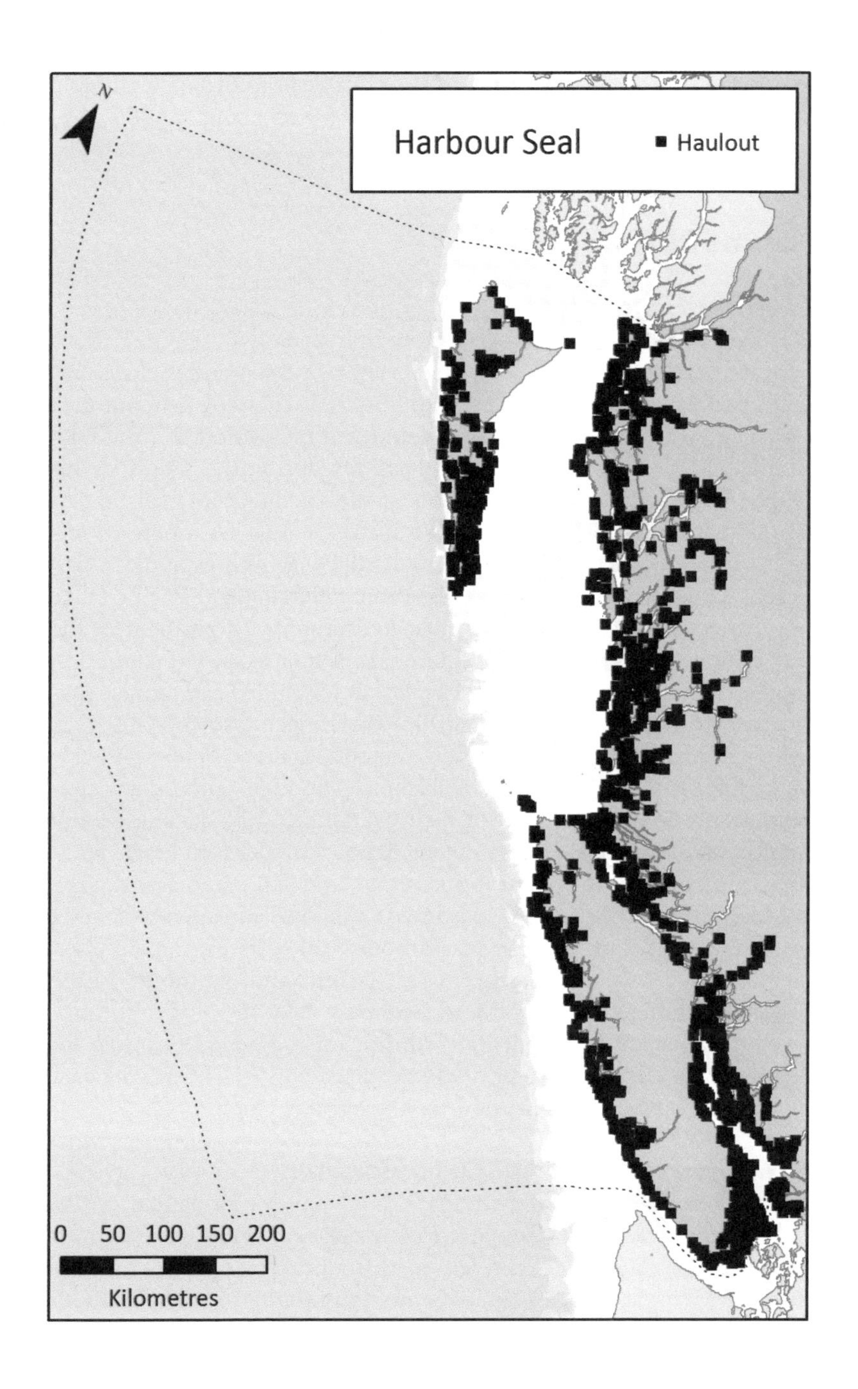

N
Harbour Seal
Haulout
0 50 100 150 200
Kilometres

log booms. The total population of Harbour Seals in British Columbia was estimated in 2008 to be 105,000, of which about 39,000 were found in the Strait of Georgia region.

Natural History

Feeding Ecology

Harbour Seals are generalist predators that feed on a variety of abundant and easily attainable prey. Diets vary by season and region. This seal's preferred food is small to medium-sized schooling fish, but they also feed on invertebrates such as squid, octopuses and shrimp. Diet studies based on extensive scat collections during the 1980s by biologist Peter Olesiuk and colleagues at the Pacific Biological Station found that Harbour Seals in the Strait of Georgia fed upon 48 different fish species from 20 different families, as well as squid, octopuses, echinoderms and even birds. Pacific Herring and Pacific Hake were the dominant prey, together accounting for 75 per cent of the diet both in terms of energy and biomass. Other fishes were found in smaller numbers (<5 per cent of scats) and included salmonids (species undetermined), Plainfin Midshipman, Lingcod, surfperches, flatfishes, sculpins, rockfishes, Pacific Tomcod, Walleye Pollock, Pacific Sand Lance, Pacific Cod and Eulachon. Although salmonids represented only about 4 per cent of their overall diet, pre-spawning adult salmon in estuaries and rivers can increase to 65 per cent of the seals' diet in certain localities during summer and fall. Studies by UBC graduate student Austen Thomas in 2012 showed considerably higher occurrences of Pink Salmon in Harbour Seal diets during summer, particularly around the southern Gulf Islands and the mouth of the Fraser River (figure 99). Harbour Seal pups feed mostly on bottom-dwelling crustaceans, particularly shrimp of the genus *Crangon*, for 1.5–3 months after weaning. Food requirements of individuals vary by age and sex but average about 1.9 kg per day.

Behaviour and Social Organization

Harbour Seals form loosely gregarious groups when hauled out to rest. These groups are composed of males and females of all ages and can range in size from a few to many hundreds of seals. In the Strait of Georgia the average haulout size is about 70 seals. Harbour Seals form social hierarchies with adult males being dominant over females and juveniles. Conflicts are mediated by aggressive displays and interactions, including snorting, growling, head butting, biting,

Figure 99. A Harbour Seal eating a salmon while another looks on. Although salmon are a seasonally important prey for Harbour Seals, they are less important than Pacific Herring, Pacific Hake and other non-salmonids. Photo: M. Malleson, near Victoria, July 2007.

fore flipper-waving and scratching. Disputes become more frequent with increased density on haulouts, as the seals compete for space on the rocks. But they do not form territories or harems on haulouts; mating takes place in the water. Harbour Seals haul out mostly at low tide for an average period of about five hours, spending only about 20 per cent of their time on land. They sleep lightly, waking frequently to scan for potential predators or other threats. Tagging studies have shown that individuals have strong fidelity to a preferred haulout site and generally forage within 10–20 km of it, although some seals have much larger ranges and move seasonally. Harbour Seals experimentally translocated from Snake Island off Nanaimo to Port Alberni returned within a few weeks to their original haulout site. In areas where there are no suitable haulout sites, Harbour Seals rest on the sea floor, surfacing at regular intervals of five to seven minutes in the same location to breathe.

When not hauled out, Harbour Seals disperse and become solitary except for mother-pup pairs during the lactation period. These seals are capable of diving to depths exceeding 500 metres, but most of their foraging dives are to less than 100 metres. They typically dive for less than 5 minutes, but can stay under for as long as 30 minutes.

Harbour Seals have a polygynous mating system, where males compete for access to receptive females. During the breeding season, males form aquatic territories in which they perform visual and acoustic displays to attract mates and for male-male competition. They make a variety of splashy, percussive actions, including lobtailing, where they throw their hindquarters vertically out of the water, then splash down on the surface loudly in somersault fashion. Males also make a distinctive underwater "roar" lasting two to four seconds as part of their display repertoire. Some interactions between males of similar size involve vigorous fighting. Prime male territories are near haulouts and along traffic corridors where they can maximize their exposure to estrous females, and males may use the same territories over at least several years. During the mating season, males spend more time in the water but less time foraging, and as a result, their body condition may decline as the season progresses.

The bond between mother and newborn pup is strong during the four-to-five-week nursing period. The mother keeps her pup close by during this time and will grab it with her fore flippers or mouth and dive to safety if threatened. She nurses her pup for about a minute every three or four hours but may leave the pup on shore for extended periods of a day or more while on a prolonged foraging excursion. Pups call to their mother with a sheep-like bleating sound, but mothers have no reciprocal contact call. Once weaned, the pup is abandoned by its mother.

Life History and Population Dynamics

Females usually have a single pup per year during a 1–2.5-month period. Pups can be born on land or in water, and can swim and dive at birth. Lactation lasts four to five weeks (average: 31 days), during which time pups double their weight to an average of 24 kg. Females come into estrus and mate about two weeks after pups are weaned, but implantation of the fertilized ovum is delayed for up to two months after mating. A gestation of about 10.5 months follows, so births take place about a year after mating. After the pupping season, seals undergo a period of moult each year, usually lasting two to three months.

The timing of pupping and mating for Harbour Seals in the eastern North Pacific varies regionally. Pupping season peaks off the west coast of Washington in late May but gets progressively earlier toward the south, with the peak off Baja California in mid March. In northern British Columbia and southeastern Alaska, pupping peaks in mid June. Interestingly, pupping in the Strait of Georgia and Puget Sound does not peak until late July to early August, more than two months later than off the outer Washington coast. It is not clear why the pupping season is so delayed in these southern inside waters.

Females reach sexual maturity at 2–5 years of age and live for an average of about 30 years. Males become sexually mature at 3–6 years and live for about 20 years. Harbour Seals have a variety of natural predators. About half of the kills we have observed by mammal-hunting Transient (Bigg's) Killer Whales in British Columbia have been of Harbour Seals. Young, naïve pups appear to be favoured by Transients, and the coastal movements of these predators are influenced by the availability of such pups. There are noticeable spikes in the occurrence of Transients in Glacier Bay, southeastern Alaska, in July and off southeastern Vancouver Island in late August and September – these periods correspond to the timing of weaning and abandonment of pups in each area. Other marine predators of Harbour Seals are Great White Sharks, Pacific Sleeper Sharks and Steller Sea Lions. Studies by the University of Victoria's Chris Darimont and colleagues have shown that Harbour Seals are also a common prey item for wolves on the central BC coast. In Pacific Rim National Park on the west coast of Vancouver Island, Harbour Seal remains were found to be the major prey items in 16 per cent of Grey Wolf scats analysed and 24 per cent of Cougar scats sampled from 2000 to 2006. The extent to which these predators actively hunt Harbour Seals as opposed to scavenging on beach-cast carcasses is not clear, but there are credible accounts of wolves hunting seals.

Exploitation

Harbour Seals were hunted for subsistence by First Nations along the British Columbia coast for thousands of years. Aboriginal hunters captured seals by clubbing them on haulouts and by harpooning or netting them in open water. While Harbour Seals were the predominant marine mammals hunted by the Coast Salish in the Strait of Georgia region, they were less important to the Nuu-chah-nulth on the west coast of Vancouver Island. At the village site of Ts'ishaa in Barkley Sound, about 10 per cent of identified marine mammal skeletal remains were from Harbour Seals, a distant second to those of Northern Fur Seals at 64 per cent.

With the arrival of Europeans to the west coast, the killing of Harbour Seals increased dramatically. Seals were hunted commercially for pelts during two periods, 1879 to 1914 and 1962 to 1968. Between these periods the government paid a bounty to cull seals, which were considered nuisances due to their depredation of salmon fisheries and the damage they caused to fishing nets. Between 1879 and 1970, when the species was given protection, a total of 172,649 pelts were sold and 114,903 bounties paid on seal snouts – required from the bounty hunters as proof of a kill before the payment (ranging from $1 to $5) was made. An average of 2913 seals were killed for bounty each year in BC from 1914 to 1964. But, when shot, Harbour Seals often sink before they can be recovered, so the records of pelts and bounty payments represent only a fraction of the total actually killed. By the late 1960s, fewer than 10,000 Harbour Seals remained from an estimated BC population of 65,000–110,000 in the late 1800s. The hunting of Harbour Seals by First Nations is currently permitted for food and social or ceremonial purposes, which likely result in annual kills of fewer than 100 seals. Harbour Seals are also killed intentionally at aquaculture sites in BC (described below).

Taxonomy and Population Structure

The subspecific structure of Harbour Seals is in a state of flux. Until recently, five subspecies were accepted: *Phoca vitulina richardii* in the eastern North Pacific, *P. v. stejnegeri* in the western North Pacific, *P. v. mellonae* in the Ungava Peninsula area of northern Quebec, *P. v. concolor* in the western North Atlantic, and *P. v. vitulina* in the eastern North Atlantic. But recent genetic data along with a lack of morphological differences indicate that there is only one subspecies in the North Atlantic, and only one subspecies, *P. v. richardii*, is currently recognized in the North Pacific (pending further assessment). The subspecies name *richardii* is often spelled *richardi*. Some researchers believe that it should be *richardsii*, as the name was given by British zoologist J.E. Gray in 1864 to honour Captain G.H. Richards, a hydrographer with the British Admiralty, and evidently Gray omitted the "s" in error.

Conservation Status and Management

After receiving protection from hunting and culling in 1970, Harbour Seals increased rapidly in British Columbia. The status of seals in the Strait of Georgia, where periodic aerial censuses have been conducted by the Pacific Biological Station since 1973, is especially well known. Harbour Seal abundance increased there at an annual rate of about 11.5 per cent during the 1970s and 1980s, slowed in the late 1980s and early 1990s, then stabilized around the mid 1990s. Only about 3900 Harbour Seals were in the Strait in 1973. This count had increased tenfold to 39,000 by 1994 and was still at about that number when the most recent survey was conducted in 2008. Surveys in selected areas on the BC coast outside of the Strait of Georgia suggest a similarly increasing trend. There are currently estimated to be about 105,000 Harbour Seals in British Columbia, a figure at or above historical levels.

Not all Harbour Seal populations in the northeastern Pacific are doing as well as in BC. Some populations in Alaska, such as in the Aleutians and Glacier Bay, have experienced sharp declines in abundance in the past 20–30 years, and they are currently depleted. It is unclear what has driven these declines, but ecosystem regime shifts, increased competition for food and increased predation by Steller Sea Lions (in the case of Glacier Bay) have been suggested as possible factors.

Many fishermen have long regarded Harbour Seals as nuisances due to their tendency to remove and damage salmon in gillnets and to take salmon off lines. Although the 50-year-long culling program in BC ended in 1964, it is still legal to obtain a "Nuisance Seal Licence" to shoot animals that are inflicting damage to fishing gear or are attempting to remove fish from salmon aquaculture facilities. From 1990 to 2011 almost 6000 seals were killed at aquaculture facilities in BC. With improved management practices and mitigation through the use of predator nets designed to keep seals from fish pens, the annual number of seals shot has decreased to fewer than 10 in 2012 and 2013. But some seals become entangled in these nets and drown. Removal of salmon from the lines of recreational fishermen is an ongoing problem in certain areas on the BC coast. It appears that only a small fraction of the Harbour Seal population acquires this habit, while most seals prey on non-salmonid fishes and may actually benefit salmon populations by consuming predators of juvenile salmon. The number of Harbour Seals illegally shot by fishermen and other mariners is not known. A management issue involving Harbour Seals that proved difficult to mitigate was their predation on out-migrating salmon smolts released from a hatchery on the Puntledge River in Courtenay on the east coast of Vancouver Island. A few dozen seals learned to use illumination from street lights on a bridge to detect and prey on smolts swimming near the surface. A directed culling program in the late 1990s removed many of these seals, but it wasn't until the lights were shielded so that they didn't illuminate the river below that the problem was finally resolved.

Remarks

People often find Harbour Seal pups on shore and mistakenly think they have been abandoned or orphaned. The pups may call repeatedly with a plaintive "maaa, maaa", which makes it hard to resist picking up the pup and "rescuing" it. While some pups are in fact abandoned, it is normal for mothers to leave their pups while they go on foraging excursions that can last several hours or sometimes longer. Pups found alone on shore should be left alone and watched from a distance to see if the mother returns. If the pup seems to be in distress or is injured, or if the mother does not return after 24 hours, people are encouraged to contact the local SPCA or the Vancouver Aquarium's Marine Mammal Rescue Centre.

Selected References: Bigg 1969 and 1981; Burg et al. 1999; Burns 2009; Cottrell et al. 2002; Darimont et al. 2008; Fisher 1952; Hayes et al. 2004; Mathews and Adkison 2010; McKechnie and Wigen 2011; McMillan et al. 2008; Olesiuk 2010; Olesiuk, Bigg and Ellis 1990a and b; Olesiuk, Bigg, Ellis et al. 1990; Stutz 1967; Sullivan 1981; Temte et al. 1991; Wilton 2007.

Sea Otter *Enhydra lutris*

Other Common Names: None.

Description

The Sea Otter is the most aquatic member of the Mustelidae (weasel family). It has a robust body with a large, rounded head and short, stocky neck. The snout is blunt with a prominent, broad and triangular nose-pad and thick whiskers, or vibrissae. The tail is short compared to other otters (less than one-fourth of overall length) and is flattened top to bottom into a paddle-like shape. The forelimbs are short with small, highly dexterous paws. The large hindlimbs are oriented backwards and flattened into flippers for swimming.

The Sea Otter's body is entirely covered with thick fur except for the eyes, nose, inside of the small earflaps and pads of the feet. The thick, luxuriant underfur is overlaid with a layer of guard hairs, and the pelage is the densest of any mammal, with more than 125,000 hairs per square centimetre. The colour of the fur is generally dark brown, though the guard hairs may be much lighter brown or blonde, particularly on the head, neck, chest and forelimbs of older animals, giving them a grizzled appearance. Newborn pups have dense, woolly fur that is a light buff colour. This is replaced by darker adult fur by 13 weeks of age.

Measurements

length (metres):

male: 1.27 (0.93–1.49) n = 30
female: 1.21 (0.98–1.37) n = 12
neonate: ~ 0.60

weight (kg):

male: 27.9 (11.8–39) n = 30
female: 22.7 (12.7–27.3) n = 12
neonate: 1.4–2.3

Dental formula

incisors: 3/2
canines: 1/1
premolars: 3/3
molars: 1/2

Identification:

Sea Otters are often confused with Northern River Otters, which live in coastal habitat in British Columbia and forage in the ocean. But the two species are quite different in appearance and behaviour and can be readily distinguished even at a distance. River Otters are much smaller and more slender than Sea Otters, and they have a much longer tail that tapers to a sharp point (see figure 1). And River Otters swim belly down, whereas Sea Otters usually float and often swim on their back at the surface. Sea Otters usually rest in tight groups of many individuals, or *rafts*, at the surface, a behaviour not exhibited by River Otters (figure 100). Sea Otters seldom haul out, particularly in areas with human habitation, and when they do so they shuffle in an ungainly manner due to their flipper-like hind limbs. River Otters, in contrast, regularly haul out on beaches, rocks and docks, and they can run quickly on land. Because Sea Otters usually inhabit outer coast waters, any otter seen in protected inside waters, especially the Strait of Georgia, is most likely to be a Northern River Otter.

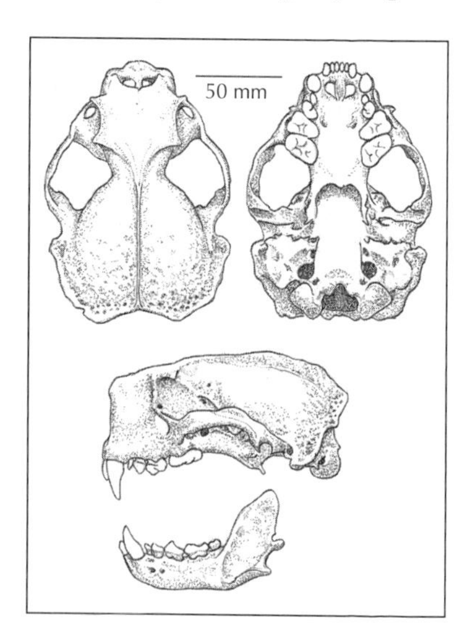

Figure 100. A raft of Sea Otters near Fingal Island on BC's central coast. Photo: G. Ellis, August 2004.

When in the water and just exposing its head, a Harbour Seal can look somewhat like a Sea Otter, but the latter has smaller eyes, a larger nose and a more furry face and head that is often light blonde in colour. Sea Otters are very wary of humans and can be difficult to approach without causing them to startle and flee. They have a good sense of smell and can detect the engine exhaust of boats approaching from upwind at a considerable distance.

Distribution and Habitat

Historically, Sea Otters were distributed in shallow, nearshore waters of the North Pacific Rim from Hokkaido, Japan, to central Baja California, Mexico. After being hunted relentlessly throughout their range during the 18th and 19th centuries, only about a dozen small remnant colonies remained, containing fewer than 2000 animals in total. Following protection in 1911, some of these surviving colonies disappeared, but others began to expand and have now reoccupied much of their former range. Sea Otters had been completely extirpated from southeastern Alaska south to northern California, but a series of translocations of otters from the Aleutians and Prince William Sound, Alaska, in the 1960s and 1970s re-established the species in several locations along this coast. But not all former Sea Otter range in these areas has yet been re-occupied.

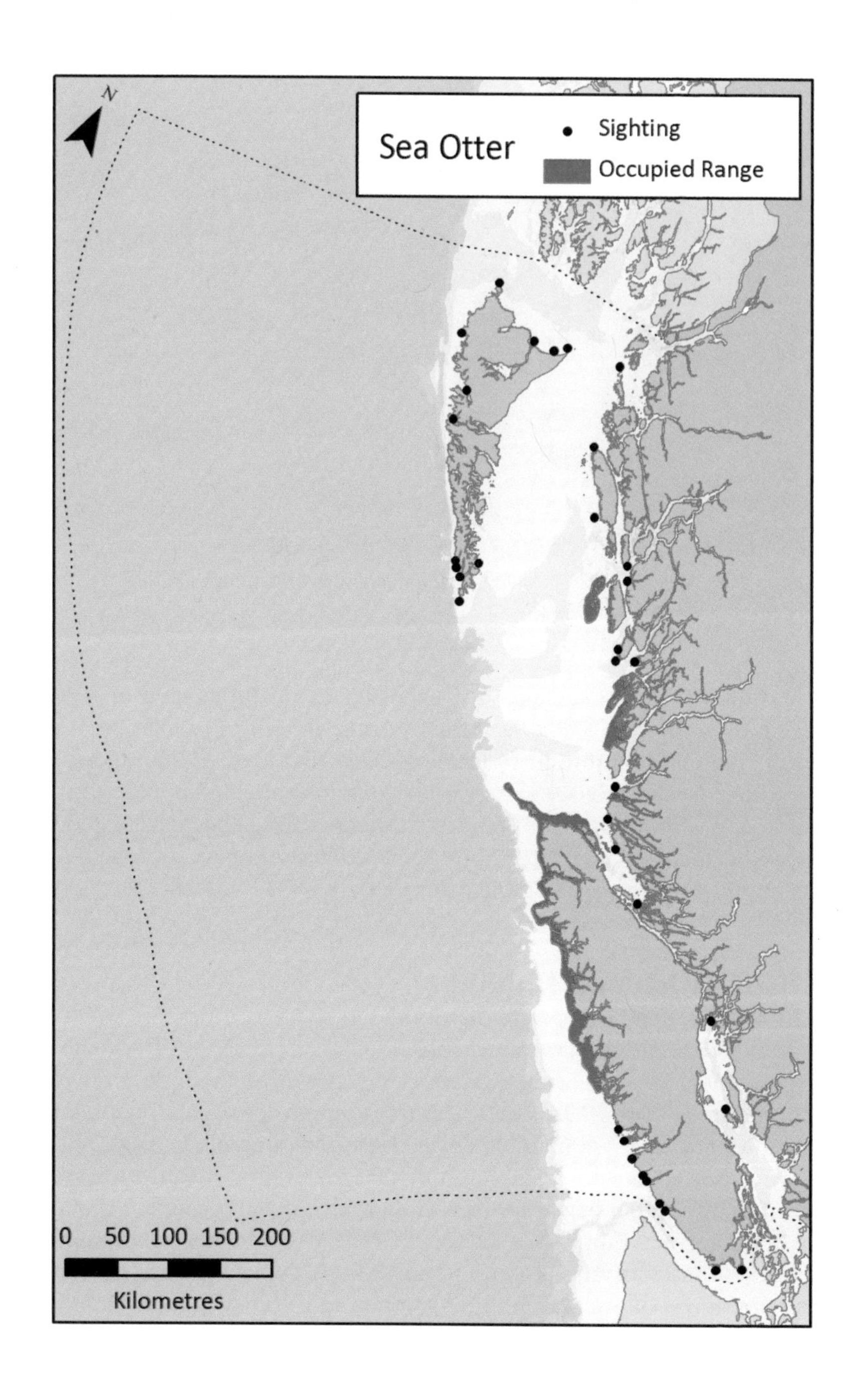
Sea Otter
Sighting
Occupied Range
N
0 50 100 150 200
Kilometres

In British Columbia, Sea Otters were reintroduced with the translocation of 89 animals captured off Amchitka Island and in Prince William Sound, Alaska. The Otters were released at Checleset Bay on the west coast of Vancouver Island on three occasions – 29 otters in 1969, 14 in 1970 and 46 in 1972. During aerial surveys in 1977 by Pacific Biological Station researchers Michael Bigg and Ian MacAskie, 55 Sea Otters were seen in Checleset Bay and another 15 off Bajo Point, Nootka Island, about 75 km to the southwest. Thereafter, regular surveys by Vancouver Island University biologist Jane Watson and colleagues documented the rapid expansion of Sea Otters along the outer coast of Vancouver Island through the 1980s and 1990s. By 1995 Sea Otters were distributed almost continuously from Estevan Point northwest to Quatsino Sound, and the population numbered more than 1500 – an impressive annual growth rate of 18–19 per cent likely due to the availability of abundant food. Since 1995 growth in the Sea Otter population along the west coast of Vancouver Island has slowed to about 8 per cent per annum.

In 1989, a small colony with about 50 otters, including mothers and pups, was reported at the Goose Group of islands and reefs off the central BC coast, about 125 km northwest of Vancouver Island. Recent genetic tests indicate this group was almost certainly founded by animals that "jumped" north from the Vancouver Island population. This northern group has also expanded its numbers and range. As of 2013 Sea Otters have been found regularly from Ucluelet to the Scott Islands off the northwestern tip of Vancouver Island, and off the northeastern coast of Vancouver Island to Port Hardy. Along the mainland BC coast, they are found from Smith Sound north to Milbanke Sound and Seaforth Channel, as well as among the mazes of islets, rocks and reefs along the west coasts of Price and Aristazabal islands as far north as the Moore Islands in eastern Hecate Strait. Lone individuals – typically males – are occasionally sighted outside these areas (as shown in accompanying map). The most recent coast-wide survey was conducted in 2013 by Pacific Biological Station biologist Linda Nichol and colleagues, who counted 5600 Sea Otters off Vancouver Island and 1200 along the central coast.

Sea Otters mostly inhabit shallow waters (<50 metres deep) within one or two kilometres of shore, though they may occasionally be found further offshore over shallow banks. They reach highest densities in complex habitat strewn with reefs, rocks and islets. Kelp beds are favoured by mothers with pups but are not essential. Habitat use patterns are affected by weather and sea state – otters can be found in open, exposed water during calm periods but will retreat to protected areas in rough storm conditions. Occurrence of Sea Otters in inlets along the west coast of Vancouver Island tends to be more common in winter than in spring and summer. They will occasionally haul out on rocky reefs but are wary and will quickly enter the water if approached.

Natural History

Feeding Ecology

Sea Otters mostly feed on a wide range of bottom-dwelling invertebrates, including crabs, clams, sea urchins, abalones and snails. Over 150 prey species have been documented, and diet varies among regions, among individuals, and according to population status. In habitat that has recently become occupied in areas of population expansion – such as off the central BC coast – Sea Otters prey on large, easy-to-catch and energetically profitable invertebrates such as Purple Sea Urchins, Red Sea Urchins, Dungeness Crabs and Northern Abalone. As abundance of such preferred prey declines, the diet diversifies to include a wider variety of invertebrates of all sizes, from Geoducks and Gumboot Chitons to small Northern Kelp Crabs and Black Turban Snails. In addition to Geoducks, other bivalve molluscs are important prey in BC, including Littleneck Clams, Horse Clams, Butter Clams, Cockles and California Mussels. Fish are common prey in the Aleutian Islands, but fish predation has not yet been observed in British Columbia. When population abundance reaches the habitat's carrying capacity, as it has in some areas along the northwestern coast of Vancouver Island, and competition for food increases, individual otters tend to specialize on particular prey species, which increases their foraging efficiency. Because Sea Otters have a high metabolic rate – as much as three times greater than terrestrial mammals – they must consume about a quarter of their body weight in food per day to meet their energy requirements.

Behaviour and Social Organization

Sea Otters are non-migratory, and individuals have strong fidelity to small home ranges that vary in size from a few to tens of kilometres of coastline. But some seasonal movements and long-distance movements of individuals may take place. Sea Otters usually segregate by sex, with males and females occupying distinct areas. They may occur singly, as mother-pup pairs, or in resting groups known as rafts. Rafts often form in predictable locations and can contain 200 or more individuals (as shown in figure 100).

Figure 101. Sea Otters eat a diversity of marine invertebrates. This male is finishing a Red Sea Cucumber before eating the Red Sea Urchin lying on its belly. Photo: E. Rechsteiner, Calvert Island, December 2013.

Otters usually forage alone, although several animals may gather in good feeding locations. Dives to obtain prey typically last only one to three minutes and reach depths of 5–60 metres, although dives of seven minutes to depths of 100 metres have been recorded. Mothers leave their young pups at the surface when on foraging dives, and re-establish contact with the pup's shrill, high-pitched screams that are loud enough to be heard a kilometre away. Sea Otters capture prey with their forepaws and can carry it along with rocks or other hard objects – which are used as tools to break open shelled prey – in loose folds of skin under the forelimbs as they swim. Sea Otters are among the few mammals, other than humans, known to use tools. A Sea Otter eats its prey at the surface while floating on its back and using its chest like a table as it breaks its meal apart (figures 101 and 102). Although they often forage in rocky reef areas, Sea Otters also search for prey on soft bottoms, and can excavate considerable amounts of sand and mud to reach buried clams such as Geoducks. They forage for as long as 12 hours each day, typically making 150–250 dives to obtain food. Because they have no blubber and depend on their dense fur to stay warm in the cold ocean, Sea Otters spend up to 10 per cent of their time cleaning and grooming their fur to maintain its insulative properties. Insulation is provided by air trapped in the fluffy underfur;

Figure 102. Sea Otters often carry rocks to the surface to help break open hard-shelled prey, in this case Red Turban Snails.
Photo: E. Rechsteiner, Queen's Sound, central BC coast, February 2014.

the otter rolls at the surface and blows into the fur while vigorously grooming to replenish the layer of air.

Sea Otters have a polygynous mating system, with breeding males establishing territories that may overlap with the home ranges of a group of females with which they will mate. Mating is a vigorous process, with the male biting the nose of the female to position her as they roll at the surface. This behaviour often results in wounds that can leave distinctive scars on the females' nose.

Life History and Population Dynamics

Female Sea Otters reach sexual maturity at three to five years. Males begin mating at five or six years, though they may reach sexual maturity at a younger age. Mating and pupping can occur throughout the year, although in British Columbia there is a distinct peak in pupping during the spring. A single pup is born, usually in the water, after a gestation period of about six or seven months, which includes a two- to three-month period of delayed implantation. Pups are cared for entirely by the female. Born with thick, woolly fur, pups are well insulated from the cold water but are unable to swim on their own and are carried on the mother's abdomen for the first few weeks of life. After nursing for about one month, pups begin to eat some solid food provided by their mother. By four months of age, the pup feeds almost exclusively on prey delivered by its mother, and by five months the pup can dive, as well as capture and break open prey on its own. Pups become independent at six to eight months, at which point the mother mates again. Mortality of pre-weaned pups can be high – as much as 60–78 per cent – in areas where populations have reached carrying capacity and food is limited. Pup deaths are mostly due to abandonment and starvation. But in expanding populations, such as those along the central BC coast, pup mortality can be as low as 10 per cent. Young pups are also vulnerable to predation by Bald Eagles when left at the surface while their mother forages for food. Females continue to reproduce throughout the remainder of their 15- to 20-year lifespan. Longevity is shorter in males, at 10–15 years.

Mammal-eating Killer Whales and Great White Sharks are known predators of Sea Otters. But despite anecdotal reports of Killer Whales pursuing Sea Otters in BC, there are no documented cases of observed predation. Lacking blubber, Sea Otters are not particularly profitable prey for Killer Whales from an energy standpoint. Sea Otter fur has been found in Grey Wolf scats on the west coast of Vancouver Island, but this is likely from scavenging rather than predation.

Exploitation

Sea Otters were hunted by First Nations in British Columbia for millennia. Remains have been found in archaeological sites as old as 5000 years along the west coast of Vancouver Island, in Haida Gwaii, on the central mainland coast and, rarely, in the Strait of Georgia area. Luxuriant Sea Otter pelts were recognized as symbols of wealth and use in chiefly regalia, but it is doubtful that these animals were significant for First Nations as food. When Europeans arrived on the west coast in the late 18th century, the hunting of Sea Otters expanded dramatically. The Europeans were eager to obtain valuable otter pelts to sell in Asia, and First Nations hunters were keen to exchange pelts for highly valued trade goods. Between 1785 and 1809, about 55,000 pelts were landed in BC, although it is not clear what proportion may have been taken in adjacent waters of Washington and southeastern Alaska. At least 6000 are known to have come from the west coast of Vancouver Island. From 1787 to 1797 at least 11,000 were obtained in trade in Haida Gwaii. This intensive Sea Otter hunt was clearly unsustainable, and the species was commercially extinct by 1850. As their scarcity increased, so too did the value of their pelts. For example, on September 1, 1904, Victoria's *British Colonist* newspaper reported that four pelts had been brought south from Masset, Haida Gwaii, and were sold for $2000, which was about the value of a large house in Victoria at the time. Despite legal protection in 1911, Sea Otters continued to be killed in British Columbia. The last reported Sea Otter kills in BC were single animals taken off the northwestern coast of Vancouver Island in 1929 and 1931. There were two reported sightings after this date – one near Langara Island in 1939 and another in Checleset Bay around 1954. Sea Otters were not seen again in the province until the first translocations from Alaska began in 1969.

Taxonomy and Population Structure

On the basis of skull morphology, three subspecies of Sea Otters are recognized – *Enhydra lutris lutris* in Russia, *E. l. kenyoni* in the region from the Aleutian Islands to Washington, including BC, and *E. l. nereis* in California. The extent to which human exploitation may have contributed to the genetic and morphological distinctions among these modern subspecies is unknown, but current differences in mitochondrial DNA suggest that there was at least some genetic differentiation among the others in these areas before human-caused population declines and fragmentation.

Conservation Status and Management

Estimates of overall Sea Otter abundance in the North Pacific before the maritime fur trade in the 18th and 19th centuries are crude and uncertain, but they range from 150,000 to 300,000. Over-exploitation drove the species to the brink of extinction by the mid 1800s, and only 1000 to 2000 animals in 13 remnant colonies survived into the early 20th century. The species finally received some protection in the early 1910s, with the banning of Sea Otter hunting in US waters and with the signing of the international Fur Seal Treaty by the United Kingdom (for Canada), the US, Japan and Russia, which included an article banning the commercial exploitation of Sea Otters in international waters (more than three nautical miles [5.5 km] from shore, where Sea Otters are seldom found). Hunting of Sea Otters was finally banned in BC with the Game Act of 1931, the year the last otter was killed in the province.

Following the end of hunting, remnant colonies of Sea Otters in western and central Alaska and California began to recover, and translocations re-established populations in southeastern Alaska, BC and Washington. By the early 1980s Sea Otter numbers around the North Pacific Rim totalled about 165,000, within the range of population estimates before the fur trade. But in the late 1980s, a precipitous and unexplained decline of Sea Otters began in the Aleutian Islands, home to as much as 40 per cent of the North Pacific population, and by 2005 the Aleutian population had been reduced by 90 per cent. Recent surveys indicate that the decline has ended and a growth trend has begun. One hypothesis is that the decline was driven by mammal-eating Killer Whales, which could have switched to hunting Sea Otters following declines in the abundance of their preferred prey, mainly Harbour Seals and Steller Sea Lions, in the region. While increased Killer Whale predation may have been at least partly responsible, the hypothesis is based on limited evidence, and other factors such as disease, contaminants and predation by sharks cannot be ruled out.

The Sea Otter population in British Columbia has shown strong growth since becoming re-established in the 1970s. After increasing at an annual rate of almost 19 per cent along the west coast of Vancouver Island between 1977 and 1995, growth has slowed to about 8 per cent in recent years as the population approaches equilibrium with available food resources. Sea Otter abundance on the central BC coast is increasing at about 11 per cent a year, and the range is expanding northward. The total population in BC was about 6800 animals in 2013. Although there have been scattered sightings of single animals on most parts of the BC coast, Sea Otters have yet to reoccupy formerly important habitat such as Haida Gwaii and the northern mainland coast. In time, these areas too will likely see the return of Sea Otters. Although the historical abundance of the species in British Columbia is not known, the carrying capacity of potential Sea Otter habitat in BC is estimated to be about 52,000 animals.

Sea Otters are vulnerable to mortality from a variety of factors related to human activities. Because they rely on clean fur for insulation, fouling from spilled oil is a serious threat as it destroys the fur's water-repellent properties. The 1989 *Exxon Valdez* oil spill in Prince William Sound, Alaska, killed as many as 3900 Sea Otters through hypothermia or toxic poisoning from ingestion of oil as the animals tried to clean their fur. Almost 25 years later Sea Otters in Prince William Sound are still being exposed to oil buried in sediments when they dig for clams and other invertebrates. Other threats to Sea Otters include entanglement in fishing nets and crab pots, strikes from fast-moving vessels, environmental contaminants, disease and illegal killing. Sea Otters are highly efficient predators of valuable shellfish such as Dungeness Crab, Geoduck and sea urchins, and they can have a significant effect on commercial and recreational fisheries. As the range of Sea Otters expands, concerns about their impacts on invertebrate harvesting have intensified, with some calling for government intervention to regulate or control population growth.

Sea Otters are considered Endangered on the IUCN's Red List because of their vulnerability to large-scale population declines. In Canada the species was listed by COSEWIC as Endangered in 1978, then downlisted to Threatened in 1996 due to population growth and range expansion beyond the west coast of Vancouver Island. With continued robust growth, Sea Otters were again downlisted to Special Concern in 2007, which removes protection of the species and its critical habitat under the Species at Risk Act. But Sea Otters are still protected under the Marine Mammal Regulations of the Fisheries Act. A limited take of Sea Otters by the Nuu-chah-nulth First Nation on the west coast of Vancouver Island for ceremonial purposes is likely to be authorized in the near future. Sea Otters are on the BC Blue List.

Remarks

Considered a "keystone species", the Sea Otter can have a major influence on the structure and function of nearshore marine ecosystems. Extirpation of Sea Otters by the maritime fur trade allowed grazing sea urchins to proliferate, which virtually eliminated the once-extensive kelp beds along the outer BC coast, creating what are known as "urchin barrens". The reintroduction and expansion of Sea Otters triggered an "otter-urchin-algae trophic cascade", where otter predation rapidly removed most urchins and allowed the re-establishment of kelp and other macroalgae. In a classic field study, marine ecologist Jane Watson of Vancouver Island University documented this cascade in painstaking detail over a 23-year period beginning in 1987, as Sea Otters expanded into new unoccupied habitat off the west coast of Vancouver Island. And as UBC graduate student Russ Markel found in his studies, the re-establishment of kelp forests has had a highly beneficial effect on the diversity and productivity of rocky reef ecosystems, including the recruitment and growth rates of rockfishes.

Selected References: Bigg and MacAskie 1978, Bodkin et al. 2012, COSEWIC 2007, Doroff and Burdin 2011, Estes et al. 2009, Gregr et al. 2008, Hatler et al. 2008, Kuker and Barrett-Lennard 2010, McKechnie and Wigen 2011, Nichol et al. 2009, Raum-Suryan et al. 2004, Sea Otter Recovery Team 2007, Szpak et al. 2012, Watson et al. 1997, Watson and Estes 2011.

APPENDIX 1

Scientific Names of Other Organisms Mentioned in this Book

Fishes

Arrowtooth Flounder	*Atheresthes stomias*
Blackbelly Eelpout	*Lycodes pacificus*
Blackfin Sculpin	*Malacocottus kincaidi*
Blue Shark	*Prionace glauca*
Bonitos	*Sarda* spp.
Capelin	*Mallotus villosus*
Chinook Salmon	*Oncorhynchus tshawytscha*
Chum Salmon	*Oncorhynchus keta*
Coho Salmon	*Oncorhynchus kisutch*
Cookiecutter Shark	*Isistius brasiliensis*
Dover Sole	*Microstomus pacificus*
Eulachon	*Thaleichthys pacificus*
Great White Shark	*Carcharodon carcharias*
High Cockscomb	*Anoplarchus purpurescens*
Jack Mackerel	*Trachurus symmetricus*
Japanese Anchovy	*Engraulis japonicus*
Lingcod	*Ophiodon elongatus*
Longnose Skate	*Raja rhina*
Mahi-mahi	*Coryphaena hippurus*
Opah	*Lampris guttatus*
Northern Anchovy	*Engraulis mordax*
Northern Sculpin	*Icelinus borealis*
Northern Smoothtongue	*Leuroglossus schmidti*
Pacific Cod	*Gadus macrocephalus*
Pacific Hagfish	*Eptatretus stoutii*

Pacific Hake	*Merluccius productus*
Pacific Halibut	*Hippoglossus stenolepis*
Pacific Herring	*Clupea pallasii*
Pacific Lamprey	*Lampetra tridentata*
Pacific Mackerel	*Scomber japonicus*
Pacific Sanddab	*Citharichthys sordidus*
Pacific Sand Lance	*Ammodytes hexapterus*
Pacific Sardine	*Sardinops sagax*
Pacific Saury	*Cololabis saira*
Pacific Sleeper Shark	*Somniosus pacificus*
Pacific Spiny Dogfish	*Squalus suckleyi*
Pacific Tomcod	*Microgadus proximus*
Pacific Viperfish	*Chauliodus macouni*
Pink Salmon	*Oncorhynchus gorbuscha*
Plainfin Midshipman	*Porichthys notatus*
Ragfish	*Icosteus aenigmaticus* (formerly *Acrotus willoughbyi*)
Sablefish	*Anoplopoma fimbria*
Saithe	*Pollachius virens*
Shiner Perch	*Cymatogaster aggregata*
Sockeye Salmon	*Oncorhynchus nerka*
Steelhead Salmon	*Oncorhynchus mykiss*
Walleye Pollock	*Theragra chalcogramma*
Yelloweye Rockfish	*Sebastes ruberrimus*
Yellowfin Tuna	*Thunnus albacares*

Invertebrates

Bay Ghost Shrimp	*Neotrypaea californiensis*
Berry Armhook Squid	*Gonatus berryi*
Black Turban Snail	*Chlorostoma funebralis*
Boreal Clubhook Squid	*Onychoteuthis borealijaponica*
Boreopacific Armhook Squid	*Gonatopsis borealis*
Butter Clam	*Saxidomus gigantea*
California Mussel	*Mytilus californianus*
Cockle	*Clinocardium nuttalli*
Dungeness Crab	*Metacarcinus magister* (formerly *Cancer*)
Flying Squid	*Ommastrephes bartrami*
Geoduck	*Panopea generosa*
Giant Squid	*Architeuthis* spp.
Gumboot Chiton	*Cryptochiton stelleri*
Horse Clam	*Tresus capax* or *T. nuttallii*

Humboldt Squid	*Dosidicus gigas*
Littleneck Clam	*Protothaca staminea*
Magister Armhook Squid	*Berryteuthis magister*
Mussel Worm	*Nereis vexillosa*
Neon Flying Squid	*Ommastrephes bartrami*
Northern Abalone	*Haliotis kamtschatkana*
Northern Kelp Crab	*Pugettia producta*
Opalescent Inshore Squid	*Doryteuthis opalescens* (formerly *Loligo*)
Parasitic copepod	*Pennella balaenoptera*
Porcelain crabs	*Pachycheles* spp.
Purple Sea Urchin	*Strongylocentrotus purpuratus*
Red Sea Cucumber	*Cucumaria miniata*
Red Sea Urchin	*Strongylocentrotus franciscanus*
Red Turban Snail	*Pomaulax gibberosus*
Robust Clubhook Squid	*Onykia robusta* (formerly *Moroteuthis*)
Sea Butterfly	*Limacina helicina*

Birds

Bald Eagle	*Haliaeetus leucocephalus*
Common Murre	*Uria aalge*
Marbled Murrelet	*Brachyramphus marmoratus*
Rhinoceros Auklet	*Cerorhinca monocerata*
Surf Scoter	*Melanitta perspicillata*

Mammals

African Elephant	*Loxodonta africana*
American Black Bear	*Ursus americanus*
Arnoux's Beaked Whale	*Berardius arnuxii*
Beluga	*Delphinapterus leucas*
Bowhead Whale	*Balaena mysticetus*
Bryde's Whale	*Balaenoptera edeni*
Burmeister's Porpoise	*Phocoena spinipinnis*
Common Bottlenose Dolphin	*Tursiops truncatus*
Cougar	*Puma concolor*
Grey Wolf	*Canis lupus*
Melon-headed Whale	*Peponocephala electra*
Moose	*Alces alces*
Narwhal	*Monodon monoceros*
North Atlantic Right Whale	*Eubalaena glacialis*
Northern River Otter	*Lontra canadensis*

Pantropical Spotted Dolphin	*Stenella attenuata*
Polar Bear	*Ursus maritimus*
Pygmy Killer Whale	*Feresa attenuata*
Sitka Black-tailed Deer	*Odocoileus hemionus sitkensis*
Southern Right Whale	*Eubalaena australis*
Spinner Dolphin	*Stenella longirostris*
Vaquita	*Phocoena sinus*
Walrus	*Odobenus rosmarus*

APPENDIX 2

Strandings, Bycatches and Post-Whaling Sightings of Rare or Uncommon Cetaceans in British Columbia Waters

Abbreviations: HG = Haida Gwaii; VI = Vancouver Island; CWS = Canadian Wildlife Service; CRP, PBS = Cetacean Research Program, Pacific Biological Station; RBCM = Royal BC Museum; DFO MMRP Pacific = Department of Fisheries and Oceans Marine Mammal Response Program, Pacific Region; BCCSN = British Columbia Cetacean Sightings Network.

Date	Locality	Latitude (°N)	Longitude (°W)	Record type	No.	Sex (M/F)	Length (cm)	Specimen/ Record no.	Source
North Pacific Right Whale									
9-13 Jun 2013	Off Port Louis, W Graham I., HG	53°37′	133° 17.2′	Sighting	1	F			CRP, PBS
25-26 Oct 2013	Swiftsure Bank, SW VI	48°31.3′	124° 52.5′	Sighting	1				CRP, PBS
Sei Whale									
2 Aug 2012	~300 km W of HG	51°58′	135°58′	Sighting	1				CRP, PBS
3 Aug 2012	~290 km SW of HG	50°44′	135°02′	Sighting	3				CRP, PBS
Pygmy Sperm Whale									
31 Jul 2003	Hanson I., NE VI	50°34.3′	126°46.5′	Stranding	1	M	330	DFO 1684	CRP, PBS

Date	Locality	Latitude (°N)	Longitude (°W)	Record type	No.	Sex (M/F)	Length (cm)	Specimen/ Record no.	Source
Dwarf Sperm Whale									
24 Sep 1981	Pachena Bay, SW VI	48°47′	125°08′	Live stranding	1	F	230	RBCM 10400; DFO 3725	Nagorsen and Stewart 1983
Baird's Beaked Whale									
3 Jul 1987	230 km W of Cape Beale, W VI	48°49.1′	128°40.8′	Sighting					Willis and Baird 1998a
Mar 1992	Fife Pt, NE Graham I., HG	54°06′	131°40′	Stranding	1		~600	DFO 5024	Willis and Baird 1998a
5 Jun 1995	42 km W of N VI	50°29.2′	128°53.1′	Sighting	21				Willis and Baird 1998a
14 May 1996	54 km W of Tofino, W VI	48°54.8′	126°35.8′	Sighting					Willis and Baird 1998a
19 Feb 1997	Cox Bay, W VI	49°06′	125°53′	Stranding	1	M	953	DFO 3563	CRP, PBS
7 May 2000	Father Charles Canyon, offshore VI	48°36.8′	126°40.4′	Sighting	12				CWS
7 May 2000	Father Charles Canyon, offshore VI	48°37.6′	126°40.2′	Sighting	13				CWS
7 May 2000	Father Charles Canyon, offshore VI	48°38.4′	126°40.0′	Sighting	10				CWS
10 May 2000	30 km SW of Quatsino Sound, W VI	50°14.6′	128°22.0′	Sighting	4				CWS
2 Jun 2002	420 km W of Cape St James, Queen Charlotte Sound	51°45.6′	137°13.6′	Sighting	2				CWS

12 Aug 2002	50 km SE of Cape St James, Queen Charlotte Sound	51°33.4′	130°31.8′	Sighting	6				CWS
24 Oct 2002	16 km NW of Solander I., VI	50°08.4′	128°13.2′	Sighting	2				BCCSN
2 Jun 2003	20 km S of Cape St James, Queen Charlotte Sound	51°42.1′	130°53.6′	Sighting	7				CRP, PBS
19 Mar 2004	Long Beach W VI	49°02′	125°42′	Stranding	1	M	980	DFO 1858	CRP, PBS
2 Jun 2004	90 km SW of Tofino, W VI	48°41.3′	127°07.6′	Sighting	8				BCCSN
11 May 2006	Cape Scott, N VI	50°47.1′	128°24.5′	Sighting	2				CRP, PBS
2 May 2007	Cape St James, Moresby I., S HG	51°57.3′	131°08.2′	Sighting	4				CRP, PBS
9 May 2007	55 km SE of Cape St James, Queen Charlotte Sound	51°33.2′	130°26.9′	Sighting	12				CRP, PBS
5 Aug 2007	SW Moresby I., HG	52°03.8′	131°23.5′	Sighting	3				CRP, PBS
18 Aug 2008	Queen Charlotte Sound	51°06.5′	130°21.6′	Sighting	5				CRP, PBS
14 Jul 2009	Queen Charlotte Sound	51°27.6′	130°25.5′	Sighting	10				CRP, PBS
15 Jul 2009	Queen Charlotte Sound	51°25.6′	130°20.5′	Sighting	1				CRP, PBS
19 Jul 2009	Queen Charlotte Sound	51°16.6′	130°20.1′	Sighting	15				CRP, PBS

Date	Locality	Latitude (°N)	Longitude (°W)	Record type	No.	Sex (M/F)	Length (cm)	Specimen/ Record no.	Source
Baird's Beaked Whale (cont.)									
19 Jul 2009	Queen Charlotte Sound	51°14.7′	130°08.4′	Sighting	2				CRP, PBS
21 Aug 2010	250 km W of Ucluelet, VI	48°50.6′	129°0.5′	Sighting	6				CWS
30 Jul 2012	100 km SW of Barkley Sound, W VI	48°07.7′	126°13.2′	Sighting	6				CRP, PBS
26 Nov 2012	160 km W of Nootka Sound, W VI	49°18.0′	128°44.9′	Sighting	2				CRP, PBS
Hubbs' Beaked Whale									
Pre-1977	unknown BC			Stranding	1	M	~ 500	RBCM 7721; DFO 5025	Mead et al. 1982
16 Dec 1962	Prince Rupert, N BC	54°17′	130°22′	Stranding	1	M	472	DFO 4335	Pike & MacAskie 1969
3 Jul 1963	Long Beach, W VI	49°02′	125°44′	Stranding	1	M	274	UBC 9037; DFO 4102	Pike & MacAskie 1969
Jul 1965	Skull Cove, Rivers Inlet, Central Coast	51°03′	127°33′	Stranding	1	M		UBC 9036; DFO 4787	Pike & MacAskie 1969
22 Aug 1967	Florencia Bay, W VI	49°00′	125°38′	Stranding	1	F	~ 530	UBC 9360; DFO 4146	Campbell and Stirling 1971
Jan 1969	Long Beach, W VI	49°03′	125°43′	Live stranding	1	F	526	UBC 9416; DFO 4137	Campbell and Stirling 1971
2 Oct 1992	0.5 km S of Pachena Pt, W VI	48°43′	125°06′	Stranding	1	F	502	DFO 4529	Willis and Baird 1998a

Stejneger's Beaked Whale									
Aug 1953	Port McNeill, N VI	50°35′	127°06′	Stranding	1	F		UBC 4501; DFO 3779	Cowan and Guiguet 1965
May 1959	Long Beach, W VI	49°02′	125°43′	Stranding	1	F	500	UBC 7433; DFO 4023	Moore 1963
16 Apr 1971	Frederick I., W Graham I., HG	53°56′	133°10′	Stranding	1	M	491	RBCM 7408; DFO 3641	Willis and Baird 1998a
9 Jul 1973	SE Moresby I., HG	52°00′	131°00′	Stranding	1	F		DFO 3640	Willis and Baird 1998a
5 Aug 1979	Departure Bay, SE VI	49°12′	123°58	Live stranding	1	F	~ 450	DFO 4800	CRP, PBS
4 Apr 1982	Point No Point, S VI	48°23′	123°58′	Stranding	1	M	> 400	RBCM 11821; DFO 4422	Willis and Baird 1998a
17 Apr 1998	1 km N of Cape Ball River, Graham I., HG	53°42′	131°52′	Stranding	1	M	492	DFO 4792	CRP, PBS
1 Jun 1998	2 km N of Halibut Bight, SE Graham I., HG	53°24′	131°55′	Stranding	1	F	421	DFO 3576	CRP, PBS
Cuvier's Beaked Whale									
Unknown	Buck Ch, Moresby I., HG	53°06′	132°29′	Stranding	1			RBCM 12580	RBCM
Pre-1952	Victoria, S VI	48°26′	123°22′	Stranding	1			RBCM 5645; DFO 3697	Cowan and Guiguet 1952
1900	Bella Bella, Central Coast	52°08′	128°07′	Stranding	1			UBC 3691; DFO 3726	Cowan and Guiguet 1952
1937	Fisherman Bay, near Cape Scott, N VI	50°47′	128°19′	Stranding	1			RBCM 4557; DFO 4566	Cowan and Hatter 1940

Date	Locality	Latitude (°N)	Longitude (°W)	Record type	No.	Sex (M/F)	Length (cm)	Specimen/ Record no.	Source
Cuvier's Beaked Whale (cont.)									
25 May 1941	3 km N of Estevan Pt, W VI	49°23′	126°33.6′	Stranding	1	M	~ 600–650	DFO 3696	Cowan 1945
Jan 1954	1.6 km S of Jordan River, SW VI	48°25′	124°03′	Stranding	1	M	~ 548	RBCM 6416; DFO 4827	Mitchell 1968
Winter 1959-60	Near Tow Hill, N Graham I., HG	54°10′	131°48′	Stranding	1	M	564	UBC 7999; DFO 4826	Mitchell 1968
Feb 1961	Sandspit, E HG	53°16′	131°48′	Stranding	1	M	545	UBC 8325; DFO 4131	Mitchell 1968
Jul 1968	HG			Stranding	1			DFO 4523	Willis and Baird 1998a
Apr 1969	N of Lawn Pt, SE Graham I., HG	53°26′	131°55′	Stranding	1		~ 590	DFO 4522	Willis and Baird 1998a
4 Jul 1971	Swan Bay, Skincuttle Inlet, E HG	52°20′	131°18′	Stranding	1			DFO 3850	Willis and Baird 1998a
Mar 1978	Swan Bay, Skincuttle Inlet, E HG	52°20′	131°18′	Stranding	1	M		DFO 4804	Willis and Baird 1998a
1 Sep 1986	300 km WSW of Moresby I., W HG	51°00′	135°20.1′	Bycatch	1		~ 700	DFO 3730	CRP, PBS
Jul 1987	Side Bay, Brooks Bay, NW VI	50°20′	127°53′	Live-stranding	1			DFO 3868	Willis and Baird 1998a
28 Aug 1988	Balcom Inlet, Kunghit I., S HG	52°06′	131°00′	Stranding	1	F	~ 580	DFO 3552	CRP, PBS

19 May 1989	Beresford Bay, NW Graham I., HG	54°05′	131°05′	Live-stranding	1			DFO 3836	Willis and Baird 1998a
01 Jun 1989	Off Skedans, E HG	52°57′	131°34′	Sighting	1				Willis and Baird 1998a
Sep 1989	Estevan Point, W VI	49°23′	126°30′	Stranding	1			DFO 4711	Willis and Baird 1998a
10 Jul 1991	N Langara I., N HG	54°15′	133°02′	Stranding	1	F	~ 470	DFO 3976	CRP, PBS
17 May 1993	Gregoire Pt, SW VI	49°54′	127°11′	Stranding	1	F	590	DFO 4710	Willis and Baird 1998a
15 Jul 1994	Tian Head, Graham I., E HG	53°47′	133°07′	Stranding	1	M		DFO 4447	Willis and Baird 1998a
3 Aug 2003	Off NW VI	50°20.1′	128°05.5′	Sighting	1				Ford, Abernethy et al. 2010
4 Aug 2004	Off NW VI	50°29.7′	128°23.5′	Sighting	1				Ford, Abernethy et al. 2010
24 Aug 2005	E Moresby I., HG	52°19.8′	131°01.2′	Sighting	1				Ford, Abernethy et al. 2010
16 May 2010	San Josef Bay, NW VI	50°39′	128°19′	Stranding	1	F	~ 550	DFO 5671	DFO MMRP Pacific
14 Jan 2014	110 km SW of Tofino, W VI	48°33.5′	127°19.4′	Sighting	2				CRP, PBS
Unidentified beaked whale*									
1987	Pyramid Hill, Graham I., HG	54°01′	133°04′	Stranding	1		~200	DFO 3509	Willis and Baird 1998a
1987	White Pt, Graham I., HG	54°00′	133°07′	Stranding	1			DFO 4799	Willis and Baird 1998a

* All are unidentified species of the genus Mesoplodon except the sighting on Apr 29, 2004, which was probably a Cuvier's Beaked Whale.

Date	Locality	Latitude (°N)	Longitude (°W)	Record type	No.	Sex (M/F)	Length (cm)	Specimen/ Record no.	Source
Unidentified beaked whale (cont.)									
17 Apr 1991	5 km S of Green Point, VI	49°03	125°43	Stranding	1			DFO 3968	Willis and Baird 1998a
10 Jun 1997	Woodruff Bay, Kunghit I., HG	51°58′	131°02′	Stranding	1	F	600	DFO 4473	CRP, PBS
29 Apr 2004	S of Triangle Island	50° 38.2′	128°51.9′	Sighting	1	M		DFO 2004	CRP, PBS
8 Apr 2013	Bajo Point, Nootka I.	49° 36.8	126° 47.9	Stranding	1	F	496	DFO 10673	CRP, PBS
Short-beaked Common Dolphin									
8 Apr 1953	Victoria, S VI	48°24.5′	123°19′	Stranding	1	M	179	RBCM 5792; DFO 3727	Guiguet 1954
25 Nov 1994	Nuchatlitz, W VI	49°46′	126°58′	Stranding	1	F	176	DFO 4037	CRP, PBS
Jul 2012	Skedans, E HG	52°57′	131°34′	Stranding	1			DFO 10672	CRP, PBS
Long-beaked Common Dolphin									
2 Feb 1993	Horton Bay, Mayne I., S BC	48°51′	123°18′	Live stranding	1	M	203	DFO 3872	Ford 2005
21 Aug 2002	Port McNeill, NE VI	50°35′	127°05′	Sighting	1				Ford 2005
Sep-Oct 2002	Burrard Inlet, Vancouver	49°17.5′	123°05′	Sightings (multiple)	2				Ford 2005
1 Nov 2002	Goose Bay, Rivers Inlet, Central Coast	51°22′	127°39′	Sighting	4				Ford 2005
23 Sep 2003	Victoria, S VI	48°25′	123°23′	Sighting	2				Ford 2005
7 Oct 2003	Alberni Inlet, W VI	49°10.6′	124°49.1′	Bycatch	2	M	187	DFO 1551	Ford 2005

28 Oct 2003	Alberni Inlet, W VI	49°09.5′	124°48.2′	Sighting	1				Ford 2005
Striped Dolphin									
Winter 1948	Muchalat Arm, Nootka Sound, W VI	49°38′	126°26′	Stranding	1			UBC 2886; DFO 5215	Cowan and Guiguet 1952
Jun 1960	Long Beach, W VI	49°03′	125°42.8′	Stranding	1			RBCM 6665; DFO 9378	RBCM
22 Apr 1961	Lawn Pt, near Cape Cook, E VI	50°02′	125°14′	Stranding	1			UBC 8011; DFO 5216	Pike and MacAskie 1969
23 Apr 1963	11 km from Kyuquot, W VI	49°51′	127°21′	Bycatch	1			UBC 9236; DFO 4795	Pike and MacAskie 1969
5 Mar 1972	Florencia Bay, W VI	49°00′	125°38′	Stranding	1		226	UBC 9470; DFO 9379	Hatler 1972
13 Mar 1972	Long Beach, W VI	49°04′	125°44′	Stranding	1	M		RBCM 7525; DFO 9380	Hatler 1972
1975	Victoria, S VI	48°27.5′	123°17.7′	Stranding	1			RBCM 8875; DFO 9381	Baird et al. 1993
17 Feb 1983	Long Beach, W VI	49°04.2′	125°45.2′	Stranding	1	M	204	RBCM 11394; DFO 9382	RBCM
10 Aug 1987	360 km W of Ucluelet, W VI	48°44.3′	131°07′	Bycatch	1			DFO 6416	Jamieson and Heritage 1988
17 Dec 1995	Wickaninnish Beach, W VI	49°01′	125°40′	Stranding	1	F	205	DFO 4563	CRP, PBS
20 Nov 2002	Green Pt, Long Beach, W VI	49°3.06′	125°43.2′	Stranding	1	F	191	DFO 6265	CRP, PBS
31 May 2005	Carmanah Point, W VI	48°35.5′	124°42.15′	Stranding	1		230	DFO 2141	CRP, PBS

Date	Locality	Latitude (°N)	Longitude (°W)	Record type	No.	Sex (M/F)	Length (cm)	Specimen/ Record no.	Source
Striped Dolphin (cont.)									
4 Jan 2007	Long Beach, W VI	49°04.4′	125°46.2′	Stranding	1	M	220	DFO 2715	CRP, PBS
29 Jan 2011	Schooner Cove, W VI	49°04.1′	125°48.0′	Stranding	1		~200	DFO 6335	DFO MMRP Pacific
Short-finned Pilot Whale*									
23 Nov 1977	Metchosin, S VI	48°22.2′	123°31.8′	Stranding	1	F	452	RBCM 9775; DFO 4363	CRP, PBS
18 Jul 1986	Offshore VI	47°27′	129°18′	Sighting	12-15				CRP, PBS
14 Aug 1986	Offshore VI	50°25′	132°20′	Bycatch	4			DFO 3558	Jamieson & Heritage 1987
Aug 1986	Offshore BC			Bycatch	2			DFO 4331	Jamieson & Heritage 1987
23 Aug 1986	Offshore HG	51°40′	136°0′	Bycatch	1			DFO 4471	Jamieson & Heritage 1987
23 Aug 1986	Offshore HG	51°40′	136°0′	Sighting	150				CRP, PBS
15 Sep 1986	Offshore VI	48°10′	129°15′	Sighting	10				CRP, PBS
13 Jul 1987	Offshore VI	47°27′	130°05′	Bycatch	1			DFO 3624	Jamieson & Heritage 1988
30 Aug 1988	Offshore VI	48°05′	128°10′	Sighting	9				CWS
28 Aug 2001	Offshore VI	48°56.2′	130°20.4′	Sighting	5				CWS

* Some early sighting reports of Short-finned Pilot Whales in Pike and MacAskie (1969) are not included as the species identification is questionable.

APPENDIX 3

Whale Catches in Canadian Pacific Waters, 1905–75.

Table A3.1.

Whale catches from shore-based whaling stations in British Columbia, 1905–67. Data from the BC Historical Whaling Database (Nichol et al. 2002) and Merilees (1985). Note that totals may differ from those in Pike and MacAskie (1969) and Nichol et al. (2002) because additional data sources were used and catches outside Canadian waters are excluded.

Whaling stations:

- S = Sechart, Barkley Sound, western Vancouver Island
- PL = Page's Lagoon, Nanaimo (now Piper's Lagoon)
- C = Cachalot, Kyuquot Sound, northwestern Vancouver Island
- NH = Naden Harbour, northern Graham Island, Haida Gwaii
- RH = Rose Harbour, Kunghit Island, southern Haida Gwaii
- CH = Coal Harbour, Quatsino Sound, northwestern Vancouver Island

Species listed:

- BBW = Baird's Beaked Whale
- BW = Blue Whale
- FW = Fin Whale
- GW = Grey Whale
- HW = Humpback Whale
- CMW = Common Minke Whale
- RW = North Pacific Right Whale
- SPW = Sperm Whale
- SW = Sei Whale

Year	Stations	Species								Total
		BBW	BW	FW	GW	HW	CMW	RW	SPW	SW	
1905	S										NA
1906	S										NA
1907*	S, PL					72					72
1908	C, S, PL		98	26		483			2		609
1909	C, S, PL		61	13		597			1		672
1910	C, R, S, PL		111	75		741			31		958
1911	C, NH, RH, S		203	364	1	1022			34		1624
1912	C, NH, RH, S		167	395		528			19		1109
1913	C, NH, RH, S		38	215		437			15		705
1914	C, NH, RH, S		82	322		134		1	25		564
1915	C, NH	1	17	131		79			1		229
1916	C, NH, RH	1	18	260		104			14	7	404
1917	C, RH, S	1	19	104		105			23	126	378
1918	C, NH, RH	1	31	214		89		1	34	130	500
1919	C, NH, RH	4	37	217		65			35	74	432
1920	C, NH, RH	1	26	148		98			56	164	493
1921	*none*										0
1922	C, NH		4	94		49			38	2	187

* Catch data in 1907 only available for Page's Lagoon. **<why is this footnote necessary? Just list "PL" in 1907, above.>**

1923	C, NH, RH	2	62	165		78			94	54	455
1924	C, NH, RH	1	49	117	1	46		1	72	100	387
1925	C, NH, RH	3	28	128		39			75	68	341
1926	NH, RH		14	118		23		1	79	26	261
1927	NH, RH		8	135		21			77	7	248
1928	NH, RH	1	49	140		21			83	13	307
1929	NH, RH		16	168		10		1	146	67	408
1930	NH, RH		10	53		12			88	88	251
1931–32	*none*										0
1933	RH		1	17					174		192
1934	NH, RH			70		13			252		335
1935	RH		5	25		1			175		206
1936	NH, RH		3	48		14			311		376
1937	NH, RH		1	44		7			265		317
1938	NH, RH		3	49		4			222		278
1939	*none*										0
1940	RH		1	92		1			120		214
1941	NH, RH		1	85		9			233		328
1942	RH		1	16		16			130		163
1943	RH			17		5			69		91
1944–47	*none*										0
1948	CH			38		119			28	3	188

Year	Stations	Species									Total
		BBW	BW	FW	GW	HW	CMW	RW	SPW	SW	
1949	CH		2	105		76			69	3	255
1950	CH	1	4	150		95			40	24	314
1951	CH	1	9	216	1	51		1	153	5	437
1952	CH		16	240		61			126	22	465
1953	CH	4	8	181	10	47			275	14	539
1954	CH	3	11	150		105			222	134	625
1955	CH	3	11	120		37			320	139	630
1956	CH	1	15	167		28			127	37	375
1957	CH	4	15	284		49			190	93	635
1958	CH	2	8	573		40			112	39	774
1959	CH		28	372		28			256	185	869
1960–61	*none*										0
1962	CH	1	26	158		16			172	340	713
1963	CH	3	30	219		24			147	154	577
1964	CH		12	140		10			106	614	882
1965	CH		9	83		18			151	601	862
1966	CH	2		134					229	355	720
1967	CH			102			1		303	87	493
Species totals		**41**	**1368**	**7497**	**13**	**5727**	**1**	**6**	**6019**	**3775**	**24447**

Table A3.2.

Whale catches in Canadian waters from pelagic whaling ships operated by Japan from 1964 to 1975 and by the Soviet Union in 1974. Data provided by the International Whaling Commission. The IWC found Soviet catch data from 1964 to 1972 to be falsified, so these are not included in this table.

Active ships:

DV = *Dalnij Vostok* (Soviet Union)
KIN = *Kinjyo Maru* (Japan)
KM = *Kyokuyo Maru* (Japan)
KM3 = *Kyokuyo Maru III* (Japan)
NIC = *Nichiei Maru* (Japan)
NM3 = *Nisshin Maru III* (Japan)
TM = *Tonan Maru* (Japan)
TM2 = *Tonan Maru II* (Japan)

Species listed:

FW = Fin Whale
HW = Humpback Whale
CMW = Common Minke Whale
SPW = Sperm Whale
SW = Sei Whale

Year	Active ships	Species					Total
		FW	HW	CMW	SPW	SW	
1964	KIN, KM	38			4	39	81
1965	KM, NIC	29	1		18	10	58
1966	–						0
1967	–						0
1968	–						0
1969	TM				13		13
1970	NM3, TM	24			90	226	340
1971	NM3	16		2	70	26	114
1972	NM3, TM2	1			148	1	150
1973	–						0
1974	NM3, KM3, DV	78			131	30	239
1975	NM3	15				3	18
Species totals		**201**	**1**	**2**	**474**	**335**	**1013**

GLOSSARY

Auditory bullae – the inflated bony capsules that enclose the middle and inner ears.

Basicranium – referring to the ventral side of the skull.

Bathypelagic – the deepwater portion of the pelagic zone (open ocean) that extends from 1000 to 4000 metres below the ocean surface.

Bony nares – nasal openings through the skull leading to the nostrils or blowhole(s).

Bony symphysis – bony attachment of the two sides of the mandible at their anterior tip.

Caudal peduncle – the tail stock of cetaceans, or tapered posterior part of the body to which the flukes are attached.

Condylobasal length (CBL) – the length of a skull, measured from the anterior point of the rostrum to the posterior surfaces of the occipital condyles.

COSEWIC – Committee on the Status of Endangered Wildlife in Canada.

Cyclostomes – the jawless fishes (hagfishes and lampreys).

DFO – the well-established abbreviation for the Canadian Department of Fisheries and Oceans, now called Fisheries and Oceans Canada.

EEZ – Exclusive Economic Zone, which in western Canada extends 200 nautical miles (370 km) offshore.

Equilibrium – in populations at equilibrium, birth rates and mortality rates are in balance, leading to a stable abundance.

Elasmobranchs – cartilaginous fishes including sharks, rays and skates.

Extant – a taxon group (e.g., species, genus or family) that currently exists or is alive (as opposed to extinct).

Extirpated – when a species ceases to exist in a geographic area that it formerly occupied, though it continues to exist elsewhere.

Extralimital – occurrence of a species outside of its normal range.

Falcate – hooked or curved like a sickle.

Foramen magnum – the hole in the posterior of the skull through which the spinal cord passes.

Frontal – the anteriormost pair of bones covering the brain, situated between the orbits, actually paired but often fusing together at an early age.

Gape – the width of the widely opened mouth of a vertebrate.

Heterodont – having teeth differentiated into various types; e.g., incisors, canines, premolars and molars.

Homodont – having teeth that are all essentially similar, usually simple cones.

IUCN – International Union for Conservation of Nature.

Mandible – the entire lower jaw; the two separate jaw bones are known as the dentary bones.

Maxillary – the bone in the upper jaw that bears the canines, premolars and molars (teeth undifferentiated, reduced or unerupted in toothed whales, absent in baleen whales).

Mesopelagic – the part of the open (pelagic) ocean that extends from a depth of 200 to 1000 metres below the surface.

Nares – the nostrils.

Nasal – the anteriormost pair of mid-dorsal bones forming the roof of the nasal passage.

Nasofrontal process – the area where the nasal and frontal join.

Occipital condyles – rounded knobs on the occipital bone at the base or back of the skull that forms a joint with the first cervical vertebra, enabling the head to move relative to the neck.

Pelage – the fur or hair of a mammal.

Pelagic – referring to the open ocean, neither close to the bottom nor near the coast.

Polygynous – a breeding system where one male mates with more than one female while each female mates with only one male.

Premaxillary – paired bones in the front of the upper jaw that bear the incisor teeth (in pinnipeds and sea otter).

Rorqual – a baleen whale of the family Balaenopteridae. The name is from the Norwegian word *røyrkval*, meaning “furrow whale”, referring to the series of longitudinal pleats or grooves below the mouth that continue along the body’s underside.

Rostrum – the beak-like projection or anterior prolongation of the head of a cetacean.

SARA – the Canadian government's 2002 Species At Risk Act provides for the protection of COSEWIC-ranked Endangered and Threatened species and their habitat in Canada; see page 63 for more details.

Sagittal crest – a longitudinal median bony ridge dorsal to the braincase.

Site fidelity – the tendency for individual animals to remain in or return to particular geographic locations, often on a small spatial scale.

Spyhopping – raising the head vertically above the surface to expose the eyes to air, presumably to better observe things above water.

Supraoccipital – a bone on the dorsal side of the foramen magnum, or hole in the rear of the skull where the spinal cord emerges.

Supraorbital processes – small bony "shelves" that project off the sides of the frontal bones above the orbit (eye socket).

Symphysis – a fibrocartilaginous fusion between two bones.

Throat pleats – longitudinal parallel furrows or grooves in the throat and anterior portion of the thorax of baleen whales that allow the throat to expand when engulfing water to capture zooplankton.

Unicuspid – teeth with a single cusp or projection on the crown.

Vertex – area on the dorsal surface of the skull where the nasal, maxillary, premaxillary, frontal and supraoccipital bones converge.

Zygomatic arch – the arch of bone that forms the lateral border of the orbit and temporal fossa.

Zygomatic width – width of the skull at the zygomatic arches.

SELECTED REFERENCES

Acheson, S., and R.J. Wigen. 2002. Evidence for a prehistoric whaling tradition among the Haida. *Journal of Northwest Anthropology* 36:155–68.

Aguilar, A. 2009. Fin Whale *Balaenoptera physalus*. In Perrin, Würsig and Thewissen 2009.

Ainley, D.G., R.E. Jones, R. Satallcup, D.J. Long, G.W. Page, L.T. Jones, L.E. Stenzel, R.L. LeValley and L.B. Spear. 1994. *Beached Marine Birds and Mammals of the North American West Coast: A Revised Guide to their Census and Identification, with Supplemental Keys to Beached Sea Turtles and Sharks.* San Francisco: National Oceanic and Atmospheric Administration, Gulf of the Farallones National Marine Sanctuary.

Allen, B.M., and R.P. Angliss. 2011. *Alaska Marine Mammal Stock Assessments, 2010.* NOAA Technical Memorandum NMFS-AFSC-223. Seattle: National Oceanic and Atmospheric Administration, National Marine Fisheries Service.

Andrews, R.C. 1916. Monographs of the Pacific Cetacea II. The Sei Whale (*Balaenoptera borealis* Lesson). 1. History, habits, external anatomy, osteology, and relationship. *Memoirs of the American Museum of Natural History (New Series)* 1:289–388.

Andrews, R.D., J.M. Straley, G.S. Schorr, A. Thode, J. Calambokidis, C.R. Lunsford and V. O'Connell. 2010. Satellite tracking Eastern Gulf of Alaska Sperm Whales: local movements around the shelf-edge contrast with rapid, long-distance migrations across stock boundaries. *Alaska Marine Science Symposium*, Anchorage, p. 73.

Archer, F.I., and W.F. Perrin. 1999. *Stenella coeruleoalba. Mammalian Species* 603:1–9.

Arima, E.Y. 1988. Notes on Nootkan sea mammal hunting. *Arctic Anthropology* 25:16–27.

Arndt, U.M. 2011. Ancient DNA analysis of northeast Pacific Humpback Whale (*Megaptera novaeangliae*). PhD thesis, Simon Fraser University, Burnaby, BC.

Ashe, E. 2007. Survival and abundance of Pacific White-sided Dolphins (*Lagenorhynchus obliquidens*) in the inshore coastal waters of British Columbia, Canada. Masters of Research thesis, University of St Andrews.

Bailey, H., B.R. Mate, D.M. Palacios, L. Irvine, S.J. Bograd and D.P. Costa. 2009. Behavioural estimation of Blue Whale movements in the Northeast Pacific from state-space model analysis of satellite tracks. *Endangered Species Research* 10:93–106.

Baird, R.W. 2009. False Killer Whale *Pseudorca crassidens*. In Perrin, Würsig and Thewissen 2009.

Baird, R.W., K. Langelier and P.J. Stacey. 1989. First records of False Killer Whales (*Pseudorca crassidens*) in Canada. *The Canadian Field-Naturalist* 103:368–71.

Baird, R.W., D. Nelson, J. Lien and D.W. Nagorsen. 1996. The status of the Pygmy Sperm Whale, *Kogia breviceps*, in Canada. *The Canadian Field-Naturalist* 110:525–32.

Baird, R.W., and P.J. Stacey. 1991a. Status of Risso's Dolphin, *Grampus griseus*, in Canada. *The Canadian Field-Naturalist* 105:233–42.

———. 1991b. Status of the Northern Right Whale Dolphin, *Lissodelphis borealis*, in Canada. *The Canadian Field-Naturalist* 105:243–50.

———. 1993. Sightings, strandings, and incidental catches of Short-finned Pilot Whales, *Globicephala macrorhynchus*, off the British Columbia coast. *Report of the International Whaling Commission* Special issue 14:475–79.

Baird, R.W., P.J. Stacey and H. Whitehead. 1993. Status of the Striped Dolphin, *Stenella coeruleoalba*, in Canada. *The Canadian Field-Naturalist* 107:455–65.

Baird, R.W., D.L. Webster, D.J. McSweeney, A.D. Ligon, G.S. Schorr and J. Barlow. 2006. Diving behaviour of Cuvier's (*Ziphius cavirostris*) and Blainville's (*Mesoplodon densirostris*) Beaked Whales in Hawai'i. *Canadian Journal of Zoology* 84:1120–28.

Balcomb, K.C. III. 1989. Baird's Beaked Whale *Berardius bairdii* Stejneger, 1883: Arnoux's Beaked Whale *Berardius arnuxii* Duvernoy, 1851. In *Handbook of Marine Mammals: Volume 4. River Dolphins and the Larger Toothed Whales*. New York: Academic Press.

Barlow, J. 2010. Cetacean abundance in the California Current estimated from a 2008 ship-based line transect survey. NOAA Technical Memorandum NMFS-SWFSC-456. Seattle: National Oceanic and Atmospheric Administration, National Marine Fisheries Service.

Barlow, J., J. Calambokidis, E.A. Falcone, C.S. Baker, A.M. Burdin, P.J. Clapham, J.K.B. Ford, C.M. Gabriele, R. LeDuc, D.K. Mattila, T.J.I. Quinn, L. Rojas-Bracho, J.M. Straley, B.L. Taylor, J. Urbán R., P. Wade, D. Weller, B.H. Witteveen and M. Yamaguchi. 2011. Humpback Whale abundance in the North Pacific estimated by photographic capture-recapture with bias correction from simulation studies. *Marine Mammal Science* 27:793–818.

Barlow, J., and K.A. Forney. 2007. Abundance and population density of cetaceans in the California Current ecosystem. *Fishery Bulletin* 105:509–26.

Barrett-Lennard, L.G., J.K.B. Ford and K.A. Heise. 1996. The mixed blessing of echolocation: differences in sonar use by fish-eating and mammal-eating Killer Whales. *Animal Behaviour* 51:553–65.

BC Marine Conservation Analysis. 2011. *Marine Atlas of Pacific Canada: a product of the British Columbia Marine Conservation Analysis (BCMCA)*. http://www.bcmca.ca.

Bearzi, M. 2006. California Sea Lions use dolphins to locate food. *Journal of Mammalogy* 87:606–17.

Berta, A. 2009. Pinniped evolution. In Perrin, Würsig and Thewissen 2009.

Berta, A., and T.A. Deméré. 2009. Mysticetes, evolution. In Perrin, Würsig and Thewissen 2009.

Berta, A., and J.L. Sumich. 1999. *Marine Mammals: Evolutionary Biology*. San Diego: Academic Press.

Bigg, M.A. 1969. The Harbour Seal in British Columbia. *Bulletin of the Fisheries Research Board of Canada* 172:33.

———. 1981. Harbour Seal *Phoca vitulina* Linnaeus, 1758 and *Phoca largha* Pallas, 1811. In *Handbook of Marine Mammals*, vol. 2: *Seals* edited by S.H. Ridgway and R.J. Harrison. London: Academic Press.

———. 1985. Status of the Steller Sea Lion (*Eumetopias jubatus*) and California Sea Lion (*Zalophus californianus*) in British Columbia. *Canadian Special Publication of Fisheries and Aquatic Sciences* 77.

———. 1988. Status of the Steller Sea Lion, *Eumetopias jubatus*, in Canada. *The Canadian Field-Naturalist* 102:315–36.

———. 1990. Migration of Northern Fur Seals (*Callorhinus ursinus*) off western North America. *Canadian Technical Report of Fisheries and Aquatic Sciences* 1764.

Bigg, M.A., and I.B. MacAskie. 1978. Sea Otters reestablished in British Columbia. *Journal of Mammalogy* 59:874–76.

Bigg, M.A., P.F. Olesiuk, G.M. Ellis, J.K.B. Ford and K.C. Balcomb III. 1990. Social organization and genealogy of Resident Killer Whales (*Orcinus orca*) in the coastal waters of British Columbia and Washington. *Report of the International Whaling Commission* Special Issue 12:383–405.

Bigg, M.A., and A.A. Wolman. 1975. Live-capture Killer Whale (*Orcinus orca*) fishery, British Columbia and Washington, 1962–73. *Journal of the Fisheries Research Board of Canada* 32:1213–21.

Bjørge, A., and K.A. Tolley. 2009. Harbour Porpoise *Phocoena phocoena*. In Perrin, Würsig and Thewissen 2009.

Black, N.A. 1994. Behavior and ecology of Pacific White-sided Dolphins (*Lagenorhynchus obliquidens*) in Monterey Bay, California. MSc thesis, San Francisco State University.

Boas, F. 1909. The Kwakiutl of Vancouver Island. In *The Jesup North Pacific Expedition, Memoir of the American Museum of Natural History*. New York: G.E. Stechert.

Bodkin, J.L., B.E. Ballachey, H.A. Coletti, G.G. Esslinger, K.A. Kloecker, S.D. Rice, J.A. Reed and D.H. Monson. 2012. Long-term effects of the Exxon *Valdez* oil spill: Sea Otter foraging in the intertidal as a pathway of exposure to lingering oil. *Marine Ecology Progress Series* 447:273–87.

Boness, D.J., P.J. Clapham and S.L. Mesnick. 2002. Life history and reproductive strategies. In Hoelzel 2002.
Bortolotti, D. 2009. *Wild Blue: A Natural History of the World's Largest Animal*. Toronto Thomas Allen Publishers.
Bowen, W.D., C.A. Beck and D.A. Austin. 2009. Pinniped ecology. In Perrin, Würsig and Thewissen 2009.
Bowen, W.D., and D.B. Siniff. 1999. Distribution, population biology and feeding ecology of marine mammals. In Reynolds and Rommel 1999.
Boyd, I.L., W.D. Bowen and S.J. Iverson. 2010. *Marine Mammal Ecology and Conservation. A Handbook of Techniques*. Oxford: Oxford University Press.
Boyd, I.L., C. Lockyer and H.D. Marsh. 1999. Reproduction in marine mammals. In Reynolds and Rommel 1999.
Brownell, R.L., L. Robert, P.J. Clapham, T. Miyashita and T. Kasuya. 2001. Conservation status of North Pacific Right Whales. *Journal of Cetacean Research and Management* Special Issue 2:269–86.
Burg, T.M., A.W. Trites and M.J. Smith. 1999. Mitochondrial and microsatellite DNA analyses of Harbour Seal population structure in the northeast Pacific Ocean. *Canadian Journal of Zoology* 77:930–43.
Burns, J.J. 2009. Harbor Seal and Spotted Seal *Phoca vitulina* and *P. largha*. In Perrin, Würsig and Thewissen 2009.
Burtenshaw, J.C., E.M. Oleson, J.A. Hildebrand, M.A. McDonald, R.K. Andrew, B.M. Howe and J.A. Mercer. 2004. Acoustic and satellite remote sensing of Blue Whale seasonality and habitat in the Northeast Pacific. *Deep-Sea Research II* 51:967–86.
Calambokidis, J., J. Barlow, J.K.B. Ford, T.E. Chandler and A.B. Douglas. 2009. Insights into the population structure of Blue Whales in the eastern North Pacific from recent sightings and photographic identification. *Marine Mammal Science* 25:816–32.
Calambokidis, J., E.A. Falcone, T.J. Quinn, A.M. Burdin, P.J. Clapham, J.K.B. Ford, C.M. Gabriele, R. LeDuc, D. Mattila, L. Rojas-Bracho, J.M. Straley, B.L. Taylor, J. Urbán R., D. Weller, B.H. Witteveen, M. Yamaguchi, A. Bendlin, D. Camacho, K. Flynn, A. Havron, J. Huggins and N. Maloney. 2008. SPLASH: Structure of Populations, Levels of Abundance and Status of Humpback Whales in the North Pacific. Report. Cascadia Research Collective, Olympia, Washington.
Campbell, R.W., and D. Stirling. 1971. A photoduplicate file for British Columbia vertebrate records. *Syesis*, 4:217–22.
Carl, G.C. 1946. A school of Killer Whales stranded at Estevan Point. *BC Provincial Museum Natural History and Anthropology* Report (1945):21–28.
Carretta, J.V., K. A. Forney, E. Oleson, K. Martien, M.M. Muto, M.S. Lowry, J. Barlow, J. Baker, B. Hanson, D. Lynch, L. Carswell, R.L. Brownell Jr, J. Robbins, D.K. Mattila, L. Ralls and M.C. Hill. 2011. *US Pacific Marine Mammal Stock Assessments: 2010*. NOAA Technical Memorandum NMFS-SWFSC-476. NOAA Fisheries, Southwest Fisheries Science Center, La Jolla, California.
Cavanagh, D.M. 1983. Northwest Coast Whaling: A New Perspective. MA thesis, University of British Columbia, Vancouver.

Chivers, S.J., A.E. Dizon, P.J. Gearin and K.M. Roberston. 2002. Small-scale population structure of eastern North Pacific Harbour Porpoises (*Phocoena phocoena*) indicated by molecular genetic analyses. *Journal of Cetacean Research and Management* 4:111–222.
Clapham, P., C. Good, S. Quinn, R. Reeves, J. Scarff and R.L. Brownell, Jr. 2004. Distribution of North Pacific Right Whales (*Eubalaena japonica*) as shown by 19th- and 20th-century whaling catch and sighting records. *Journal of Cetacean Research and Management* 6:1–6.
Clapham, P.J. 2009. Humpback Whale *Megaptera novaeangliae*. In Perrin, Würsig and Thewissen 2009.
Clapham, P.J., and J.G. Mead. 1999. *Megaptera novaeangliae*. *Mammalian Species*:1–9.
Connor, R.C. 2002. Ecology of group living and social behaviour. In Hoelzel 2002.
COSEWIC. 2003. COSEWIC assessment and status report on the Sei Whale *Balaenoptera borealis* in Canada. Committee on the Status of Endangered Wildlife in Canada, Ottawa.
———. 2006. COSEWIC status report on Common Minke Whale *Balaenoptera acutorostrata*. Committee on the Status of Wildlife in Canada, Ottawa.
———. 2007. COSEWIC assessment and update status report on the Sea Otter *Enhydra lutris* in Canada. Committee on the Status of Endangered Wildlife in Canada, Ottawa.
———. 2010. COSEWIC assessment and status report on the Northern Fur Seal *Callorhinus ursinus* in Canada. Committee on the Status of Endangered Wildlife in Canada, Ottawa.
———. 2011. COSEWIC assessment and status report on the Humpback Whale *Megaptera novaeangliae* in Canada. Committee on the Status of Endangered Wildlife in Canada, Ottawa.
———. In press. COSEWIC assessment and status report on Steller Sea Lion *Eumetopias jubatus* in Canada. Committee on the Status of Endangered Wildlife in Canada, Ottawa.
Cottrell, P.E., S. Jeffries, B. Beck and P.S. Ross. 2002. Growth and development in free-ranging Harbour Seal (*Phoca vitulina*) pups from southern British Columbia, Canada. *Marine Mammal Science* 18:721–33.
Cowan, I.M. 1939. The Sharp-headed Finner Whale of the eastern Pacific. *Journal of Mammalogy* 20:215–25.
———. 1944. The Dall Porpoise, *Phocoenoides dalli* (True), of the northern Pacific Ocean. *Journal of Mammalogy* 25:295–306.
———. 1945. A beaked whale stranded on the coast of British Columbia. *Journal of Mammalogy* 26:93–94.
Cowan, I.M., and G.C. Carl. 1945. The Northern Elephant Seal (*Mirounga angustirostris*) in British Columbia waters and vicinity. *The Canadian Field-Naturalist* 59:170–71.
Cowan, I.M., and C.J. Guiguet. 1952. Three cetacean records from British Columbia. *Murrelet* 33:10–11.
———. 1965. *The Mammals of British Columbia*. Victoria: British Columbia Provincial Museum.

Cowan, I.M., and J. Hatter. 1940. Two mammals new to the known fauna of British Columbia. *Murrelet* 21:9.

Crockford, S.J., S.G. Frederick and R.J. Wigen. 2002. The Cape Flattery fur seal: an extinct species of *Callorhinus* in the eastern North Pacific? *Canadian Journal of Archaeology* 26:152–74.

Crossman, C.A. 2012. Population structure in Harbour Porpoises (*Phocoena phocoena*) of British Columbia and widespread hybridization in cetaceans. MSc thesis, University of British Columbia, Vancouver.

Dahlheim, M.E., and J.E. Heyning. 1999. Killer Whale *Orcinus orca* (Linnaeus, 1758). In *Handbook of Marine Mammals: Volume 6: The Second Book of Dolphins and the Porpoises* edited by S.H. Ridgway and R.S. Harrison. San Diego: Academic Press.

Darimont, C.T., P.C. Paquet and T.E. Reimchen. 2008. Spawning salmon disrupt trophic coupling between wolves and ungulate prey in coastal British Columbia. *BMC Ecology* 8:14.

Darling, J.D. 1984. Gray Whales off Vancouver Island, British Columbia. In *The Gray Whale* Eschrichtius robustus edited by M.L. Jones, S.L. Swartz and S. Leatherwood. New York: Academic Press.

Darling, J.D., K.E. Keogh and T.E. Steeves. 1998. Gray Whale (*Eschrichtius robustus*) habitat utilization and prey species off Vancouver Island, BC. *Marine Mammal Science* 14:692–720.

Dawson, S., J. Barlow and D. Ljungblad. 1998. Sounds recorded from Baird's Beaked Whale, *Berardius bairdii. Marine Mammal Science* 14:335–44.

Deecke, V.B., J.K.B. Ford and P. Slater. 2005. The vocal behaviour of mammal-eating Killer Whales: communicating with costly calls. *Animal Behaviour* 69:395–405.

Demarchi, D.A. 2011. The British Columbia ecoregion classification. Ecosystem Information Section, BC Ministry of Environment, Victoria.

Doroff, A., and A. Burdin. 2011. *Enhydra lutris*. In IUCN 2012. IUCN Red List of Threatened Species. www.iucnredlist.org.

Dorsey, E.M., S.J. Stern, A.R. Hoelzel and J. Jacobsen. 1990. Minke Whales (*Balaenoptera acutorostrata*) from the west coast of North America: individual recognition and small-scale site fidelity. *Report of the International Whaling Commission* 12:357–68.

Drucker, P. 1951. *The Northern and Central Nootkan Tribes. Bulletin of the Bureau of American Ethnology* (Smithsonian Institution) 144.

Dunham, J.S., and D.A. Duffus. 2002. Diet of Grey Whales (*Eschrichtius robustus*) in Clayoquot Sound, British Columbia, Canada. *Marine Mammal Science* 18:419–37.

Ellis, G.M., J.R. Towers and J.K.B. Ford. 2011. Northern resident killer whales of British Columbia: Photo-identification catalogue and population status to 2010. *Canadian Technical Report of Fisheries and Aquatic Sciences* 2942.

Estes, J.A., J.L. Bodkin and M. Ben-David. 2009. Otters, marine. In Perrin, Würsig and Thewissen 2009.

Falcone, E.A., G.S. Schorr, A.B. Douglas, J. Calambokidis, E. Henderson, M.F. McKenna, J. Hildebrand and D. Moretti. 2009. Sighting characteristics and photo-identification of Cuvier's Beaked Whales

(*Ziphius cavirostris*) near San Clemente Island, California: a key area for beaked whales and the military? *Marine Biology* 156:2631–40.

Ferguson, M.C., J. Barlow, S.B. Reilly and T. Gerrodette. 2006. Predicting Cuvier's (*Ziphius caviorostris*) and *Mesoplodon* beaked whale population density from habitat characteristics in the eastern tropical Pacific Ocean. *Journal of Cetacean Research and Management* 7:287–99.

Fish, F.E., P.W. Weber, M.M. Murray and L.E. Howle. 2011. The tubercles on Humpback Whales' flippers: application of bio-inspired technology. *Integrative and Comparative Biology* 51:203–13.

Fisher, H.D. 1952. The status of the Harbour Seal in British Columbia, with particular reference to the Skeena River. *Bulletin of the Fisheries Research Board of Canada* 93:58.

Fisheries and Oceans Canada. 2009. Management Plan for the Pacific Harbour Porpoise (*Phocoena phocoena*) in Canada. Fisheries and Oceans Canada, Ottawa.

———. 2010. Management Plan for the Eastern Pacific Grey Whale (*Eschrichtius robustus*) in Canada. Fisheries and Oceans Canada, Ottawa.

———. 2011. Recovery Strategy for the North Pacific Right Whale (*Eubalaena japonica*) in Pacific Canadian Waters. Fisheries and Oceans Canada, Ottawa.

Fisheries Branch. 1918. Fiftieth Annual Report of the Fisheries Branch of the Department of Naval Service, for 1916–17 (Ottawa). *Sessional Paper* 38.

Flinn, R.D., A.W. Trites, E.J. Gregr and R.I. Perry. 2002. Diets of Fin, Sei and Sperm whales in British Columbia: an analysis of commercial whaling records, 1963–1967. *Marine Mammal Science* 18:663–79.

Ford, J.K.B. 1989. Acoustic behaviour of resident Killer Whales (*Orcinus orca*) off Vancouver Island, British Columbia. *Canadian Journal of Zoology* 67:727–45.

———. 1991. Vocal traditions among resident Killer Whales (*Orcinus orca*) in coastal waters of British Columbia. *Canadian Journal of Zoology* 69:1454–83.

———. 2005. First records of Long-beaked Common Dolphins, *Delphinus capensis*, in Canadian waters. *The Canadian Field-Naturalist* 119:110–13.

———. 2009. Killer Whales *Orcinus orca*. In Perrin, Würsig and Thewissen 2009.

Ford, J.K.B., R.M. Abernethy, A.V. Phillips, J. Calambokidis, G.M. Ellis and L.M. Nichol. 2010. Distribution and relative abundance of cetaceans in western Canadian waters from ship surveys, 2002–2008. *Canadian Technical Report of Fisheries and Aquatic Sciences* 2913.

Ford, J.K.B., J.W. Durban, G.M. Ellis, J.R. Towers, J.F. Pilkington, L.G. Barrett-Lennard and R.D. Andrews. 2013. New insights into the northward migration route of Gray Whales between Vancouver Island, British Columbia and southeastern Alaska. *Marine Mammal Science* 29:325–37.

Ford, J.K.B., and G.M. Ellis. 1999. *Transients: Mammal-Hunting Killer Whales of British Columbia, Washington and Southeastern Alaska.* Vancouver: UBC Press.

Ford, J.K.B., and G.M. Ellis. 2006. Selective foraging by fish-eating Killer Whales *Orcinus orca* in British Columbia. *Marine Ecology Progress Series* 316:185–99.

Ford, J.K.B., G.M. Ellis and K.C. Balcomb. 2000. *Killer Whales: The Natural History and Genealogy of* Orcinus orca *in the Waters of British Columbia and Washington*, 2nd edition. Vancouver: UBC Press; Seattle: University of Washington Press.

Ford, J.K.B., G.M. Ellis, L.G. Barrett-Lennard, A.B. Morton, R.S. Palm and K.C. Balcomb III. 1998. Dietary specialization in two sympatric populations of Killer Whales (*Orcinus orca*) in coastal British Columbia and adjacent waters. *Canadian Journal of Zoology* 76:1456–71.

Ford, J.K.B., G. Ellis, C.O. Matkin, M.H. Wetklo, L.G. Barrett-Lennard and R.E. Withler. 2011. Shark predation and tooth wear in a population of northeastern Pacific Killer Whales. *Aquatic Biology* 11:213–24.

Ford, J.K.B., G.M. Ellis, D.R. Matkin, K.C. Balcomb, D. Briggs and A.B. Morton. 2005. Killer Whale attacks on Minke Whales: prey capture and antipredator tactics. *Marine Mammal Science* 21:603–18.

Ford, J.K.B., B. Koot, S. Vagle, N. Hall-Patch and G. Kamitakahara. 2010. Passive acoustic monitoring of large whales in offshore waters of British Columbia. *Canadian Technical Report of Fisheries and Aquatic Sciences* 2898.

Ford, J.K.B., A.L. Rambeau, R.M. Abernethy, M.D. Boogaards, L.M. Nichol and L.D. Spaven. 2009. An assessment of the potential for recovery of Humpback Whales off the Pacific coast of Canada. DFO Canadian Science Advisory Secretariat, Research Document 2009/015.

Ford, J.K.B., and R.R. Reeves. 2008. Fight or flight: antipredator strategies of baleen whales. *Mammal Review* 38:50–86.

Ford, J.K.B., E.H. Stredulinsky, J.R. Towers and G.M. Ellis. 2013. Information in support of the identification of critical habitat for Transient Killer Whales (*Orcinus orca*) off the west coast of Canada. DFO Canadian Science Advisory Secretariat Research Document 2012/155.

Fordyce, R.E. 2009. Cetacean evolution. In Perrin, Würsig and Thewissen 2009.

Forney, K.A., and P.R. Wade. 2006. Worldwide distribution and abundance of Killer Whales. In Whales, Whaling, and Ocean Ecosystems, edited by J.A. Estes, D.P. Demaster, D.F. Doak, T.M. Williams and R.L. Brownell Jr. Berkeley: University of California Press.

Fortune, S.M., A.W. Trites, W.L. Perryman, M.J. Moore, H.M. Pettis and M.S. Lynn. 2012. Growth and rapid early development of North Atlantic Right Whales (*Eubalaena glacialis*). *Journal of Mammalogy* 93:1342–54.

Frasier, T.R., S.M. Koroscil, B.N. White and J.D. Darling. 2011. Assessment of population substructure in relation to summer feeding ground use in the eastern North Pacific Gray Whale. *Endangered Species Research* 14:39–48.

Gabriele, C.M., J.M. Straley, L.M. Herman and R.J. Coleman. 1996. Fastest documented migration of a North Pacific Humpback Whale. *Marine Mammal Science* 12:457–64.
Gaskin, D.E. 1982. *The Ecology of Whales and Dolphins*. London: Heinemann Educational Books.
Gentry, R.L. 2009. Northern Fur Seal *Callorhinus ursinus*. In Perrin, Würsig and Thewissen 2009.
Gentry, R.L., and D.E. Withrow. 1986. Steller Sea Lion. In *Marine Mammals*, 2nd edition, revised, edited by D. Haley. Seattle: Pacific Search Press.
Gibson, J.R. 1992. *Otter Skins, Boston Ships and China Goods: The Maritime Fur Trade of the Northwest Coast, 1785–1841*. Montreal: McGill-Queen's University Press.
Goddard, J. 1997. *A Window on Whaling in British Columbia*. Victoria: Jonah Publications.
Goldbogen, J.A. 2010. The ultimate mouthful: lunge feeding in rorqual whales. *American Scientist* 98:124–31.
Goldbogen, J.A., J. Calambokidis, A.S. Friedlaender, J. Francis, S.L. DeRuiter, A.K. Stimpert, E. Falcone and B.L. Southall. 2012. Underwater acrobatics by the world's largest predator: 360° rolling manoeuvres by lunge-feeding Blue Whales. *Biology Letters* 9.
Goldbogen, J.A., J. Calambokidis, E. Oleson, J. Potvin, N.D. Pyenson, G. Schorr and R.E. Shadwick. 2011. Mechanics, hydrodynamics and energetics of Blue Whale lunge feeding: efficiency dependence on krill density. *Journal of Experimental Biology* 214:131–46.
Gregr, E.J. 2011. Insights into North Pacific Right Whale *Eubalaena japonica* habitat from historic whaling records. *Endangered Species Research* 15:223–39.
Gregr, E.J., J. Calambokidis, L. Convey, J.K.B. Ford, R.I. Perry, L. Spaven and M. Zacharias. 2006. Recovery strategy for Blue, Fin, and Sei whales (*Balaenoptera musculus*, *B. physalus* and *B. borealis*) in Pacific Canadian waters. Species at Risk Act Recovery Strategy Series, Fisheries and Oceans Canada, Vancouver, BC.
Gregr, E.J., and K.O. Coyle. 2009. The biogeography of the North Pacific Right Whale (*Eubalaena japonica*). *Progress in Oceanography* 80:188–98.
Gregr, E.J., L. Nichol, J.K.B. Ford, G. Ellis and A.W. Trites. 2000. Migration and population structure of northeastern Pacific whales off coastal British Columbia: an analysis of commercial whaling records from 1908–1967. *Marine Mammal Science* 16:699–727.
Gregr, E.J., L.M. Nichol, J.C. Watson, J.K.B. Ford and G.M. Ellis. 2008. Estimating carrying capacity for Sea Otters in British Columbia. *Journal of Wildlife Management* 72:382–88.
Gregr, E.J., and A.W. Trites. 2001. Predictions of critical habitat for five whale species in the waters of coastal British Columbia. *Canadian Journal of Fisheries and Aquatic Sciences* 58:1265–85.
Guiguet, C.J. 1953. California Sea Lion (*Zalophus californianus*) in British Columbia. *The Canadian Field-Naturalist* 67:140.

Guiguet, C.J. 1954. A record of Baird's dolphin (*Delphinus bairdii* Dall) in British Columbia. *The Canadian Field-Naturalist* 68:136.
Guiguet, C.J., and G.C. Pike. 1965. First specimen record of the Gray Grampus or Risso Dolphin, *Grampus griseus* (Cuvier) from British Columbia. *Murrelet* 46:16.
Guiguet, C.J., and W.J. Schick. 1970. First record of Right Whale Dolphin, *Lissodelphis borealis* (Peale) from British Columbia. *Syesis* 3:188.
Hall, A.M. 2004. Seasonal abundance, distribution and prey species of Harbour Porpoise (*Phocoena phocoena*) in southern Vancouver Island waters. MSc thesis, University of British Columbia, Vancouver.
———. 2011. Foraging behaviour and reproductive season habitat selection of northeast Pacific porpoises. PhD thesis, University of British Columbia, Vancouver.
Hammond, P.S., G. Bearzi, A. Bjørge, K. Forney, L. Karczmarski, T. Kasuya, W.F. Perrin, M.D. Scott, J.Y. Wang, R.S. Wells and B. Wilson. 2008a. *Delphinus capensis*. IUCN Red List of Threatened Species, version 2011.1. www.iucnredlist.org.
———. 2008b. *Stenella coeruleoalba*. IUCN Red List of Threatened Species, version 2011.1. www.iucnredlist.org.
Hatler, D.F. 1971. A Canadian specimen of Risso's Dolphin. *The Canadian Field-Naturalist* 85:188–89.
———. 1972. The mammals of Pacific Rim National Park. Unpublished report, National and Historic Parks Branch, Western Region, Calgary, Alberta.
Hatler, D.F., and J.D. Darling. 1974. Recent observations of the Gray Whale in British Columbia. *The Canadian Field-Naturalist* 88:449–59.
Hatler, D.F., D.W. Nagorsen and A.M. Beal. 2008. *Carnivores of British Columbia*. Victoria: Royal BC Museum.
Hayes, S.A., D.P. Costa, J.T. Harvey and B.J. Le Boeuf. 2004. Aquatic mating strategies of the male Pacific Harbor Seal (*Phoca vitulina richardii*): are males defending the hotspot? *Marine Mammal Science* 20:639–56.
Heath, C.B., and W.F. Perrin. 2009. California, Galapagos and Japanese sea lions *Zalophus californianus*, *Z. wollebaeki* and *Z. japonicus*. In Perrin, Würsig and Thewissen 2009.
Heise, K. 1997a. Diet and feeding behaviour of Pacific White-sided Dolphins (*Lagenorhynchus obliquidens*) as revealed through the collection of prey fragments and stomach content analyses. *Report of the International Whaling Commission* 47:807–15.
———. 1997b. Life history and population parameters of Pacific White-sided Dolphins (*Lagenorhynchus obliquidens*). *Report of the International Whaling Commission* 47:817–25.
Heyning, J.E. 1989. Cuvier's Beaked Whale *Ziphius cavirostris* G. Cuvier, 1823. In *Handbook of marine mammals. Volume 4: River Dolphins and the Larger Toothed Whales*, edited by S.H. Ridgway and S.R. Harrison. New York: Academic Press.
Heyning, J.E., and G.M. Lento. 2002. The evolution of marine mammals. In *Marine mammal Biology: An Evolutionary Approach*, edited by A.R.

Hoelzel. Oxford: Blackwell Science.
Heyning, J.E., and J.G. Mead. 1996. Suction feeding in beaked whales: morphological and observational evidence. *Contributions in Science* (Natural History Museum of Los Angeles County) 464.
———. 2009. Cuvier's Beaked Whale *Ziphius cavirostris*. In Perrin, Würsig and Thewissen 2009.
Heyning, J.E., and W.F. Perrin. 1994. Evidence for two species of Common Dolphins (genus *Delphinus*) from the eastern North Pacific. *Contributions in Science* (Natural History Museum of Los Angeles County) 442.
Hoelzel, A.R., ed. 2002. *Marine Mammal Biology: An Evolutionary Approach*. Oxford: Blackwell Publishing.
Horwood, J. 2009. Sei Whale *Balaenoptera borealis*. In Perrin, Würsig and Thewissen 2009.
Houston, J. 1990a. Status of Hubbs' Beaked Whale, *Mesoplodon carlhubbsi*, in Canada. *The Canadian Field-Naturalist* 104:121–24.
———. 1990b. Status of Stejneger's Beaked Whale, *Mesoplodon stejnegeri*, in Canada. *The Canadian Field-Naturalist* 104:131–34.
———. 1991. Status of Cuvier's Beaked Whale, *Ziphius cavirostris*, in Canada. *The Canadian Field-Naturalist* 105:215–18.
Insley, S.J. 2001. Mother-offspring vocal recognition in Northern Fur Seals is mutual but asymmetrical. *Animal Behaviour* 61:129–37.
Ivashchenko, Y.V., R.L. Brownell Jr and P.J. Clapham. 2013. Soviet whaling in the North Pacific: revised catch totals. *Journal of Cetacean Research and Management* 13:59–71.
Ivashchenko, Y.V., and P.J. Clapham. 2012. Soviet catches of Bowhead (*Balaena mysticetus*) and Right (*Eubalaena japonica*) whales in the North Pacific and Okhotsk Sea. *Endangered Species Research* 18:201–17.
Jamieson, G.S., and G.D. Heritage. 1987. Offshore squid (drift gillnet fishery). *Canadian Technical Report of Fisheries and Aquatic Sciences* 1576.
———. 1988. Experimental flying squid fishery off British Columbia. 1987. *Canadian Industry Report of Fisheries and Aquatic Sciences* 186.
Jefferson, T.A. 1987. A study of the behavior of Dall's Porpoise (*Phocoenoides dalli*) in Johnstone Strait, British Columbia. *Canadian Journal of Zoology* 65:736–44.
———. 1988. *Phocoenoides dalli*. *Mammalian Species* 319.
———. 1990. Status of Dall's porpoise, *Phocoenoides dalli*, in Canada. *The Canadian Field-Naturalist* 104:112–16.
Jefferson, T.A., S. Leatherwood and M.A. Webber. 1993. *Marine Mammals of the World*. FAO species identification guide. Rome: FAO.
Jefferson, T.A., M.W. Newcomber, S. Leatherwood and K. van Waerebeek. 1994. Right whale dolphins, *Lissodelphis borealis* (Peale, 1848) and *Lissodelphis peronii* (Lacepede, 1804). In *Handbook of Marine Mammals*, vol. 5: *The First Book of Dolphins*, edited by S.H. Ridgway and S.R. Harrison. New York: Academic Press.
Jefferson, T.A., M.A. Webber and R.L. Pitman. 2008. *Marine Mammals of the World. A Comprehensive Guide to their Identification*. San Diego: Academic Press.

Jones, M.L., and S.L. Swartz. 2009. Gray Whale *Eschrichtius robustus*. In Perrin, Würsig and Thewissen 2009.

Josephson, E., T.D. Smith and R.R. Reeves. 2008a. Historical distribution of Right Whales in the North Pacific. *Fish and Fisheries* 9:155–68.

———. 2008b. Depletion within a decade: the American 19th-century North Pacific Right Whale fishery. In *Oceans Past: Management Insights From the History of Marine Animal Populations*, edited by D.J. Starkey, P. Holm and M. Barnard. London: Earthscan.

Kasuya, T. 2009. Giant beaked whales *Berardius bairdii* and *B. arnuxii*. In Perrin, Würsig and Thewissen 2009.

Kenney, R.D. 2009. North Atlantic, North Pacific and Southern Right Whales – *Eubalaena glacialis*, *E. japonica* and *E. australis*. In Perrin, Würsig and Thewissen 2009.

Keple, A.R. 2002. Seasonal abundance and distribution of marine mammals in the southern Strait of Georgia, British Columbia. MSc thesis, University of BC, Vancouver.

Kooyman, G.L. 2009. Diving physiology. In Perrin, Würsig and Thewissen 2009.

Kuker, K., and L. Barrett-Lennard. 2010. A re-evaluation of the role of Killer Whales *Orcinus orca* in a population decline of Sea Otters *Enhydra lutris* in the Aleutian Islands and a review of alternative hypotheses. *Mammal Review* 40:103–24.

Leatherwood, S., W.F. Perrin, V.L. Kirby, C.L. Hubbs and M. Dahlheim. 1980. Distribution and movements of Risso's Dolphin, *Grampus griseus*, in the eastern North Pacific. *Fishery Bulletin* 77:951–64.

Leatherwood, S., and W.A. Walker. 1979. The Northern Right Whale Dolphin *Lissodelphis borealis* Peale in the eastern North Pacific. In *Behavior of Marine Animals*, vol. 3: *Cetaceans*, edited by H.E. Winn and B.L. Olla. New York: Plenum.

Le Boeuf, B.J. 1974. Male-male competition and reproductive success in elephant seals. *American Zoologist* 14:163–76.

Le Boeuf, B.J., D.E. Crocker, D.P. Costa, S.B. Blackwell, P.M. Webb and D.S. Houser. 2000. Foraging ecology of Northern Elephant Seals. *Ecological Monographs* 70:353–82.

Le Boeuf, B.J., and R.M. Laws. 1994. *Elephant Seals: Population Ecology, Behavior and Physiology*. Berkeley: University of California Press.

Loughlin, T.R. 2009. Steller Sea Lion *Eumetopias jubatus*. In Perrin, Würsig and Thewissen 2009.

Loughlin, T.R., C.H. Fiscus, A.M. Johnson and D.J. Rugh. 1982. Observations of *Mesoplodon stejnegeri* (Ziphiidae) in the central Aleutian Islands, Alaska. *Journal of Mammalogy* 63:697–700.

Loughlin, T.R., and M.A. Perez. 1985. *Mesoplodon stejnegeri*. *Mammalian Species* 250.

Lux, C.A., A.S. Costa and A.E. Dizon. 1997. Mitochondrial DNA population structure of the Pacific White-sided Dolphin. *Report of the International Whaling Commission* 47:645–52.

Mate, B., A.L. Bradford, G. Tsidulko, V. Vertyankin and V. Ilyashenko. 2011.

Late-feeding season movements of a western North Pacific Grey Whale off Sakhalin Island, Russia, and subsequent migration into the Eastern North Pacific. Paper SC/63/BRG23 presented to the International Whaling Commission Scientific Committee. www.iwcoffice.org.

Mate, B.R., and D.P. DeMaster. 1986. California Sea Lion. In *Marine Mammals of Eastern Pacific and Arctic Waters*, edited by D. Haley. Seattle: Pacific Search Press.

Mathews, E.A., and M.D. Adkison. 2010. The role of Steller Sea Lions in a large population decline of Harbour Seals. *Marine Mammal Science* 26:803–36.

McAlpine, D.F. 2009. Pygmy and Dwarf sperm whales *Kogia breviceps* and *K. sima.* In Perrin, Würsig and Thewissen 2009.

McGinnis, S.M., and R.J. Schusterman. 1981. Northern Elephant Seal *Mirounga angustirostris* Gill 1866. In *Handbook of Marine Mammals*, vol. 2: *Seals*, edited by S.H. Ridgway and R.J. Harrison. London: Academic Press.

McKechnie, I., and R.J. Wigen. 2011. Toward a historical ecology of pinniped and Sea Otter hunting traditions on the coast of southern British Columbia. In *Human Impacts on Seals, Sea Lions and Sea Otters: Integrating Archaeology and Ecology in the Northeast Pacific*, edited by T.J. Braje and T.C. Rick. Berkeley: University of California Press.

McMillan, A.D., I. McKechnie, D.E. St Claire and S.G. Frederick. 2008. Exploring variability in maritime resource use on the Northwest Coast: a case study from Barkley Sound, western Vancouver Island. *Canadian Journal of Archaeology* 32:214–38.

McSweeney, D.J., R.W. Baird and S.D. Mahaffy. 2007. Site fidelity, associations and movements of Cuvier's (*Ziphius cavirostris*) and Blainville's (*Mesoplodon densirostris*) beaked whales off the island of Hawaii. *Marine Mammal Science* 23:666–87.

Mead, J.G., W.A. Walker and W.J. Houck. 1982. Biological observations on *Mesoplodon carlhubbsi* (Cetacea: Ziphiidae). *Smithsonian Contributions to Zoology* 344.

Mellinger, D.K., K.M. Stafford and C.G. Fox. 2004. Seasonal occurrence of Sperm Whale (*Physeter macrocephalus*) sounds in the Gulf of Alaska, 1999–2001. *Marine Mammal Science* 20:48–62.

Merilees, B. 1985. The Humpback Whales of Georgia Strait. *Waters* (Journal of the Vancouver Aquarium) 8:7–24.

Minamikawa, S., T. Iwasaki and T. Kishiro. 2007. Diving behaviour of a Baird's Beaked Whale, *Berardius bairdii*, in the slope water region of the western North Pacific: first dive records using a data logger. *Fisheries Oceanography* 16:573–77.

Mitchell, E.D. 1968. Northeast Pacific stranding distribution and seasonality of Cuvier's Beaked Whale, *Ziphius cavirostris*. *Canadian Journal of Zoology* 46:265–79.

Mizroch, S.A., D.W. Rice, D. Zweifelhoffer, J. Waite and W.L. Perryman. 2009. Distribution and movements of Fin Whales in the North Pacific Ocean. *Mammal Review* 39:193–227.

Monks, G.G., A.D. McMillan and D.E. St Claire. 2001. Nuu-chah-nulth whaling: archaeological insights into antiquity, species preferences and cultural importance. *Arctic Anthropology* 38:60–81.
Moore, J.C. 1963. Recognizing certain species of beaked whales of the Pacific Ocean. *American Midland Naturalist* 70:396–428.
Moore, J.E., and J.P. Barlow. 2013. Declining abundance of beaked whales (family Ziphiidae) in the California Current large marine ecosystem. *PLOS ONE* 8:1:e52770.
Morin, P.A., F.I. Archer, A.D. Foote, J. Vilstrup, E.E. Allen, P. Wade, J. Durban, K. Parsons, R. Pitman, L. Li, P. Bouffard, S.C. Abel Nielsen, M. Rasmussen, E. Willerslev, M.T.P. Gilbert and T. Harkins. 2010. Complete mitochondrial genome phylogeographic analysis of Killer Whales (*Orcinus orca*) indicates multiple species. *Genome Research* 20:908–16.
Morton, A.B. 1990. A quantitative comparison of the behavior of Resident and Transient forms of the Killer Whale off the central British Columbia coast. *Report of the International Whaling Commission* Special Issue 12:245–48.
———. 2000. Occurrence, photo-identification and prey of Pacific White-sided Dolphins (*Lagenorhynchus obliquidens*) in the Broughton Archipelago, Canada, 1984–1998. *Marine Mammal Science* 16:80–93.
Moss, M.L., D. Yang, S.D. Newsome, C.F. Speller, I. McKechnie, A.D. MacMillan, R.J. Losey and P.L. Koch. 2006. Historical ecology and biogeography of North Pacific pinnipeds: isotopes and ancient DNA from three archaeological assemblages. *Journal of Island and Coastal Archaeology* 1:165–90.
Murie, D.J. 1981. The migration of the Northern Fur Seal, *Callorhinus ursinus* Linnaeus 1758, in the eastern North Pacific Ocean and eastern Bering Sea: an analysis of pelagic sealing logs of the years 1886 to 1911. BSc thesis, University of Victoria, Victoria.
Nagorsen, D.W. 1985. *Kogia simus*. *Mammalian Species* 239.
———. 1990. *The Mammals of British Columbia: A Taxonomic Catalogue*. Victoria: Royal BC Museum.
Nagorsen, D.W., and G.E. Stewart. 1983. A Dwarf Sperm Whale (*Kogia simus*) from the Pacific coast of Canada. *Journal of Mammalogy* 64:505–06.
Nichol, L.M., R. Abernethy, L. Flostrand, T.S. Lee and J.K.B. Ford. 2010. Information relevant for the identification of critical habitats of North Pacific Humpback Whales (*Megaptera novaeangliae*) in British Columbia. DFO Canadian Science Advisory Secretariat Research Document 2009/116.
Nichol, L.M., M.D. Boogaards and R.M. Abernethy. 2009. Recent trends in the abundance and distribution of Sea Otters (*Enhydra lutris*) in British Columbia. DFO Canadian Science Advisory Secretariat Research Document 2009/016.
Nichol, L.M., E.J. Gregr, R. Flinn, J.K.B. Ford, R. Gurney, L. Michaluk and A. Peacock. 2002. British Columbia commercial whaling catch data 1908 to 1967: a detailed description of the BC historical whaling database. *Canadian Technical Report of Fisheries and Aquatic Sciences* 2396.

Nichol, L.M., A.M. Hall, G.M. Ellis, E. Stredulinsky, M. Boogaards and J.K.B. Ford. 2013. Dietary overlap and niche partitioning of sympatric Harbour Porpoises and Dall's Porpoises in the Salish Sea. *Progress in Oceanography* 115:202–10.
Odell, D.K., and K.M. McClune. 1999. False Killer Whale *Pseudorca crassidens* (Owen, 1946). In *Handbook of Marine Mammals*, vol. 6: *The Second Book of Dolphins and the Porpoises*, edited by S.H. Ridgway and S.R. Harrison. New York: Academic Press.
Olesiuk, P.F. 2008. Preliminary assessment of the recovery potential of Northern Fur Seals (*Callorhinus ursinus*) in British Columbia. DFO Canadian Science Advisory Secretariat Research Document 2007/076.
———. 2010. *An Assessment of Population Trends and Abundance of Harbour Seals (*Phoca vitulina*) in British Columbia*. Ottawa: Fisheries and Oceans Canada.
———. 2011. Abundance of Steller Sea Lions in (*Eumetopias jubatus*) in British Columbia. DFO Canadian Science Advisory Secretariat Research Document 2010/000.
Olesiuk, P.F., and M.A. Bigg. 1988. *Seals and Sea Lions on the British Columbia Coast.* Nanaimo, BC: Pacific Biological Station, Department of Fisheries and Oceans.
Olesiuk, P.F., M.A. Bigg and G.M. Ellis. 1990a. Life history and population dynamics of Resident Killer Whales (*Orcinus orca*) in the coastal waters of British Columbia and Washington state. *Report of the International Whaling Commission* Special Issue 12:209–42.
———. 1990b. Recent trends in the abundance of Harbour Seals, *Phoca vitulina*, in British Columbia. *Canadian Journal of Fisheries and Aquatic Sciences* 47:992–1003.
Olesiuk, P.F., M.A. Bigg, G.M. Ellis, S.J. Crockford and R.J. Wigen. 1990. An assessment of the feeding habits of Harbour Seals (*Phoca vitulina*) in the Strait of Georgia, British Columbia, based on scat analysis. *Canadian Technical Report of Fisheries and Aquatic Sciences* 1730.
Olesiuk, P.F., G.M. Ellis and J.K.B. Ford. 2005. Life history and population dynamics of northern Resident Killer Whales (*Orcinus orca*) in British Columbia. DFO Canadian Science Advisory Secretariat Research Document 2005/045.
Oleson, E.M., J. Calambokidis, W.C. Burgess, M.A. McDonald, C.A. LeDuc and J.A. Hildebarnd. 2007. Behavioral context of call production by eastern North Pacific Blue Whales. *Marine Ecology Progress Series* 330:269–84.
Olsen, P.A. 2009. Pilot whales *Globicephala melas* and *G. macrorhynchus*. In Perrin, Würsig and Thewissen 2009.
Osborne, R., J. Calambokidis and E.M. Dorsey. 1988. *A Guide to Marine Mammals of Greater Puget Sound, Including a Catalog of Individual Orca and Minke Whales.* Anacortes, Washington: Island Publishers.
Perez, M.A., and M.A. Bigg. 1986. Diet of Northern Fur Seals, *Callorhinus ursinus*, off western North America. *Fishery Bulletin* 84:957–71.
Perrin, W.F. 2009. Common Dolphins *Delphinus delphis* and *D. capensis*. In Perrin, Würsig and Thewissen 2009.

Perrin, W.F., and R.L. Brownell Jr. 2009. Minke Whales *Balaenoptera acutorostrata* and *B. bonaerensis*. In Perrin, Würsig and Thewissen 2009.
Perrin, W.F., B. Würsig and J.G.M. Thewissen, eds. 2009. *Encyclopedia of Marine Mammals: Second Edition*. San Diego: Academic Press.
Pike, G.C. 1950. Stomach contents of whales caught off the coast of BC. *Pacific Progress Report* (Fisheries Research Board of Canada) 83:27–28.
———. 1953a. Preliminary report on the growth of Finback Whales from the coast of British Columbia. *Norsk Hvalfangst-Tidende* 1:11–15.
———. 1953b. Two records of *Berardius bairdi* from the coast of British Columbia. *Journal of Mammalogy* 34:98–104.
———. 1960. Pacific Striped Dolphin, *Lagenorhynchus obliquidens*, off the coast of British Columbia. *Journal of the Fisheries Research Board of Canada* 17:123–24.
———. 1962. Migration and feeding of the Gray Whale, *Eschrichtius gibbosus*. *Bulletin of the Fisheries Research Board of Canada* 19:815–37.
Pike, G.C., and I.B. MacAskie. 1969. Marine mammals of British Columbia. *Bulletin of the Fisheries Research Board of Canada* 71.
Pitman, R.L., and P. Ensor. 2003. Three forms of Killer Whales (*Orcinus orca*) in Antarctic waters. *Journal of Cetacean Research and Management* 5:131–39.
Rambeau, A.L. 2008. Determining abundance and stock structure for widespread migratory animals: the case of Humpback Whales (*Megaptera novaeangliae*) in British Columbia, Canada. MSc thesis, University of British Columbia, Vancouver.
Rankin, S., J. Oswald, J. Barlow and M. Lammers. 2007. Patterned burst-pulse vocalizations of the Northern Right Whale Dolphin, *Lissodelphis borealis*. *Journal of the Acoustical Society of America* 121:1213–18.
Raum-Suryan, K., K. Pitcher and R. Lamy. 2004. Sea Otter, *Enhydra lutris*, sightings off Haida Gwaii / Queen Charlotte Islands, British Columbia, 1972-2002. *The Canadian Field-Naturalist* 118:270–72.
Reeves, R.R. 2009. Conservation efforts. In Perrin, Würsig and Thewissen 2009.
Reeves, R.R., and E.D. Mitchell. 1993. Status of Baird's Beaked Whale, *Berardius bairdii*. *The Canadian Field-Naturalist* 107:509–23.
Reeves, R.R., T.D. Smith, J.N. Lund, S.A. Lebo and E.A. Josephson. 2010. Nineteenth-century ship-based catches of Gray Whales, *Eschrichtius robustus*, in the eastern North Pacific. *Marine Fisheries Review* 72:1:26–65.
Reeves, R.R., B.S. Stewart, P.J. Clapham and J.A. Powell, with P. Folkens. 2002. *National Audubon Society Guide to Marine Mammals of the World*. New York: Alfred A. Knopf.
Reeves, R.R., and H. Whitehead. 1997. Status of the Sperm Whale, *Physeter macrocephalus*, in Canada. *The Canadian Field-Naturalist* 111:16.
Reimchen, T. 1980. Sightings of Risso's Dolphins (*Grampus griseus*) off Queen Charlotte Islands, British Columbia. *Murrelet* 61:44–45.
Reynolds III, J.E., and S.A. Rommel. 1999. *Biology of Marine Mammals*. Washington: Smithsonian Institution Press.
Rice, D.W. 1974. Whales and whale research in the eastern North Pacific. In

The Whale Problem, edited by W.E. Schevill. Cambridge, Mass.: Harvard University Press.
———. 1977. Synopsis of biological data on the Sei Whale and Bryde's Whale in the eastern North Pacific. *Report of the International Whaling Commission* Special Issue 1:92–97.
———. 1998. *Marine Mammals of the World: Systematics and Distribution*. Special Publication 4. San Francisco: The Society for Marine Mammalogy.
Rice, D.W., and A.A. Wolman. 1971. *The Life History and Ecology of the Gray Whale* (Eschrichtius robustus). Special Publication. Lawrence, Kansas: American Society of Mammalogists.
Robbins, J., L. Dalla Rosa, J.M. Allen, D.K. Mattila, E.R. Secchi, A.S. Friedlaender, P.T. Stevick, D.P. Nowacek and D. Steel. 2011. Return movement of a Humpback Whale between the Antarctic Peninsula and American Samoa: a seasonal migration record. *Endangered Species Research* 13:117–21.
Robinson, P.W., D.P. Costa, D.E. Crocker et al. 2012. Foraging behavior and success of a mesopelagic predator in the northeast Pacific Ocean: Insights from a data-rich species, the Northern Elephant Seal. *PLOS ONE* 7:e36728.
Ross, P.S., G.M. Ellis, M.G. Ikonomou, L.G. Barrett-Lennard and R.F. Addison. 2000. High PCB concentrations in free-ranging Pacific Killer Whales, *Orcinus orca*: effects of age, sex and dietary preference. *Marine Pollution Bulletin* 40:504–15.
Scammon, C.M. 1869. On the cetaceans of the western coast of North America. *Proceedings of the Academy of Natural Sciences of Philadelphia* 21:13–63.
———. 1874. *The Marine Mammals of the Northwestern Coast of North America, Together with an Account of the American Whale-fishery*. New York: Dover Publications.
Scheffer, V.B., and J.W. Slipp. 1948. The whales and dolphins of Washington State. *American Midland Naturalist* 39:257–337.
Schorr, G.S., E.A. Falcone, D.J. Moretti and R.D Andrews. 2014. First long-term behavioral records from Cuvier's Beaked Whales (*Ziphius cavirostris*) reveal record-breaking dives. *PLOS ONE* 9:e92633.
Sea Otter Recovery Team. 2007. Recovery Strategy for the Sea Otter (*Enhydra lutris*) in Canada. Fisheries and Oceans Canada, Vancouver.
Sears, R., and J. Calambokidis. 2002. COSEWIC status report on the Blue Whale from Atlantic and Pacific waters (*Balaenoptera musculus*). Committee on the Status of Endangered Wildlife in Canada, Ottawa.
Sears, R., and W.F. Perrin. 2009. Blue Whale *Balaenoptera musculus*. In Perrin, Würsig and Thewissen 2009.
Stacey, P.J., and R.W. Baird. 1991a. Status of the False Killer Whale, *Pseudorca crassidens*, in Canada. *The Canadian Field-Naturalist* 105:189–97.
———. 1991b. Status of the Pacific White-sided Dolphin, *Lagenorhynchus obliquidens*, in Canada. *The Canadian Field-Naturalist* 105:219–32.

Stacey, P.J., and R.W. Baird. 1993. Status of the Short-finned Pilot Whale, *Globicephala macrorhynchus*, in Canada. *The Canadian Field-Naturalist* 107:481–89.
Stafford, K.M., J.J. Citta, S.E. Moore, M.A. Daher and J.E. George. 2009. Environmental correlates of Blue and Fin whale call detections in the North Pacific Ocean from 1997 to 2002. *Marine Ecology Progress Series* 395:37–53.
Stafford, K.M., S.L. Nieukirk and C.G. Fox. 2001. Geographic and seasonal variation of Blue Whale calls in the North Pacific. *Journal of Cetacean Research and Management* 3:65–76.
Stelle, L.L. 2001. Behavioral ecology of summer resident Gray Whales (*Eschrichtius robustus*) feeding on mysids in British Columbia, Canada. PhD thesis, University of California, Los Angeles.
Stelle, L.L., W.M. Megill and M.R. Kinzel. 2008. Activity budgets and diving behavior of Gray Whales (*Eschrichtius robustus*) in feeding grounds off coastal British Columbia. *Marine Mammal Science* 24:462–78.
Stewart, B.S., and S. Leatherwood. 1985. Minke Whale *Balaenoptera acutorostrata* Lacépède, 1804. In *Handbook of Marine Mammals*, vol. 3: *The Sirenians and Baleen Whales*, edited by S.H. Ridgway and S.R. Harrison. London: Academic Press.
Stroud, R.K. 1968. Risso Dolphin in Washington state. *Journal of Mammalogy* 49:347–48.
Stutz, S.S. 1967. Pelage patterns and population distributions of the Pacific Harbour Seal (*Phoca vitulina richardsi*). *Journal of the Fisheries Research Board of Canada* 24:451–55.
Sullivan, R.M. 1981. Agonistic behavior and dominance relationships in the Harbor Seal, *Phoca vitulina*. *Journal of Mammalogy* 63:554–69.
Suttles, W.P. 1987. Notes on Coast Salish sea mammal hunting, in *Coast Salish Essays*. Seattle: University of Washington Press.
Swartz, S.L., B.L. Taylor and D.J. Rugh. 2006. Gray Whale *Eschrichtius robustus* population and stock identity. *Mammal Review* 36:66–84.
Szpak, P., T.J. Orchard, I. McKechnie and D.R. Gröcke. 2012. Historical ecology of late Holocene Sea Otters (*Enhydra lutris*) from northern British Columbia: isotopic and zooarchaeological perspectives. *Journal of Archaeological Science* 39:1553–71.
Taylor, B.L., R. Baird, J. Barlow, S.M. Dawson, J. Ford, J.G. Mead, G. Notarbartolo Di Sciara, P. Wade and R.L. Pitman. 2008. *Globicephala macrorhynchus*. IUCN Red List of Threatened Species, version 2011.1. www.iucnredlist.org.
Temte, J.L., M.A. Bigg and O. Wiig. 1991. Clines revisited: the timing of pupping in the Harbour Seal (*Phoca vitulina*). *Journal of Zoology* (London) 224:617–32.
Thewissen, J.G.M. 2009. Sensory biology: overview. In Perrin, Würsig and Thewissen 2009.
Thomson, R.E. 1981. Oceanography of the British Columbia coast. *Canadian Special Publication of Fisheries and Aquatic Sciences* 56.
Towers, J.R. 2011. *Minke Whales of the Straits off Northeastern Vancouver*

Island, 2nd edition. Alert Bay, BC: Marine Education and Research Society.
Towers, J.R., C.J. McMillan, M. Malleson, J. Hildering, J.K.B. Ford and G. M. Ellis. In press. Seasonal movements and ecological markers as evidence for migration of Common Minke Whales photo-identified in the eastern North Pacific. *Journal of Cetacean Research and Management.*
Townsend, C.H. 1935. The distribution of certain whales as shown by logbook records of American whaleships. *Zoologica* 19:1:1–50.
Trites, A.W., and M.A. Bigg. 1996. Physical growth of Northern Fur Seals (*Callorhinus ursinus*): seasonal fluctuations and migratory influences. *Journal of Zoology* (London) 238:459–82.
True, F.W. 1885. On a new species of porpoise, *Phocaena dalli,* from Alaska. *Proceedings of the United States National Museum* 8:95–98.
Tyack, P.L., M. Johnson, N. Aguilar Soto, A. Sturlese and P.T. Madsen. 2006. Extreme diving of beaked whales. *Journal of Experimental Biology* 209:4238–53.
Tyack, P.L., and E.H. Miller. 2002. Vocal anatomy, acoustic communication and echolocation. In Hoelzel 2002.
Wade, P.R., A. De Robertis, K.R. Hough, R. Booth, A. Kennedy, R.G. LeDuc, L. Munger, J. Napp, K.E.W. Shelden, S. Rankin, O. Vasques and C. Wilson. 2011. Rare detections of North Pacific Right Whales in the Gulf of Alaska, with observations of their potential prey. *Endangered Species Research* 13:99–109.
Wailes, G.H., and W.A. Newcombe. 1929. Sea lions. *Museum Art Notes* 4:43–50.
Walker, W.A., and M.B. Hanson. 1999. Biological observations on Stejneger's Beaked Whale, *Mesoplodon stejnegeri*, from strandings on Adak Island, Alaska. *Marine Mammal Science* 15:1314–29.
Walker, W.A., M.B. Hanson, R.W. Baird and T.J. Guenther. 1998. Food habits of the Harbor Porpoise, *Phocoena phocoena*, and the Dall's Porpoise, *Phocoenoides dalli*, in the inland waters of British Columbia and Washington. NOAA National Marine Fisheries Service AFSC Processed Report, 98-10:63–75
Walker, W.A., J.G. Mead and R.L. Brownell Jr. 2002. Diets of Baird's Beaked Whales, *Berardius bairdii*, in the southern Sea of Okhotsk and off the Pacific coast of Honshu, Japan. *Marine Mammal Science* 18:902–19.
Ware, D.M., and G.A. McFarlane. 1989. Fisheries production domains in the northeast Pacific Ocean. In *Effects of Ocean Variability on Recruitment and an Evaluation of Parameters Used in Stock Assessment Models*, edited by R.J. Beamish and G.A. McFarlane. *Canadian Special Publication of Fisheries and Aquatic Science* 108.
Wartzok, D., and D.R. Ketten. 1999. Marine mammal sensory systems. In Reynolds and Rommel 1999.
Watson, J., G. Ellis, T. Smith and J. Ford. 1997. Updated Status of the Sea Otter, *Enhydra lutris*, in Canada. *The Canadian Field-Naturalist* 111:10.
Watson, J., and J.A. Estes. 2011. Stability, resilience, and phase shifts in rocky subtidal communities along the west coast of Vancouver Island, Canada. *Ecological Monographs* 81:215–39.

Webb, R.L. 1988. *On the Northwest: Commercial Whaling in the Pacific Northwest, 1790-1967.* Vancouver: UBC Press.

Weller, D.W., A. Klimek, A.L. Bradford, J. Calambokidis, A.R. Lang, B. Gisborne, A.M. Burdin, W. Szaniszlo, J. Urbán, A. Gómez-Gallardo Unzueta, S. Swartz, R.L. Brownell Jr. 2012. Movements of Gray Whales between the western and eastern North Pacific. *Endangered Species Research* 18:193–99.

Wells, R.S., D.J. Boness and G.B. Rathbun. 1999. Behavior. In Reynolds and Rommel 1999.

Whitehead, H. 2009. Sperm Whale *Physeter macrocephalus*. In Perrin, Würsig and Thewissen 2009.

Williams, R., and L. Thomas. 2007. Distribution and abundance of marine mammals in the coastal waters of British Columbia, Canada. *Journal of Cetacean Research and Management* 9:15–28.

Williams, T.M., and G.A.J. Worthy. 2002. Anatomy and physiology: the challenge of aquatic living. In Hoelzel 2002.

Willis, P.M., and R.W. Baird. 1998a. Sightings and strandings of beaked whales on the west coast of Canada. *Aquatic Mammals* 24:21–25.

———. 1998b. Status of the Dwarf Sperm Whale (*Kogia simus*) with special reference to Canada. *The Canadian Field Naturalist* 112:114–25.

Willis, P.M., B.J. Crespi, L.M. Dill, R.W. Baird and B.L. Hanson. 2004. Natural hybridization between Dall's Porpoises (*Phocoenoides dalli*) and Harbour Porpoises (*Phocoena phocoena*). *Canadian Journal of Zoology* 82:828–34.

Wilton, W. 2007. *Wild Coast Project: Carnivore Scat Composition Analysis.* Long Beach, BC: Pacific Rim National Park Reserve.

Würsig, B., T.A. Jefferson and D.J. Schmidly. 2000. *The Marine Mammals of the Gulf of Mexico.* College Station: Texas A&M University Press.

Zacharias, M.A., D.E. Howes, J.R. Harper and P. Wainwright. 1998. The British Columbia marine ecosystem classification: rationale, development and verification. *Coastal Management* 26:105–24.

ACKNOWLEDGEMENTS

I owe much to my mentors for encouragement, guidance and opportunities as my interest in marine mammals of British Columbia grew: Michael A. Bigg (to whom this book is dedicated), H. Dean Fisher, Ian McTaggart Cowan, Murray Newman and Ian MacAskie. I thank my wife, Bev Ford, who has assisted on many forays into the field, provided steadfast support and kept the home fires burning during my extended absences doing field work and writing. My children Michael and Katie completed the "Ford family research team" and have supported and endured my focus on the book.

I am especially grateful to my long-term colleagues and friends, with whom I've shared so many great experiences out on the water observing marine mammals along the BC coast. Graeme Ellis deserves very special mention for almost four decades of research collaboration and camaraderie in the field. Others who also go back more years than many of us would care to admit include: Ken Balcomb, Lance Barrett-Lennard, Jim Borrowman, Jim Darling, Volker Deecke, Brian Falconer, Karen Hansen, Kathy Heise, Stan Hutchings, Craig Matkin, Bill Mackay, Rod MacVicar, John McCulloch, Alexandra Morton, Linda Nichol, Rod Palm, Paul Spong, Helena Symonds, Andrew Trites and Jane Watson.

I thank the current and past members of the Cetacean Research Program team at the Pacific Biological Station for their outstanding efforts in the field and in the lab: Robin Abernethy, Melissa Boogaards, Hitomi Kimura, Barbara Koot, Tatiana Lee, Chantal Levesque, Christie McMillan, Linda Nichol, Miriam O, Kristy O'Brien, Mayuko Otsuki, Alana Phillips, James Pilkington, Andrea Rambeau (Ahrens),

Silvia Scali, Hawsun Sohn, Lisa Spaven, Chelsea Stanley, Tara Stevens, Eva Stredulinksy, Jared Towers and Brianna Wright.

Others who I've had the good fortune to work with at various times over the years include: David Bain, Robin Baird, Lynne Barre, Ron Bates, Michelle Bigg, Caitlin Birdsall, Nancy Black, David Briggs, Randy Burke, Doug Burles, Rob Butler, John Calambokidis, Deb Cavanagh, Paul Cottrell, Marilyn Dahlheim, Luciano Dalla Rosa, Doug Davis, Nic Dedeluk, Mike deRoos, Dave Duffus, John Durban, Dave Ellifrit, Fred Felleman, Wayne Garton, Brian Gisborne, Kyla Graham, Annely Greene, Ed Gregr, Christophe Guinet, Anna Hall, Brad Hanson, Marilyn Hargreaves, Marty Haulena, Gil Hewlett, Jackie Hildering, Erich Hoyt, Dave Huff, Steve Insley, Jeff Jacobsen, Steve Jeffries, Mark Jonah, Alison Keple, Ron Lewis, Shayne MacLellan, John Mair, Mark Malleson, Dena Matkin, Murray McGregor, Bill McIntyre, Hermann Meuter, Dave Myers, Hiromi Naito, Flip Nicklin, John Nightingale, Brent Norberg, Peter Olesiuk, Rich Osborne, Bruce Paterson, Chris Picard, Meg Pocklington, Stephen Raverty, Erin Rechsteiner, Rhonda Reidy, Amalis Riera, Rudy Riesch, Peter Ross, Doug Sandilands, Eva Saulitis, Leah Saville, Alisa Schulman-Janiger, Ari Shapiro, Fred Sharpe, Norm Sloan, Jon Stern, Jan Straley, Wendy Szaniszlo, Robin Taylor, Frank Thomsen, Leah Thorpe, Chris Tulloch, Svein Vagle, Scott Wallace, Jody Weir, Rob Williams, Janie Wray, Clint Wright, and Harald Yurk. My thanks to you all.

Field research on marine mammals can be logistically difficult and expensive, and I am grateful to the following organizations for their support: Fisheries and Oceans Canada's Species-at-Risk Program, the Vancouver Aquarium and its Killer Whale Adoption Program, Langara Fishing Adventures, Stubbs Island Whale Watching, and the University of British Columbia. I also thank the officers and crews of the Canadian Coast Guard ships John P. Tully, Vector, W.E. Ricker, Arrow Post, Tanu and Gordon Reid, who have been so accommodating and helpful during our surveys at sea.

I thank Donald Gunn for his painstaking attention to detail and his great patience in preparing the excellent skull illustrations. I am also very grateful to Uko Gorter for providing the fine illustrations of cetaceans used in the species accounts and for happily making the occasional "tweaks" to ensure they accurately reflect the appearance of whales in BC waters. Alastair Denbigh prepared the pinniped and Sea Otter illustrations while working as illustrator at the Pacific Biological Station. Robin Abernethy deserves special thanks for preparing the numerous maps with such good humour and forbearance, despite my frequent requests for revision and updating.

Many thanks to the following people and organizations for allowing me to include their excellent photographs or illustrations: Robin Abernethy, Jim Borrowman, Carl Buell, John Calambokidis, Anne Carson, Diane Claridge and Charlotte Dunn of the Bahamas Marine Mammal Research Organisation, Brian Collen, Doug Davis, Graeme Ellis, Brian Falconer, Brian Gisborne, Jackie Hildering, Brad Hill, Mark Hobson, Jacob Joslin, Barb Koot, Derrek Lundy and the Gulf Islands Driftwood, Roger McDonell, Mark Malleson, Sheena Majewski, Christie McMillan, Raisa Mirza, Sally Mizroch, Linda Nichol, Cornelia Oedekoven, Stef Olcen, Pacific Rim National Park Service, Rod Palm, James Pilkington, Bob Pitman, Erin Rechsteiner, Royal BC Museum, Val Shore, Steve Suddes, Wendy Szaniszlo, Charlotte Tarver, Leah Thorpe, Jared Towers and Steve Wischniowski.

I also thank those who kindly helped by providing specialized information or other kinds of assistance in the preparation of this book: Arlene Armstrong, Mary Borrowman, Jenn Broom, Wayne Campbell, Phil Clapham, Paul Cottrell, Susan Crockford, Chris Darimont, Hugh Denbigh, Garry Fletcher, Jeremy Goldbogen, Dave Hatler, Mark Hipfner, Margaret Horsfield, David Janiger, Tom Jefferson, Lesley Kennes, Jeff Laake, Heather Lord, Misty MacDuffee, Andree Mackay, Bruce Mate, Colin MacLeod, Jim Mead, Ken Morgan, Debra Murie, Dave Nagorsen, Bill Perrin, Nick Pyenson, Kotoe Sasamori, Norm Sloan, Doug Stewart, Rick Thomson, Bill Walker and Becky Wigen.

I am very grateful to my colleagues Randall Reeves and Jane Watson for their detailed technical reviews of the manuscript and their many valuable suggestions and insightful comments. I also thank Val Shore, Gavin Hanke, and my PBS colleagues Linda Nichol, Eva Stredulinksy, Jared Towers and Sheena Majewski for providing many helpful comments. I am especially grateful to Gerry Truscott of the Royal BC Museum for his encouragement, for keeping the faith that a manuscript would eventually be delivered, and for his careful edits when it finally did arrive.

Marine Mammals of British Columbia

Editing by Amy Reiswig and Gerry Truscott, with assistance from Alex Van Tol.

Design and layout by Gerry Truscott. Typeset in Plantin Std 10/12 (body) and Optima LT Std 9/11 (captions).

Cover design by Jenny McCleery.

Front cover photograph by Mark Malleson (© Mark Malleson).

Back cover photographs by Jared Towers (Common Minke Whale, also on page 131), Brian Gisborne (Risso's Dolphins, also on page 263, and Sea Otter) and Raisa Mirza (California Sea Lions, also on page 369). (All © photographers.)

Skull drawings by Donald Gunn (© Royal BC Museum).

Whole-animal drawings of cetaceans by Uko Gorter (© Uko Gorter).

Whole-animal drawings of carnivores by Alastair Denbigh (© Fisheries and Oceans Canada).

Maps by Robin Abernethy (© Fisheries and Oceans Canada).

Printed and bound by Friesens.

The complete Mammals of British Columbia series:

1. *Bats of British Columbia* by David W. Nagorsen and R. Mark Brigham (1993, copublished with UBC Press).
2. *Opossums, Shrews and Moles of British Columbia* by David W. Nagorsen (1996, copublished with UBC Press).
3. *Hoofed Mammals of British Columbia* by David Shackleton (1999; revised and updated in 2013).
4. *Rodents and Lagomorphs of British Columbia* by David W. Nagorsen (2005).
5. *Carnivores of British Columbia* by David F. Hatler, David W. Nagorsen and Alison M. Beal (2008).
6. *Marine Mammals of British Columbia* by John K.B. Ford (2014).

Go to www.royalbcmuseum.bc.ca/Publications for details.

INDEX